Cognitive Studies in Literature and Performance

This series offers cognitive approaches to understanding perception, emotions, imagination, meaning-making, and the many other activities that constitute both the production and reception of literary texts and embodied performances.

More information about this series at
https://link.springer.com/bookseries/14903

Howard Mancing · Jennifer Marston William

Restoring the Human Context to Literary and Performance Studies

Voices in Everything

palgrave
macmillan

Howard Mancing
Purdue University
West Lafayette, IN, USA

Jennifer Marston William
Purdue University
West Lafayette, IN, USA

Cognitive Studies in Literature and Performance
ISBN 978-3-030-89077-3   ISBN 978-3-030-89078-0 (eBook)
https://doi.org/10.1007/978-3-030-89078-0

Cover illustration: Oliver Flick

This Palgrave Macmillan imprint is published by the registered company Springer Nature Switzerland AG
The registered company address is: Gewerbestrasse 11, 6330 Cham, Switzerland

We dedicate this book to our children:
to Catherine and Christina (Howard), and to Aidan and Kai (Jen),
whose voices are in everything we do.

Acknowledgements

Very special thanks to Elena Coda, Catherine Connor, Elaine Francis, Charles Ganelin, Madeleine Henry, Norm Holland, Paula Leverage, and Maren Linett for their ongoing strong support throughout the writing process. Many thanks also go to Fritz Breithaupt, Tom Broden, Mark Bruhn, Merlin Donald, Tyler Gabbard-Rocha, Richard Gerrig, Sarah Gretter, Tato Gyulamiryan, Elizabeth Hart, Frank Hakemulder, David Herman, Bruce McConachie, Keith Oatley, Alan Palmer, Isabel Jaén Portillo, Ryan Schneider, Rich Schweickert, Julien Simon, Barbara Smirka, Mark Turner, Blakey Vermeule, Christina Weiler, Lisa Zunshine, and all the wonderfully insightful students who took the various graduate and undergraduate classes we offered in recent years as this book was coming together. We are indebted to all of the above for their expertise, collaborative spirit, and friendship over the years.

Contents

About the Authors

Howard Mancing Professor Emeritus of Spanish at Purdue University, is a renowned expert on Cervantes and early modern Spanish literature as well as cognitive literary studies. He has published two previous monographs: *The Chivalric World of Don Quixote: Style, Structure, and Narrative* (University of Missouri Press, 1982) and *Miguel de Cervantes' 'Don Quixote': A Reference Guide* (Greenwood Press, 2006), and has co-edited six scholarly essay volumes. He also authored the two-volume *Cervantes Encyclopedia* (Greenwood Press, 2004). He has published journal articles and chapters on cognitive studies topics such as embodied cognition, narrative and affect, and Theory of Mind representations in literature.

Jennifer Marston William is Professor of German with specializations in twentieth and twenty-first-century literature, culture, and film at Purdue University, where she currently serves as Head of the School of Languages and Cultures. She is author of two previous monographs: *Killing Time: Waiting Hierarchies in the Twentieth-Century German Novel* (Bucknell University Press, 2010) and *Cognitive Approaches to German Historical Film: Seeing is Not Believing* (Palgrave Macmillan, 2017), and has co-edited three scholarly essay volumes. She has published articles and chapters on conceptual metaphor and on analyzing literature in the context of Theory of Mind.

Both Mancing and William, along with their colleague in French literature at Purdue University, Dr. Paula Leverage (author of *Reception and Memory: A Cognitive Approach to the Chansons de Geste*, Rodopi, 2010), were co-founders of the Center for Cognitive Literary Studies at Purdue, which has existed since 2008 and was recently renamed the Center for NeuroHumanities.

CHAPTER 1

Introduction: The Human Contexts of Literary and Performance Studies

Context is the key.
—Robert Frenay

Context is, of course, the key.
—David Lodge

We are evolved animals, with an imaginative mind-brain that enables us to be unique, individual, and contextualized agents who use language as a tool in our complex social relationships. Our aim in this book is to describe this idea's relevance for literary and performance studies. This stands in opposition to the view still implicit in some contemporary humanistic and social science theories, that we are socially constructed (and thus debiologized) subjects, born with a blank slate mind and brought into being by social forces such as language, ideology, and power/knowledge relationships. This introductory chapter will lay out the problems—linguistic, psychological, and biological—presented by some forms of literary theory; sketch out our proposal for an alternative approach to understanding and appreciating literature and performance; and outline the human contexts of literary and performance studies as examined throughout this book.

H. Mancing and J. Marston William, *Restoring the Human Context to Literary and Performance Studies*, Cognitive Studies in Literature and Performance,
https://doi.org/10.1007/978-3-030-89078-0_1

1 Outlining the Problem: Contemporary Literary Studies' Incompatibility with Modern Science

Works of literature consist of only one thing: language. Therefore, all literary theory must be based, explicitly or implicitly, on a theory—or at least some concept—of language. The varieties of language theory that have dominated literary studies during the final decades of the twentieth century and into the twenty-first century so far have been firmly grounded in, and depend vitally upon, the linguistic program sketched out a century ago by Ferdinand de Saussure. Yet linguistics as a discipline has itself virtually abandoned the Saussurean paradigm. Thus, the curious situation has emerged in which some literary scholars still assume a concept of language which linguists no longer find useful. One of our main goals in this book is to trace the rift between linguistics and literary theory and to suggest reconciliation.

This seemingly modest project requires an excavation into the ways fundamentally inaccurate ideas about humans and language have become entrenched in literary studies. All literary theory rests, whether explicitly or implicitly, on a psychological theory. It is impossible to talk or write of authors, readers, characters, and/or narrators without having some idea about what constitute these entities. And since these entities are always human beings (or anthropomorphized biological organisms or machines), it is impossible to teach literature, write criticism of specific literary works, or engage in theorizing about literature, without understanding what a human being is, what the mind and psyche consist of, and what the nature of cognitive processes might be. Since language is arguably the single most distinctive feature of the human species, language and psychology cannot be considered separately. Therefore, any attempt to trace the rift between linguistic theory and literary theory also necessarily implies tracing relationships between psychology and language. The psychology informing current literary theory turns out to be just as divergent from contemporary psychological theory as the linguistic theory informing current literary theory is from contemporary linguistic theory.

Any psychological theory rests on some stance with respect to biology. What is the relationship between mind and body, nature and nurture, genes and culture? What is the role of evolution in human cognition? Much literary theory ignores not only the latest advances in genetics and neuroscience, but also the solid foundation of evolution that has supported all biological study since the middle of the nineteenth century. The biological assumptions that underlie today's theory are not shared by the majority of biologists any more than are the linguistic and psychological theories of theory widely accepted by linguists and psychologists. Literary theory as such does not need to disappear, but we assert that it needs some serious reconsideration and reconceptualization. While most disciplines reject Cartesian dualism today, prevailing theories in the humanities often posit a socially constructed subject about whom nothing biological is of significance. The general assumption in many theories about literature in particular as well as about social/cultural contexts in general in recent decades is that nature (genes, biology, innateness, the material body in any significant way) is not a factor in human subjectivity, while nurture (culture, social or environmental forces, ideology, and, above all, language) is what exclusively determines the human subject.

The linguistic, psychological, and biological assumptions that have dominated literary theory and cultural studies since the ascendance of poststructuralism—that is, since about the 1970s, with its origins in France in the decade before—have been, for the most part, shared by scholars throughout the humanities and many of the social sciences. How did it happen that literary studies in the United States became dominated by linguistic and psychological theories that are so out of date? And how can literary studies continue to be dominated by such, even when the term "poststructuralism" may no longer be in vogue, as it is said we are living in something like a post-poststructuralist era?[1] This book will explore some possible answer to these puzzling questions, but more pressingly will propose alternatives to being stuck in an outdated mode of literary analysis.

Today, the study of linguistics is dominated not by Saussure and his followers but by Chomsky and those who have come in his wake. The dominant paradigm in psychology is not behaviorist but cognitive. And modern biology is firmly based on the theory of evolution by natural selection as outlined by Charles Darwin in the mid-nineteenth century and rounded out by twentieth-century genetics and subsequent developments. As for neuroscience and our understanding of the human mind-brain, humans have learned more in the last few decades than we did in all of previous history.

In recent decades, there has been a rising tide of criticism within literary studies against the reigning brand(s) of theory based on the linguistics of Saussure and best exemplified in the work of Derrida,

Foucault, Lacan, and Althusser. The feeling that "Theory is dead" echoes through the lecture halls and in the pages of professional journals, and yet seldom is anything proposed to take its place. While those who label themselves "postmodernists" or "deconstructionists" are increasingly rare, the literary scholarship that is practiced is often still very much in line with these ways of thinking.[2] Brian Boyd asked fifteen years ago, "Are reports of the death of Theory exaggerated? Or is Theory dead in a special way, one of the Undead, a zombie or a vampire?" (2006, 290). Theory is most definitely *not* dead, but continues to underlie much literary criticism published in scholarly journals today, while approaches to literature that are more empirical in nature are often dismissed.

In critiquing literary and cultural theory, we have in mind the specific brand of theory that rests on premises that contradict modern psychology and biology and that resists (usually unknowingly) what science now understands about embodied cognition. A main goal of this book is to make the most important principles of these fields known to humanities scholars so that they can incorporate them as appropriate into their work and think critically about theoretical and other work that doesn't account for them. In the context of the 2020s in which we write, we maintain that critical race theory, feminist theory, critical disability theory, affect theory, and others are crucial in the fight against unconscious bias and prejudice, if they don't undermine themselves by ignoring scientific developments. Cognitive studies are perfectly compatible with these humanistic approaches. Throughout this book and in other works, we cite, endorse, and rely on numerous theorists and philosophers who are not directly involved with cognition studies. Some of them expand on or clarify traditional types of theory (what we'll be calling THEORY to distinguish it from sound theoretical work that aligns with modern science), updating them for the twenty-first century by closely considering the embodied human mind and its dialogic exchange with culture and society.[3]

Theory has become the unmentioned but still dominant set of concepts that form the background of much of what is written about literature in the first decades of the twenty-first century. Theory is a classic case of what historian Daniel Lord Smail calls "ghost theories," which he defines as "old ideas that continue to structure our thinking without our being fully aware of their controlling presence" (2008, 3). The following chapters draw attention to this implicit lingering theory in literary studies, while contrasting it with alternate ways of thinking in modern scientific writing and in scholarship that takes various empirical and cognitive approaches to the humanities.

2 An Alternative: Reconciling Literature and Science Through Context

Context is the single most important and unifying concept in this book. Whether in evolution, brain science, biology, psychology, philosophy, or linguistics, the thinkers whom we take as positive models from these and other fields call specific attention to the role of context in all thought and action. It is significant that the two intellectual trends of which we are most critical—the structuralist-semiotic-poststructuralist mainstream of literary theory and the mind-as-computer approach to artificial intelligence—are both characterized by decontextualization. In contrast, no one in the twentieth century placed more emphasis on context in language, aesthetics, and philosophy than Mikhail M. Bakhtin. In this, as well as in his better-known concept of dialogism, Bakhtin is exemplary and provides a framework for our considerations here.[4] We return to Bakhtin throughout the work to show how cognitive linguists, language philosophers, and others carried further the foundations that Bakhtin had laid for considering utterances in context.

The human context necessarily involves biological, psychological, and social (cultural, historical, ideological, linguistic, and individual) factors. We call particular attention to the biological aspect, not because it is more important, but because it is the element most often overlooked or even denied in the discourse of theory, where social forces are considered to be all-powerful. In the twenty-first century, no serious scholar can afford to disregard or deny biology, including evolutionary theory. But this does not mean that biology trumps culture. Quite the contrary, everything is both biological and social at the same time. Daniel G. Freedman (1979) put it aptly: "Our species is biocultural—100% biological and 100% cultural" (108). Dualisms such as the classic Cartesian mind–body split or the nature-nurture debate become incoherent ideas that cannot be meaningfully addressed within contextualized—and contextualizing—discourse.

Science reveals that we need to conceive of things in terms of continua, interrelationships, mutually defining processes, and complex interactive systems, rather than in simplistic binaries and dualisms. What we are, what happens in the world, and how we understand and act upon what is said and done can never be described as simple biological necessity or equally simple (and simplistic) social construction. Thus, *binary thought* and *determinism* are among the major targets of our ongoing critique. Fruitful explanations do not emerge out of an insistence upon describing the world in either-or terms or a reliance upon some version of Cartesian (or any other) dualism. The monocausal assumptions of determinism, whether biological (that our genes determine us) or social (that language or ideology determines us) similarly oversimplify, offering little explanation beyond chilling implications of manipulation. The concerns expressed by feminist psychologist and ethicist Carol Gilligan nearly three decades ago in 1982 are still relevant today:

> I find the question of whether gender differences are biologically determined or socially constructed to be deeply disturbing. This way of posing the question implies that people, women and men alike, are either genetically determined or a product of socialization—that there is no voice—and without voice, there is no possibility for resistance, for creativity, or for a change whose wellsprings are psychological. (1993, xix)

More recently, sexual neuroscientist Debra W. Soh (2017) has argued that some feminists—namely transgender feminists and gender feminists, which she distinguishes from "traditional equity feminists"—are ignoring contemporary scientific developments and thereby undermining their arguments. The gender feminists, as Soh observes, embrace the blank slate theory and disregard proven inherent sex differences in the brain in favor of social constructivism, while at the other extreme, transgender feminists tend to see sex differences as inherently biological and assert that gender identity is stable from an early age. Neurological science does not support these views but rather points to something more in between the poles of all-nature and all-nurture when it comes to gender identity development. We agree with Soh's viewpoint in her article's concluding statements:

> Both the gender feminist and transgender movements are operating with good intentions: the desire to obtain the dignity women and transgender people rightly deserve. But it's never a good idea to dismiss scientific nuances in the name of a compelling argument or an honourable cause. We must allow science to speak for itself.

Arguments advancing diversity, equity, and inclusion should and must continue to be made loudly, particularly in the current era of social conservatism, and humanists will play a vital role in doing exactly this. These arguments need not, and should not, fly in the face of science. Working together with like-minded scientists, humanists can reframe the current arguments so that they are scientifically sound and thus advance social justice more credibly and effectively.

Not only humanists have neglected to consider the ways in which biological sex differences matter. 2017 saw the publication of a special issue of the *Journal of Neuroscience Research* on the topic titled "An Issue Whose Time Has Come: Sex/Gender Influences on Nervous System Function."[5] Eric M. Prager, the journal's Editor-in-Chief, opens his editor's column in the issue with this statement:

> Neuroscience today relies on the overwhelming belief that biological sex does not matter and can be safely ignored in preclinical research. Common practice within neuroscientific research is that findings in one sex (usually males) can be generalized to the other sex (usually females). Authors will even take the extreme approach of developing questionable methods to "prove" that sex differences are not present in the brain.

The journal, following the National Institutes of Health's mandate to include biological sex as a variable in studies, has developed a policy requiring its authors to do this as well, stating unequivocally that "sex fundamentally influences the brain" (Prager). The point is not to try to prove inherent superiority or inferiority of either the female or the male brain, but to ensure that studies are as beneficial as possible to all sexes. A neuroscientific study that bases its results on male brains and extrapolates those results to everyone may not be telling the whole story, thereby disadvantaging women greatly—or in extreme cases may even endanger them when it comes to health care. Advances in neuroimaging provide evidence supporting differing male and female brain structures, while at the same time affirming "the interaction of neurobiological sex and sociocultural gender is beyond simplicity" (Pavlova 237). Neurological conditions like autism seem linked to biological sex rather than gender, with males more frequently considered to be on the autism spectrum than females, but new research points to social and behavioral differences that result in girls more often being able to "mask" their autism.[6] Does that tendency have its basis primarily in the biological sex, or are the behaviors also socially influenced? Pavlova also cites work on "gender-specific susceptibility to negative information" (237) that point to differences in coping between women and men that seem to be influenced by culturally determined and reinforced gender conceptions.[7] But when it comes to gender-stereotyping research, which studies the effects of positive and negative gender stereotype messages on task performance, the "[b]rain mechanisms underlying gender stereotyping appear to be sex specific" (ibid.). Most importantly: "Sex-related differences in the brain do not always parallel behavior" (237). A recent large study suggested higher rates of autism, self-reported autistic traits, and other neurodevelopmental conditions in transgender and gender-diverse people than in cisgender individuals (Warrier et al., 2020), pointing to the need for enhanced support systems for these groups, while raising many questions about the roles played by both biological and social factors.

This kind of complexity that does not lend itself to an either-or position regarding nature and nurture is likewise emphasized by biopsychologist Nigel Barber in a 2016 article regarding the blank slate controversy. He points out that some personality traits are present at birth and remain across a person's development, and also that certain prenatal factors such as nutrition have been proven to have a significant effect on intelligence. At the same time, Barber notes that "the brain itself has blank-slate-like properties," concluding, "The brain may not be entirely blank at birth but it is not entirely programmed either. It is an interesting mix of script and improvisation" (Barber).

Where does all this leave humanists and social scientists, when even the neuroscientists still don't have the answers to the nature vs. nurture debate? This is precisely where the so-called softer science fields can contribute to a knowledge base in a way that complements "hard-scientific" advancements. Our intensive study of human cultural artifacts, careful considerations of literary production and reception, and deep understanding of subjectivity in relation to aesthetics comes ultimately *not* from abstract theorization but from observation of our students and ourselves as we read and discuss texts and their contexts—it is no exaggeration that these and other areas of humanistic expertise can help scientists continue to piece together exactly how the brain works and develops. But this level of contribution to the wider world of knowledge is only possible if we keep up with current science and ensure that our scholarship is consilient with rather than contradictory to it.

3 Methodology, Terminology, and Underlying Inspirations

Along with context, a number of interrelated concepts recur frequently throughout this book. A few of the most notable are *emergence* (the whole is greater than the sum of the parts), *complexity* (multiple factors are at work at the same time, so that we cannot reduce anything to a single underlying cause), and *agency* (human beings act with intentions and are neither biological automatons nor social constructs). In making the case for an alternate approach to literary and performance studies, we have undertaken certain specific strategies in the chapters that follow. Foremost among them is to present the work of dozens of linguists, biologists, psychologists, and others using two methods: summary and quotation. We have included some longer citations in the deliberate attempt to offer the reader a chance to hear voices that have generally not been heard in theory. The subtitle of this book is, after all, *Voices in Everything*, and to provide a comprehensive context, we want many voices, not only our own, to shine through.

Because contemporary linguistics, biology, and psychology are such large and complex fields, this book offers a good starting point as a primer about the most important principles of embodied cognition, so that literature and performance can be studied, interpreted, and discussed in accordance with contemporary scientific understanding. As humanistic scholars writing about evolution, the structure and function of the mind-brain, child development, artificial intelligence and robotics, and other similar subjects, we aim to present this material from the point of view of students and teachers of literature, performance, and film. Too often good writing on, say, the structure of the brain or human evolution by professionals in those fields sounds alien to the ears of many humanists. A technical or decontextualized aspect often makes them seem irrelevant or, at best, tangential to literary students and scholars. By making frequent comparisons with related concerns in literary theory and criticism, we attempt to make these chapters more relevant to the concerns of our colleagues and students in the humanities. Furthermore, since our primary concerns are literature and performance, we emphasize language: its role and relevant theories about its nature and functions.

By allowing a good sample of the major thinkers and scientists in a wide variety of fields to have a direct and largely unmediated voice in the chapters that follow, we hope to approximate in the pages of this book something like a Bakhtinian dialog. Rather than filter the unique and often idiosyncratic voices of many others through our own, more monologic, discourse, we let them speak for themselves. No voice inspires this volume more than that of Bakhtin, perhaps the most misunderstood and misappropriated major intellectual figure of the twentieth century. He is a central figure in this book, and he inspired its subtitle. Bakhtin's moving statement about voices in the very last essay he wrote suggests in a profound way the importance of letting those whom we want to introduce to literary scholars speak for themselves: "I hear *voices* in everything and dialogic relations among them" (1986, 169). Voices indicate embodied cognition and human context.

The alternative approach to the current theoretical paradigm for literary study we lay out in this book is situated in the vast interdisciplinary area comprised of post-Chomskyan linguistics and pragmatics, evolution, biology, and contextualist embodied cognitive science (which is non-Cartesian and not allied with the disembodied COGNITIVISM that compares the human brain to the computer). Throughout this book, we employ small caps for the terms COGNITIVISM (and COGNITIVIST) and THEORY (together with THEORIST and THEORETICAL) to refer not generally to all theory, but to that which implicitly or explicitly excludes the biological, the linguistic, and the psychological foundations of human experience, particularly the creation and enjoyment of literature and performance. We also write CONTEXTUALISM in small caps, but refrain from doing so with related words such as context, contextual, contextualize, and so forth. This usage calls attention to the concepts and theoretical approaches with which we take issue due to their incompatibility with contemporary science. It also calls attention to the alternative that we propose, which does not take one neat form, but diverges in many directions, with the common ground resting in conscious efforts to align with current scientific developments and to integrate the science and humanities whenever possible. Both realms are enriched as a result, and the humanists' voices are not lost in the growing sea of STEM researchers.

4 A Roadmap to Contextualism: Book Layout and Chapter Summaries

Part I, "Linguistics and the Legacy of Bakhtin's Philosophy of Language," deals with foundational theories of language in the twentieth century and consists of three chapters. Chapter 2 briefly reviews the important contributions to the study of linguistics by Ferdinand de Saussure, followed by a critique of these theories, particularly outlining the role of Russian philosopher and theorist Mikhail Bakhtin in pointing out the shortcomings of Saussure's work, namely its lack of regard for linguistic aspects such as change over time, syntax, and, most importantly for the purposes of this book, context.

Next, in Chapter 3, comes a schematic description and critique of some of the main currents in literary theory in the latter third of the twentieth century. In this third chapter, we set forth some of the foundations for documenting the profound conceptual shortcomings of contemporary THEORY. Previous critiques of the structuralist-semiotic-poststructuralist enterprise, we contend, have too often been superficial, needlessly heavy on jargon, and reliant—even if only implicitly—on the outdated linguistic theories of Saussure. This chapter also deals briefly with the major reorientation of the study of language, with its origin in Noam Chomsky's work in the 1950s and 1960s. Rejecting the structuralist and semiotic concept of language as a social entity, Chomsky grounds language in species biology and individual psychology. Normally, Chomsky is ignored by literary critics and theorists; when he is mentioned, it is most often in a distorted version of his earliest work, long since surpassed by him and others. Not infrequently, Chomsky appears merely as a footnote to Saussure, rather than recognizing him as Saussure's replacement. We assess Chomsky's major contributions to the study of language and review the work of two of the most influential successors to Chomsky who continue to develop his theories, namely Ray Jackendoff and Steven Pinker. The chapter ends with a brief note on the importance of Chomskyan and post-Chomskyan linguistics for literary scholars. We assert that if literary theory is to align with the contemporary scientific understanding of language, it must be anchored in modern biology and psychology.

Chapter 4 presents in more detail a contemporary of both Saussure and Chomsky: Mikhail Bakhtin, whose proto-pragmatism sets the scene for a consideration of the contemporary study of pragmatics—language in context. We have attempted to read the work of Bakhtin and his colleagues in a manner free of the Marxist, Christian, semiotic, and post-structuralist prejudices that have led many to misread and appropriate his stunningly original contributions to the study of language. We then turn to the rich pragmatic and cognitive trends in language study that most closely adhere to, derive from, and develop in parallel ways to Bakhtin's approaches. Particular emphasis is placed on the theory of relevance proposed by Dan Sperber and Deidre Wilson. These works provide a linguistic context for an approach to literature that owes nothing to Saussure, little to Chomsky (although much of it was made possible by the Chomskyan breakthrough), and very much (even when in ignorance of his work) to Bakhtin. It is this concept of language as a cognitive tool employed by embodied human agents that, as we suggest, is most useful to literary scholars.

Part II, "Biology, Language, and the Brain," deals with various facets of biology, the area of study that is generally least known and least understood by literary scholars, yet is most important to any contemporary discourse about things human. Thus, Chapter 5 starts from the premise of Darwin's theory of evolution by natural selection, a subject about which many humanists are considerably uninformed (as were we before undertaking this study). This chapter opens with an outline of Darwinian Theory and the modern synthesis that provides the reigning paradigm for study in evolution, including Darwin's theory of sexual selection, far less known than his theory of natural selection. We discuss some of the important work by major contemporary feminist biologists and women's role in human evolution. To round out the review of evolution, we end with a short sketch of human evolution. We then turn to the topic most relevant to scholars dealing with language and literature: the evolution of language. This is a conflicted area of research, thus here we attempt to draw together some disparate approaches and theories into a coherent account of how one species, *Homo sapiens*, acquired the capacity to speak, and, later, write and read, symbolically.

The evolution, structure, and function of the brain are the focus of Chapter 6. Literary scholars should know what the brain is and how it works, at least in general outline, to begin to understand how it processes language in both oral and written forms. We also look at emotion and sexuality as related to the brain. At the heart of Chapter 7 is a discussion of the process of Neural Darwinism, as proposed by Gerald Edelman. Edelman's theory of how the brain develops as a constant interplay between biology and environment is emblematic of one of this book's main themes: the meaninglessness of the traditional nature-nurture binary. The latest research on the impact of electronic media on brain development and function is also discussed in this chapter. After a consideration of the brain in older adults, Chapter 7 ends with a brief summary of some recent advances in neurolinguistics.

Part III, "Psychology and the Development of the 'Literary Mind,'" shifts the focus from biology to psychology and explores the concept of the mind-brain. The subject of Chapter 8 is the mind: the activity of the brain. Sometimes it is appropriate, or at least convenient, to refer to the brain, the physical structure, in opposition to the mind, the activity of the brain. Very often, however, we would do best to refer to these two interrelated and mutually defining concepts by a single term, such as the one we prefer: the mind-brain. First and foremost, we present an approach to consciousness and concepts of self, based on the latest relevant research from scientific fields as well as the humanities. Since ideas related to selfhood and self-identity are central and crucial concerns in literary studies, this section illuminates particularly the pertinence of the integrated, interdisciplinary approach that we are proposing. Equally important is the section on perceptual systems, including the fact that not all perception is mediated by language, contrary to what is assumed in most **THEORY**. The chapter proceeds to a consideration of sleep and dreams, as well as of the cognitive unconscious, and a review of current knowledge about memory, another major focal point of literary and cultural studies today. Next, we consider narrative thought, schema theory, and categorization, ending with what we, following Mark Turner, call the "literary mind," and the role of mental images and the creative imagination.

Chapter 9 provides an overview of some important issues in developmental psychology, beginning with the fact that the human mind is not, and by definition cannot be, a blank slate onto which ideology or language can inscribe itself. Recent advances in infant cognition and child development lead to conclusions that are radically at odds with the assumptions of psychoanalytic theory and other versions of developmental psychology that still inform much contemporary literary theory. Understanding how children's minds develop from the very first hours after birth as the embodied infant grows and learns is crucial to understanding adult subjectivity, identity, and sense of self. The last section of this chapter deals with language acquisition by the child and the role of language in cognitive development. In Chapter 10, we discuss what is known in cognitive psychology as Theory of Mind (ToM). This term refers to the way in which we infer what we are thinking and what we think others are thinking and why they are thinking it. An awareness of our evolved ToM enables us to understand our own cognitive processes better. We elaborate in this chapter on the major theoretical approaches to the subject, the "theory theory" and the "simulation theory," and review some common criticisms of both. The chapter ends with some observations on the role of ToM in language acquisition and comprehension, and how it relates intimately to literary and performance study. ToM is one of the richest areas that contemporary cognitive psychology opens up to literary and performance scholars, which is why we devote an entire chapter to it.

Part IV, "Context in Science and the Humanities," compares two major approaches to cognition: one, here called COGNITIVISM, posits a computer model for the mind, while the second, CONTEXTUALISM, emphasizes instead the embodied, contextualized character of cognitive processes. In Chapter 11, we describe the orientation toward computation and discuss the possibility of constructing genuinely intelligent machines and robots. We reject this mechanistic approach entirely as a framework for literary studies. Significantly, both the COGNITIVIST approach to strong artificial intelligence (brain = computer) and the cognitive assumptions that underlie much contemporary THEORY share the basic tenets of behaviorism, the approach to psychology that limits itself to empirical studies of external behavior and casts the mind as an inscrutable black box.

In Chapter 12, we explore at length the second of the theoretical approaches to cognition, CONTEXTUALISM, recommending it as the ideal framework for undertaking literary and performance studies. This chapter is the heart of this book, laying out why post-Chomskyan linguistics, cognitive science, biology, and evolutionary theory are the most appropriate contexts for literary research. We begin by describing an alternative to mechanism as an overarching worldview: the contextualist metaphor, as laid out brilliantly by psychologist Diane Gillespie. Just as Bakhtin realized that the reality of language lies in its organic and dynamic use in specific contexts, contextualist approaches similarly understand that the human context is crucial. The researchers referenced in this chapter recognize the necessity of taking context into consideration in all things human. First and foremost among these scholars are Humberto Maturana and Francisco Varela, two Chilean neurobiologists who developed the concept of *autopoiesis*, an organism's self-organization within—and its structural coupling with—its environment. This concept implies a rewriting of both evolutionary and social theory, and it is closely related to (and lays a foundation for) a series of related conceptual approaches to what is now called embodied cognition: dynamical systems theory, constructivism, ecological psychology, cultural psychology, and developmental psychology. Chapter 13 discusses evolutionary psychology (and its predecessor, sociobiology), which is the discipline that most explicitly unites biology and psychology and the discipline that is the most egregiously misunderstood (and most bitterly attacked) among the contemporary fields relevant to the contextualist framework.

Part V "Contextualism—Changing the Paradigm in Literary and Performance Studies for the Twenty-First Century" includes three summarizing and concluding chapters dealing with work compatible with the cognitive-biological framework outlined in preceding chapters. Chapters 14 and 15 offer overviews of selected contemporary published scholarly work that is grounded in a cognitive paradigm. There is no single cognitive approach to literary and performance studies: recent findings from modern linguistics, biology, and psychology collectively provide the beginnings of a paradigm, and many of these studies can be taken as models of various ways to approach the humanities more generally.

Chapter 16 recapitulates this book's major points, emphasizing that we are not providing a solitary method for analysis but rather an argument for a change in mindset that yields new and more meaningful ways to read, understand, discuss, write about, and theorize about literature, film, and performance, in a way that is compatible with the cognitive-biological framework outlined in the preceding chapters and supported by Bakhtinian dialogism. We begin this concluding chapter with a review of the various paths one can take with cognitive literary and performance studies, pointing out not only where these paths diverge but also where they converge, namely in the bridging approach of CONTEXTUALISM. Paradigm shifts always take time, but—particularly given the crisis that the humanities currently face in terms of funding support and credibility—the moment for such a major shift is now. We suggest it is time to stop talking and writing in terms of signification and information processing and to start listening to *voices*—the voices of embodied human beings in context.

Notes

1. An indication of this is that in recent years only a handful of papers at the Modern Language Association Convention directly have addressed poststructuralism, as a perusal of the convention program archives attests, and those that did have titles implying a post-poststructuralist standpoint, such as "Beyond Poststructuralism: Teaching Theory in the Digital Era" (William Stephen Davis, 2012), "What Comes after Poststructuralism? New Ecological Realisms in Contemporary Theory" (Monika Kaup, 2016), and "New Realisms

after Postmodernism and Poststructuralism" (session title, 2018). At the same time, a number of papers at the convention continue to (re)consider the thinkers of poststructuralism and deconstruction like Deleuze, Foucault, and Derrida, usually in an at least implicit assertion of the theorists' continued relevance. In a recent monograph, Johannes Angermuller (2015) interestingly reflects on the potentially useful interaction between the poststructuralism that shaped an "intellectual generation" and contemporary social theory. More importantly though, Angermuller poses a reasonable question that aligns with one of our stances in this book, namely that the application of theory for theory's sake is neither tenable nor justifiable methodology: "Is the phenomenon of poststructuralism not an example of a movement whose unity is an imaginary effect of its reception?" (20).

2. Exceptions include the journal *Derrida Today* and its related conference, which are still going strong in their attempts to maintain the relevance of deconstructionism. The journal's mission is stated as follows: "*Derrida Today* focuses on what Derrida's thought offers to contemporary debates about politics, society and global affairs. Controversies about power, violence, identity, globalisation, the resurgence of religion, economics and the role of critique all agitate public policy, media dialogue and academic debate. *Derrida Today* explores how Derridean thought and deconstruction make significant contributions to this debate, and reconsider the terms on which it takes place. *Derrida Today* invites papers that deal with the ongoing relevance of Derrida's work and deconstruction in general to contemporary issues; the way it reconfigures the academic and social protocols and languages by which such issues are defined and discussed, and innovative artistic practices that adopt a 'deconstructive' approach to how our contemporary situation can be represented" (http://www.euppublishing.com/loi/drt).
3. See for instance, Sara Ahmed's *The Cultural Politics of Emotion* (2004) and other work that focuses on affect and embodiment.
4. With his notion of dialogism, Bakhtin conceptualized human expression of all sorts as existing not in a decontextualized environment but always in dialog with past and (potential) future expressions: "... every extra-artistic prose discourse–in any of its forms, quotidian, rhetorical, scholarly–cannot fail to be oriented toward the 'already uttered,' the 'already known,' the 'common opinion' and so forth.

The dialogic orientation of discourse is a phenomenon that is, of course, a property of any discourse. It is the natural orientation of any living discourse. On all its various routes toward the object, in all its directions, the word encounters an alien word and cannot help encountering it in a living, tension-filled interaction" (Bakhtin 1981, 279).

5. Vol. 95, Issue 1–2, January/February 2017. Edited by Larry Cahill.
6. See, e.g., Dean et al. (2016) as cited in *ScienceDaily* (Leiden).
7. For example: "When diagnosed with breast cancer, women are faced with quite a lot of threatening information that may substantially hinder their cognitive abilities and decision making (e.g., during informed consent) and, eventually, result in gender-specific (and often suboptimal) coping with the disease (see, e.g., Sokolov et al. 2016)" (Pavlova 236–37).

References

Ahmed, Sara. 2004. *The Cultural Politics of Emotion*. London: Routledge.

Angermuller, Johannes. 2015. *Why There Is No Poststructuralism in France: The Making of an Intellectual Generation*. New York: Bloomsbury.

Bakhtin, Mikhail M. 1981. *The Dialogic Imagination: Four Essays*. Ed. Michael Holquist. Trans. Caryl Emerson and Michael Holquist. Austin: U of Texas P.

———. 1986. *Speech Genres and Other Late Essays*. Ed. Caryl Emerson and Michael Holquist. Trans. Vern W. McGee. Austin: U of Texas P.

Barber, Nigel. 2016. "The Blank Slate Controversy." *Psychology Today*, 21 May (Online).

Boyd, Brian. 2006. "Theory Is Dead—Like a Zombie." *Philosophy and Literature* 30: 289–98.

Dean, Michelle, Robin Harwood, and Connie Kasari. 2016. "The Art of Camouflage: Gender Differences in the Social Behaviors of Girls and Boys with Autism Spectrum Disorder." *Autism*, Special issue (29 November): 1–12.

Freedman, Daniel G. 1979. *Human Sociobiology: A Holistic Approach*. New York: Free P.

Gilligan, Carol. 1993 (1982). *In a Different Voice: Psychological Theory and Women's Development*. Cambridge: Harvard UP.

Leiden, Universiteit. 2017. "Girls Are Better at Masking Autism Than Boys." *ScienceDaily*, 3 April (Online).

Pavlova, Marina A. 2017. "Sex and Gender Affect the Social Brain: Beyond Simplicity." *Journal of Neuroscience Research* 95.1–2: 235–50.

Prager, Eric M. 2017. “Editor’s Column: Addressing Sex as a Biological Variable.” *Journal of Neuroscience Research* 95.1–2: 11.

Smail, Daniel Lord. 2008. *On Deep History and the Brain.* Berkeley: U of California P.

Soh, Debra W. 2017. “New Gender Theories Sound Good but Are at Odds with the Science.” *The Australian*, 22 April (Online).

Sokolov, Alexander N., Marina A. Pavlova, et al. 2016. “Negative Gender-Related Information Reduces Social Cognition in Breast Cancer Patients.” Program No. 464.20. 2016 Abstract Viewer/Itinerary Planner. Washington, DC: Society for Neuroscience.

Warrier, Varun, David M. Greenberg, Elizabeth Weir, Clara Buckingham, Paula Smith, Meng-Chuan Lai, Carrie Allison, and Simon Baron-Cohen. 2020. “Elevated Rates of Autism, Other Neurodevelopmental and Psychiatric Diagnoses, and Autistic Traits in Transgender and Gender-Diverse Individuals.” *Nature Communications* 11.1: 3959.

PART I

Linguistics and the Legacy of Bakhtin's Philosophy of Language

The meaning of a word is determined entirely by its context. In fact, there are as many meanings of a word as there are contexts of its usage.

—M. M. Bakhtin

Meaning is contextual and holistic, and no word exists alone.

—Johan Lehrer

CHAPTER 2

Saussurean Linguistics and Bakhtin's Critique

This means that communication must be contextualized. Context is not an optional extra.

—Roy Harris

Literary theory as practiced in the last third of the twentieth century and the first decades of the twenty-first century has been firmly grounded in the linguistic theories of Ferdinand de Saussure (1857–1913). Saussure lived the majority of his life and was intellectually formed in the second half of the nineteenth century; his contemporaries included Emile Durkheim and Sigmund Freud, with whom he had much in common. A consideration of the state of literary studies today must start with a review of Saussure's theories, which still lurk behind much THEORY. Thus, this chapter consists of three sections: a review of Saussure's theory of language; a critique of that theory; and a look at the critical assessment of Saussurean linguistics by M. M. Bakhtin and his collaborators.

1 General Linguistics

Saussure's famous 1907–1911 courses on general linguistics at the University of Geneva inspired a generation of disciples, some of whom gathered together their notes for publishing, as Saussure himself never

H. Mancing and J. Marston William, *Restoring the Human Context to Literary and Performance Studies*, Cognitive Studies in Literature and Performance,
https://doi.org/10.1007/978-3-030-89078-0_2

did, his linguistic theory as best they understood it. The *Cours de linguistique générale* (*Course in General Linguistics*) was first published in 1916, edited by Charles Bally and Albert Sechehaye. The structuralist and semiotic enterprises that arose directly out of Saussure's work had considerable influence on early- and mid-twentieth-century language study (and other fields, especially anthropology via Claude Lévy-Strauss and child development via Jean Piaget), and his influence continues to be felt in certain areas—above all, in literary theory and, to a lesser extent, in philosophy.

But Saussure's importance in the history of linguistics shifted fundamentally when Noam Chomsky appeared on the scene in the late 1950s. Since Chomsky, linguistics has moved rapidly in a series of directions (not all due directly to Chomsky, of course) in which Saussure's work plays virtually no role. Norman N. Holland made the point already thirty years ago that

> Chomsky's work rendered Saussure's linguistics, indeed much of post-Saussurean linguistics, obsolete. I am not claiming that Chomsky is right, only that Chomsky has proven that Saussure is wrong. Linguists who reject Chomsky claim to be going beyond Chomsky, or they cling to phrase-structure grammars. They are not turning back to Saussure. Saussure's views are not held, so far as I know, by modern linguists, only by literary critics and the occasional philosopher. (1992, 141)

Saussure is an important chapter in linguistic history but should not inform contemporary theoretical models. Granted, some "Neo-Saussurean" approaches have been put forth by linguists in recent years (see, e.g., Bouchard 2014 with his "sign theory of language" and Halmøy 2016 on the Norwegian nominal system), but have been challenged by others (see, e.g., Newmeyer 2017).

Above all, Saussure conceived of a language as a vast system, or structure, that exists as an independent and identifiable entity in society. It is this system—rather than the philology and historical study of the development of individual languages that had previously dominated nineteenth-century linguistic study—that is, for Saussure, the arena of the scientific study of linguistics (1986, 9). This system can be described as follows:

> Structuralism maintained that the meaning of a linguistic form is determined by the language system itself. The world out there and how people

> interact with it, how they perceive and conceptualize it, are, in the structuralist view, extra-linguistic factors which do not impinge on the language system itself. Of course, people use language to talk about, to interpret, and to manipulate the world, but language remains a self-contained system, with its own structure, its own constitutive principles, its own dynamics. Language, in a word, is autonomous. (Taylor 1995, 16)

It is this core concept of an autonomous language that best represents the discipline-changing importance of Saussure's contribution to linguistics. Thus, Saussure makes a distinction between two manifestations of language: *langue* and *parole*. *Langue* is the system—the set of fixed codes—that has an independent existence, is self-contained (8–11), and governs all actual speech events. These "speech-events" themselves—the actual utterances by specific individuals in specific contexts—to which Saussure gives the name *parole* (13–14) are but mere individual, ephemeral, manifestations of the system—"little more than the droppings of *langue*" in the sharp phrasing of Richard Harland (1993, 20, 24)—and merit little attention from the scientist who wants systematically to study language. *Langue* on the other hand, according to Saussure, is "not a function of the speaker. It is the product passively registered by the individual. It never requires premeditation, and reflexion enters into it only for the activity of classifying" (14). *Langue* is "social in its essence and independent of the individual" (19), and it is specifically this social product that linguists should study. The scientific study of language as a system is *synchronic*, dealing with the status, or structure, of a language during a specific period of time, rather than *diachronic*, dealing with language change through different epochs (Saussure 81; see 79–98). Thus, the Saussurean approach to linguistics is referred to and gives rise to a movement known as *structuralism*: It deals with the *structure* of the *system* of a language at a certain time.

To study these synchronic linguistic functions and structures, Saussure opted for the individual *word* as the basic unit of language (109). He devoted a considerable amount of attention to phonemes, the components of words, and some to sentences, of which words are components. Ultimately, somewhat reluctantly, he settled on the individual word as the primary unit of analysis. For Saussure, a "word" is best thought of as a "sign," consisting of the union of a concept (which he called a *signifié*, or *signified*) and a sound image (*signifiant*, or *signifier*) (65–67). Crucial for Saussure is the fact that these signs of language are totally arbitrary; that is, there is no motivational reason why, for example, *cane* (Italian), *cão* (Portuguese), *chien* (French), *собака* (Russian), *dog* (English), *Hund* (German), and *perro* (Spanish) should all refer to the same animal. Nothing in the thing itself necessitates the use of one sign over another, nor is there anything in the sign that links it causally to the thing (67–69).

For Saussure, all thought is linguistic in nature, and for him this is a self-evident, timeless, universally recognized truth: "In itself, thought is like a swirling cloud, where no shape is intrinsically determinate. No ideas are established in advance, and nothing is distinct, before the introduction of linguistic structure" (110). And if thought is linguistic, it is a function of that great, complex social system that is language: "A language is a system in which all the elements fit together, and in which the value of any one element depends on the simultaneous coexistence of all the others" (113). According to Saussure, words within this system convey *value* by the fact that they differ from other words.

In sum, then, *langue* consists of a very large number of arbitrary but fixed and culturally encoded signs, and the scientific study of language consists of the study of these signs. So important is the concept of *sign* to Saussure that he even goes so far as to conceive that in the future there might be a whole general "science" of signs, to which he gives the name *semiology* (15–17). It was in response to Saussure's call for a science of signs that what became known as *semiotics* developed, along with structuralism, into a subject of intense investigation, first in Eastern Europe in the 1920s and 1930s, then in France in the 1950s and 1960s, and finally throughout the world in subsequent decades.

The fact that structuralism and semiotics spread far beyond their Prague and Paris spheres of influence at about the same time that Chomsky appeared on the scene means that structuralism/semiotics gained widespread currency just when mainstream linguistics went away from the Saussure-inspired structuralist paradigm. It is no small irony that both movements spread concurrently in linguistics and literary studies. The divorce of literary study from linguistics has led to the current state of affairs in which literary theory is still at least implicitly informed by a concept of language that virtually no contemporary linguist finds useful, while linguistics proceeds without any perceptible influence on (or from) literary studies.

Saussure occupies a position of great importance in the history of the study of language. He unquestionably broke with the reigning nineteenth-century paradigms of the nature of linguistic study; the originality and influence of his thought are beyond question. What is in question, however, is the conceptual and functional validity of his linguistic theory. It is to that question that we now turn.

2 Critique

Crucial to Saussure's concept of human communication is what he calls the "speech circuit," which he illustrates graphically with a drawing of two heads connected by lines running from mouth to ear to brain. Saussure stated that his model consists of "an active part and a passive part, the former comprising everything which goes from the association centre of one individual to the ear of the other, and the latter comprising everything which goes from an individual's ear to his own association centre" (13).

Roy Harris uses the term *telementation* (or translation theory) to describe the supposedly psychological process (association) that underlies Saussure's speech circuit (1987, 205), a process that was a general assumption about language at least since the days of John Locke. Harris (213–18) suggests that Saussure may have incorporated this two-century-old model into his theory because it seemed consistent with some of the most recent technological advances in communication in the late nineteenth and early twentieth century, such as the telegraph and the telephone: "It can hardly be a coincidence that the illustration of A and B talking in the *Cours* shows them schematically linked by what look suspiciously like telephone wires" (215–16). This idea is at the heart of Saussure's understanding of communication; as Guy Cook recognizes, Saussure's "basic idea (confused and contradictory as it is) is of thought as something separate, within the individual, *encoded* and transmitted through language to the receiver, where it is *decoded*" (1994, 157).

Saussure makes no effort to explain what happens when the sound waves sent from the speaker's lips enter the listener's ear. They simply enter the black box of the human mind and somehow imprint themselves there as an exact replica of the speaker's original concept. Further, Saussure insists, listeners themselves have no role in this process; they are entirely passive. In effect, the speaker's *signifier* replicates itself in the listener's brain and becomes the *signified* for the listener. This (mysterious) process is called *signification*. Thus, according to this model, it is language—*langue* itself—that does all the work, not the human beings involved. And although *langue* is an external, social entity, in order to actually *do* things, Saussure's *langue* must have some sort of animate nature or agency. This is a behaviorist model of cognition in the most literal sense: The mind is a black box, the human is a passive subject, and the process is one of stimulus and response (Holland 1992, 130). Chomsky has described the approach of structural linguistics as one of "militant anti-psychologism," directly related to the especially narrow behaviorist brand of psychology, elaborating that "the term 'behavioral science' suggests a not-so-subtle shift of emphasis toward the evidence itself and away from the deeper underlying principles and abstract mental structures that might be illuminated by the evidence of the behavior. It is as if natural science were to be designated 'the science of meter readings'" (1972, 65).

So it is that *langue*, which exists without any relationship to the rest of reality and which cannot be influenced by anything or anyone, has the enormous power to *signify* and thus to make both human linguistic communication and subjectivity itself possible. Meaning comes into being as a function of language, not as a human psychological process. For Saussure, *langue* works because *signifiers signify*—because each signifier is different from all other signifiers. Nothing psychological or biological needs to be involved in this neat binary, mechanistic, depersonalized, and decontextualized function. In effect, what Saussure did was to psychologize the process of signification while eliminating psychology from the people involved (Holland 1992, 130–32).

Importantly, for Saussure *langue* was by no means an abstract or theoretical construct but a real entity in the world. As Harris notes, "for Saussure the claims of a subject to be a science depend essentially on whether or not what it studies really exists … If there is no such thing as *la langue*, then it will not do as the basis for any kind of science at all" (1987, 197). And there is no question that Saussure, like Marx, Freud, and Durkheim, was entirely dedicated to the *scientific* study of his subject: the first words of his book are, "The *science* which has grown up around linguistic facts are …" (1, emphasis added).

A critique of Saussure would be incomplete without recognizing the role of *binary oppositions* throughout his theory. It is no accident that everything of importance in the Saussurean construct comes in twos: *langue/parole*, signifier/signified, diachronic/synchronic, speaker/listener, active/passive, speech/writing, syntagmatic/paradigmatic, and so forth. The tendency to see everything in terms of binaries may be a universal human tendency (see Brown 1991,

90, 131–32), but is also understood to be a locally and culturally differentiated phenomenon (see, e.g., Spariosu 2004, 29).

Saussure, unfortunately, is committed by his theory to "a model of speech communication which depends essentially on being able to produce and recognize 'the same' linguistic items irrespective of the occasion" (Harris 1987, 233; see also Holland 1992, 122–23). This means that such crucial considerations as the context in which something is said cannot be a factor in this always-perfect process of understanding; that the listener's personal memories, values, concepts, or experiences cannot in any way influence what is understood; that gender, race, ethnicity, and class cannot be factors that figure in meaning and understanding. In effect, the process is a mechanical one in which the system itself functions and in which comprehension is guaranteed.

As we will see throughout this book, the comprehension of language is an active, complex process that cannot be reduced to a mechanical imprinting of meaning on the brain (see Holland 1992, 123–29, 200). Precisely those factors that Saussure eliminates from consideration—context, personal memory, experience, gender, race, socio-economic status, emotion, and so forth—are the things that *must* function in order for an individual to understand a word, a sentence, and an utterance. The hearer is not, never has been, and never could be, simply passive. The hearer (and the reader) is, always has been, and always must be, an active and creative participant in the process of understanding. It is not just that Saussure underestimated the hearer's role in a communicative situation; he basically misunderstood and therefore misrepresented that situation by reducing it to a simple, mechanical process, something that at best plays only a minimal role in linguistic communication.

Overall, the Saussurean concept of language that was so innovative and original in its day has not held up well; it fails both as a description of what a language is and how it functions. In short, as Leonard Jackson demonstrates at length, "there are quite fundamental weaknesses in the Saussurean model of language: weaknesses so severe that we can say, merely by inspecting that model, that it is logically impossible that it should provide an adequate theory of human languages" (1991, 5). Jackson refers to this fact as a "logical poverty." Saussure's concept of *langue* is, we now recognize, untenable: *langue* in the sense he conceives it does not exist, does not have the power to signify that he attributes to it, and cannot provide a legitimate foundation for a theory of language.

Saussure's speech circuit is no longer considered an acceptable explanatory model for how human beings communicate linguistically. Saussure's outmoded behaviorist psychology has been largely rejected and cast aside for nearly half a century now. Saussure's reliance on binary oppositions is too facile and mechanical to be employed so extensively. The sign, and its components the signifier and the signified, cannot function in the manner that Saussure describes.

This criticism of Saussure comes from the vantage point of more than a century of distance and decades of work in linguistics that is based on a concept of language that is radically different from Saussure's. Saussure cannot be held accountable for his lack of knowledge of our current more sophisticated appreciation of the neural and cognitive processes. Yet even on its own terms, Saussure's linguistics is inconsistent and flawed, and scholars of the first six decades of the twentieth century could and should have seen these flaws. At least one did: Mikhail M. Bakhtin.

3 An Early Critic: Bakhtin

In the 1920s, Bakhtin and a number of friends and colleagues formed a small, isolated intellectual community known as the Bakhtin Circle. In the latter half of the decade, they published four books of stunning originality: a sharp critique of Freudianism, a comparable critique of Russian Formalism, a book on language theory, and a book on Dostoevsky, which was the only volume published under Bakhtin's own name. Authorship of these books is a hotly disputed question in Bakhtin scholarship, with opinions running from one extreme, which attributes everything in all four books solely to Bakhtin, to the opposite extreme, which denies Bakhtin authorship of anything other than the Dostoevsky book. We take a middle-of-the-road stance and assume that Bakhtin, clearly the dominant intellectual figure in the Bakhtin Circle, where all of these works were discussed at great length, either wrote or was in basic sympathy with everything in the disputed texts that agrees with what he wrote in other works. The most relevant work here is the one on language theory, *Marxism and the Philosophy of Language* (1986, Russian original 1929, published under the name of V. N. Voloshinov), because it contains a perceptive, original, and still valid critical analysis of Saussurean linguistics. Stanley Aronowitz called this book "perhaps the most persuasive statement in our century of an alternative conception of language to that of Saussurian structuralism" (1994, 153). The view of language set forth in the Voloshinov book and the one assumed by Saussure are, simply, "irreconcilable" (Cook 1994, 151).

The book is divided into three sections. The first establishes a semiotic framework for the study of language, a framework that is largely ignored in the rest of the work and that is inconsistent with Bakhtin's ongoing criticism of semiotics. The substance of the second section is a critique of two "basic trends" in linguistics, which are labeled "individualistic subjectivism" and "abstract objectivism." The former refers primarily to the philological school of Karl Vossler and others and does not concern us here. The latter, the major target of the book's criticism, refers specifically to Saussure and his followers (see especially 52–82). The third section elaborates the original idea of language use in contextualized utterances, and it maintains a running critique of Saussurean abstract objectivism. The Bakhtin/Voloshinov approach criticizes the same shortcomings in the concept of *langue*, arguing that

- a synchronic, closed system of *langue* does not and cannot exist;
- the divorce of *langue* from *parole* orients language study toward a non-existent theoretical construct and away from the utterances that are the only real substance of language;
- abstract objectivism is a mechanistic, rationalistic, construct divorced from the complexity of social reality;
- Saussure's system cannot account for language change over time;
- Saussure completely ignores context, generativity, and ideology;
- understanding is never passive, but always active and participative; and
- the system ignores one of the most crucial aspects of language: syntax.

In other words, by the late 1920s virtually all the shortcomings of Saussure's approach were clear for those who wanted to see them. Bakhtin maintained this sort of criticism throughout his life. For example, in an essay written in 1952–1953, Bakhtin is even more specific in his criticism of the Saussurean speech circuit, referring to it as "a scientific fiction" (1986, 68). If Bakhtin and his colleagues were capable of perceiving these shortcomings in Saussure's theory in the 1920s, it was apparently not necessary to wait for more than half a century of subsequent research, the rise of Chomskyan linguistics, and the contributions of biology, cognitive science, and neuroscience to see that Saussure's linguistic theory was, from the start, highly problematic and debatable. That Bakhtin and his colleagues were virtually the only ones to see all this at such an early date is quite remarkable. Surprisingly, the practitioners of the two related trends that developed directly out of Saussure's work, structuralism and semiotics, showed little evidence of perceiving these errors, inconsistencies, and absurdities in their founding theory. And it is even more surprising that what grew out of structuralism and semiotics—that collection of interrelated orientations that we conveniently group under the rubric of poststructuralism and that still influences the "post-poststructuralist" theoretical world—is firmly rooted in a century-old, very inadequate linguistic theory.

References

Aronowitz, Stanley. 1994. *Dead Artists, Live Theories, and Other Cultural Problems*. New York: Routledge.

Bakhtin, Mikhail M. 1986. *Speech Genres and Other Late Essays*. Ed. Caryl Emerson and Michael Holquist. Trans. Vern W. McGee. Austin: U of Texas P.

Bouchard, Denis. 2014. *The Nature and Origin of Language*. Oxford: Oxford UP.

Brown, Donald E. 1991. *Human Universals*. New York: McGraw-Hill.

Chomsky, Noam. 1972 (1968). *Language and Mind*. New York: Harcourt Brace Jovanovich.

Cook, Guy. 1994. "Contradictory Voices: A Dialogue Between Russian and Western European Linguists." In *Literature and the New Interdisciplinarity: Poetics, Linguistics, History*. Ed. Roger D. Sell and Peter Verdonk. Amsterdam: Rodopi. 151–61.

Halmøy, Madeleine. 2016. *The Norwegian Nominal System: A Neo-Saussurean Perspective*. Berlin: de Gruyter.

Harland, Richard. 1993. *Beyond Superstructuralism: The Syntagmatic Side of Language*. London: Routledge.

Harris, Roy. 1987. *Reading Saussure: A Critical Commentary on the* Cours de linguistique générale. La Salle: Open Court.

Holland, Norman N. 1992. *The Critical I*. New York: Columbia UP.

Jackson, Leonard. 1991. *The Poverty of Structuralism: Literature and Structuralist Theory*. London: Longman.

Newmeyer, Frederick J. 2017. "Two Challenges for 'Neo-Sassurean' Approaches to Morphosyntax." In *Formal Models in the Study of Language*. Ed. Joanna Blochowiak, et al. Cham: Springer. 49–64.

Saussure, Ferdinand de. 1986 (1916). *Course in General Linguistics*. Ed. Charles Bally and Albert Sechehaye, with Albert Riedlinger. Trans. and annotated by Roy Harris. La Salle: Open Court.

Spariosu, Mihai. 2004. *Global Intelligence and Human Development: Toward an Ecology of Global Learning*. Cambridge: MIT P.

Taylor, John R. 1995 (1989). *Linguistic Categorization: Prototypes in Linguistic Theory*. Oxford: Clarendon P.

Voloshinov, V. N. 1986 (1929). *Marxism and the Philosophy of Language*. Trans. Ladislav Matejika and I. R. Titunik. Cambridge: Harvard UP.

CHAPTER 3

On Theory, Rewriting Saussure, and Chomsky

"Value" and "meaning" are both context-sensitive.
—James L. Battersby

Après Saussure, le déluge. The influence of Saussure in certain areas is so enormous that it cannot be addressed adequately in a mere chapter of a book: it deserves a book itself, and we are not writing that book. Dealing in a brief and schematic way with post-Saussurean trends, we will synthesize and generalize, with few excursions into the specific. Today's literary scholarship and cultural studies are predominantly informed by an amorphous blend of Saussurean and post-Saussurean (Derrida, de Man, Barthes) theories about language, usually in combination with Marxist thought (Althusser, Foucault) and psychoanalytic theory (Freud, Lacan). This scholarship often rests on linguistic, psychological, and biological assumptions that are unstated and taken for granted as fact. This chapter elaborates on the critique of Saussure and his followers, introduces generative linguistics and Noam Chomsky as a bridge to the contextualism we argue for as foundational for sound literary studies.

H. Mancing and J. Marston William, *Restoring the Human Context to Literary and Performance Studies*, Cognitive Studies in Literature and Performance,
https://doi.org/10.1007/978-3-030-89078-0_3

1 Theory

"Theory" could, in principle, refer to any kind of theoretical approach to literature. But for the THEORISTS discussed in this chapter, this is not the case. Jonathan Culler, for example, reserves the term "theory" exclusively for "works that succeed in challenging and reorienting thinking in fields other than those to which they ostensibly belong because their analyses of language or mind or history or culture offer novel and persuasive accounts of signification" (1987, 168). Thus, since *signification* is a necessary factor in theory, all THEORY must be Saussurean at heart. Leonard Jackson notes the "breathtaking arrogance" of elevating mere "literary theory" to a sort of metaphysical "theory" in general, adding that "[t]heory is the current metaphysics of cultural and literary criticism" (1991, 195). Even scientists such as Richard Dawkins have agreed, objecting to the way "[t]he very word 'theory' has been hijacked for some extremely narrow parochial literary purpose—as though Einstein didn't have theories; as though Darwin didn't have theories" (qtd. in Brockman 1995, 23).

Tony Hilfer writes at length about THEORY's ruthlessly "hegemonic" status in literature departments at the beginning of the twenty-first century, calling it "a monolithic ideological discourse oppressive to literary study" (2003, xiii). While our sense is that in most literature departments THEORY is not necessarily flaunted in the same way now, during the century's third decade in an allegedly post-theoretical age, its underlying hold can still be keenly felt in the field. For instance, the *Norton Anthology of Theory and Criticism* (2001 and later editions, most recently 2018) editor Vincent Leitch omits the word *Literary* from the title altogether. Commenting on this title, Brian Boyd notes that

> [w]ith the characteristic provincialism and hubris of recent literary theory, it claims 'Theory' as its empire, as if all theory were literary, as if the theories of gravity, evolution, and relativity were nugatory compared with the anti-foundationalist truths of the 'greatest generation' [i.e., the generation of theorists since the 1960s, according to Louis Menand]. (2006, 25)

Boundaries in the vast body of post-Saussurean work are fuzzy. Structuralism is not the same as semiotics, but the two are clearly intimately related. Deborah Cameron makes a helpful distinction:

> Semiology does not mean the same as structuralism. Semiology is a discipline—Saussure's 'science of signs'—while structuralism is a method you can use to do semiology. Neither semiology nor structuralism is equivalent to the study of language. Semiology studies any and all sign systems, and the structuralist method can be applied to other things besides language. (1992, 25)

Although the semiotics of the French school has dominated theoretical discourse in literary scholarship, there is an alternative. The American pragmatist Charles Sanders Peirce was developing his *semiotics* virtually contemporaneously with Saussure's *semiology*, but Peirce's theories have not led directly to a dominant position in literary theory. Indeed, Peirce is best known to a modest-sized cadre of devotees who have tried, mostly in vain, to make inroads into the French-dominated *semiology* of THEORY and (with somewhat more success) philosophy. Perhaps the single most important distinction between the Saussurean approach and the Peircean is that whereas the former is irrevocably tied to twos, the latter is equally fervently devoted to threes. Thus, for Saussure a *signifier* (1) *signifies* (2), while for Peirce *something* (1) *means something* (2) *to someone* (3). Note what is included here that is lacking in Saussure: *someone*, a human being.

However, Peirce is equivocal about this someone. The technical term he uses for this third element in the semiotic process is *interpretant*, to which he refers on the one hand as a person, a mind, or a consciousness, and on the other as nothing more than another sign. To the degree that Peirce involves human thought and understanding in the semiotic process, he approximates—in a way that Saussure and his followers cannot—developments in contemporary cognitive science. To the degree that he dehumanizes the interpretant and makes it just another sign in an infinite semiotic play of signs, he utterly divorces himself from that position. If, as Peirce has stated, the universe consists of nothing but signs, his own brand of semiotic imperialism dwarfs that of the French school, encompassing as it would not only all human activities, but everything that takes place in the entire universe. Peirce is more conceptually subtle, complicated, and intellectually appealing than Saussure, and is a more interesting figure (see Colapietro 1989; Merrell 1992). But, in literary theory, Peirce is at best a footnote, while Saussure is the primary text.

The trouble with semiotics is that when it is genuinely descriptive, it is often trivial: for example, red hexagonal signs at traffic intersections mean *stop* in some cultures. And when it is ambitious, it is false: everything in the world consists of signs and can only be known as signs. Early structuralist work, such as Roman Jakobson's analyses of literary texts (especially poetry), Claude Lévi-Strauss's many anthropological studies, and Vladimir Propp's morphological analysis of Russian folktales, all had a certain intellectual integrity to them and were properly influential. Even some examples from the late age of structuralism, such as A. J. Greimas's structural semantics with its "semiotic squares" and its "actantial" (i.e., involving actants, or agents) analysis of plot structure, are characterized by originality, honesty, and consistency. In many ways, they continue to represent the best pragmatic accomplishments of Saussure's theoretical descendants. The relative integrity and intellectual honesty of this foundational work, however, has been eclipsed by later developments in THEORY.

One of the tenets of THEORY is that language or culture at large determines what is inscribed in the blank slate of the passive human mind; we do not use language as a cognitive tool, but language uses, or "thinks," us. In THEORY, the socially constructed subject is (wholly and solely) determined by language and/or ideology. The exaggerated concept of language that is developed in THEORY emerges out of Saussure's fundamental principle that *langue* exists independently in the social sphere as a distinct entity that actively engenders signification. Language, it turns out, not only makes meaning possible, it creates meaning, and language is not only the most powerful force in society, it becomes the sole force: language *determines* reality. The word precedes the world, according to Lacan, and thus there exists something called the Symbolic Order, into which a child gains entry at a crucial stage in development.

The similarities between poststructuralism and behaviorism are impossible to ignore. In his book *Beyond Freedom and Dignity* (1971), behaviorist B. F. Skinner asserts as explicitly as any post-Saussurean ever could that "[t]he social environment is what is called a culture. It shapes and maintains the behavior of those who live in it" (143). Ultimately, for Skinner, "a person does not act upon the world, the world acts upon him" (211). The comfortable either/or way that Skinner's behaviorism works is consistent with Saussurean binaries, as is the passive nature of the person. The essential behaviorism that informs THEORY from Saussure to postmodernism and beyond is rarely acknowledged but is always present.

In poststructuralist thought, people are relegated to a caricature of humanity, the Enlightenment liberal humanist individual. The human being is dead, posits Foucault; the disembodied person becomes a subject constituted by the play of discourse systems. Personal relationships are illusory (as are persons themselves); all is ideology, power (or power/knowledge), interpellation, suturing, inscription by language. Authority depends on power, not any ability to convince logically, emotionally, or intellectually. The disembodied person becomes a subject constituted by the play of discourse systems, "a fiction and illusion *tout court* ... a mere node within self-governing technical semiotic systems" (Best and Kellner 1991, 284). The subject is, in effect, dead.

Emerging from Saussure's concept of *langue* as a social entity, linguistic determinism posits that language is externally imposed and frames language entirely as social construction. Since contemporary (non-semiotic) linguistic theory and practice reject the concept of *langue* as an entity, any philosophy or literary theory that relies on it is a shaky edifice indeed. Without Saussurean *langue*, there can be no linguistic determinism. The notion of linguistic determinism is so absurd that linguist and science fiction writer Suzette Haden Elgin, for example, calls it "science fiction" and expresses incredulity that anyone takes it seriously: "*No scholar (including Sapir and Whorf) has ever suggested—much less openly claimed—that any such hypothesis is valid*.... This idea *is* nonsense; no question about it" (1999, 52).[1] This "nonsense" is the bedrock on which much contemporary THEORY is constructed.

Another patch of shaky terrain underlying THEORY is the concept of *subjectivity*. The unified, autonomous self that purportedly characterizes liberal humanism is normally traced back to Enlightenment thought, when "man" (to mirror critically the reigning sexist terminology of that time) and his logical idealism were made a permanent part of Western metaphysics, though its roots go back to Plato and Aristotle. The crimes that THEORY pins on liberal humanism are many; it "has been accused of fostering or tolerating a tyranny of instrumental reason, a desacralization and exploitation of nature, a colonization by Europeans of the rest of the world, a patriarchal domination of women, and a class hegemony of the capitalist bourgeoisie" (Good 2001, 3; see also Leonard Jackson 2000, 182–83). Quite a record of accomplishment.

Poststructuralism, like Skinner's behaviorism, comes to bury the concept of a human being altogether in a clever and subversive reversal of the traditional hierarchical binaries (man/woman, logical/emotional, and so forth), replacing the liberal humanist individual with a radically decentered posthumanist subject. The putative distinction between the traditional (liberal, Enlightenment, male) individual and the linguistically constructed, passive, posthuman subject is central to THEORETICAL discourse. A subject is not a person (as is evidenced by the use of the pronoun "it" to refer to a "subject"), nor is the subject the human biological organism (which would entail recognition of biological reality, with all its complexity and subtlety) but rather a linguistically constructed entity unrelated to any sort of body. Daniel Chandler comments on this nonorganic subject as follows:

> In "theories of subjectivity" a distinction is made between "the subject" and "the individual." While the *individual* is an actual person, the *subject* is a set of *roles* constructed by dominant cultural and ideological values (e.g. in terms of class, age, gender and ethnicity). Ideology turns individuals into subjects. Subjects are not actual people but exist only in relation to interpretative practices and are constructed through the use of signs. (2002, 180)

This binarism begins with an abstracted concept of human beings that has never existed anywhere; in other words, it is a classic straw man. It is also a circular definition that leaves no room for any alternative. The liberal humanist/decentered posthumanist subject distinction rules out exactly the sort of embodied, contextualized, human agent studied in cognitive science and discussed throughout this book. As Jackson correctly notes, "[p]hrases like 'the decentering of the subject' have been used in literary theory as mantras, to be chanted for effect whether or not they make scientific or philosophical sense" (2000, 27). The subject seems to be nothing more than a linguistic pattern with no material existence at all, a mere figment of words (Joseph Carroll 1995, 418–24). Not, "I think, therefore I am," nor even "I speak, therefore I am," but more like "I am spoken, therefore I can be spoken of."

Much is made in THEORY of the "speaking subject." This term (first used by Saussure) is perhaps most associated with Julia Kristeva. Since she took so much of her theoretical structure from Bakhtin, and since Bakhtin uses the same term, it might be thought that Kristeva's usage is Bakhtinian. But for Bakhtin the term "speaking subject" never implies anything more than a person, a consciousness, a speaker, or an author (1986, 60–102, especially 71–76). For Bakhtin, the term does not ever imply anything even remotely like the poststructuralist linguistically determined subject. This, together with Kristeva's use of the term "intertextuality," is one of the best examples of how Bakhtin's thought can be betrayed and appropriated in attempts to make him into a semiotician, poststructuralist, or postmodernist *avant la lettre*. It is rather more consistent with Bakhtin's ideology to understand his dialogism as almost the exact opposite of Kristeva's intertextuality. The spirit of Bakhtin's work is in every way antithetical to the version presented by Kristeva. Where Kristeva sees texts in contact, Bakhtin hears voices in dialogue; for Bakhtin: "It is not works that come into contact, but people" (Bakhtin/Medvedev 1978, 152).

In his single most famous and most influential statement, Derrida has proclaimed, "There is nothing outside of the text"—or, "There is no outside-text," an alternate translation proposed by Gayatri Spivak for the French *il n'y a pas de hors-texte*—(1976, 158). There is, in other words, no context: everything is text. As Valentine Cunningham (1994, 24) reminds us, the French *hors-texte* (which has no direct English equivalent) usually refers to the introductory, prefatory, contextualizing material at the beginning of books, essentially what Lanser (1981, 122–31) calls "extrafictional" and Genette (1997) calls the "paratext." Derrida is largely jettisoning this textual context to free the text from any specific meaning (or authorial intention) that might logically be inferred by a reader.

But Derrida and his followers often claim that Derrida is, precisely, celebrating and emphasizing context; and it must be noted that at a later date, Derrida did apparently change his mind: The phrase *Il n'y a pas de hors-texte* "means nothing else: there is nothing outside context" (1988, 136). As the Derridian nemesis John Searle notes, "So the original preposterous thesis that there is nothing outside of texts is now converted into the banality that everything exists in some context or other" (1994, 665). But for Derrida, context is boundless, not recoverable, and ultimately just more text; as Culler sums it up, "Meaning is context-bound, but context is boundless" (1997, 67). In other words, Derrida pulls a con game here: text is always context—but context dissolves into nothing, or simply into more text.

The doctrine of textuality also includes the idea of *intertextuality*, according to which self-generated and exclusively self-referential relationships exist among texts, as texts confront, conflict with, rewrite, deconstruct, and otherwise have dealings with each other. Intertextuality is a concept that Julia Kristeva (1980, 66) popularized in modern THEORY based on her reading of Bakhtin. Bakhtin writes often of dialogic relationships through time, the utterance as a link in an endless chain of speech communication, one text's coming into contact with other texts, and so forth. He does not use the word "intertextuality," although that is the impression Kristeva gave her readers (1986, 37). Bakhtin never implies anything like what Kristeva and other THEORISTS mean by the term. As a result of Kristeva's implicitly attributing her concept to Bakhtin, many have erroneously come to believe that the term was originally his, a fact that Kristeva has recognized and attempted to clarify (Guberman 1996, 189). In Kristeva's defense: she was a pioneer who was initially familiar with just a portion of the Bakhtin corpus, especially the Rabelais and Dostoevsky books, so she did not have much perspective on Bakhtin's overall worldview, and naturally perceived everything through the lens of her own semiotic theory (see Morson and Emerson 1990, 4).

Karine Zbinden (1999) has shown that French translations of some of Bakhtin's major works very substantially misrepresent his thought, particularly the social dimension of key terms such as *heteroglossia* and *dialogism* that come across in these versions as reinforcing the structuralist/semiotic emphasis on textuality. Zbinden notes that in her preface to the French translation of the Dostoevsky book, Kristeva

> claims the legitimacy of contemporizing the works of past scholars, and specifically of Bakhtin.... This is how Bakhtin emerged as a semiotician before his time. Her stripping of "dialogism" down to "intertextuality" to address the preoccupations of her time rather than Bakhtin's shows just how misleading the notion of readability can be. (57)

See also Zbinden's (2006) extended study of the distortion of Bakhtin's work by Kristeva and Todorov. These early misreadings, then, create a Bakhtin—or a series of Bakhtins—that can be used to bolster readings and theories that are antithetical to the spirit of Bakhtin's actual writings.

The assumptions of linguistic determinism and textuality lead to the inevitable elimination of the author. Roland Barthes (1977) boldly proclaimed "the death of the author," which supposedly liberates texts from the yoke of the God-like creator-genius and frees texts to signify all alone. This "proclamation of the death of the author became the war cry of postmodernism" (Fokkema 1999, 39). Though Barthes affirms that a corollary of the death of the author is "the birth of the reader," this nascent reader's growth is stunted by linguistic determinism.[2] Michel

Foucault, asking rhetorically "What matter who speaks?" also posits the author is not a flesh-and-blood person but merely a "function" we conventionally ascribe to texts. This position is, in effect, "a somewhat pretentious statement of the distinction between narrator, implied author, and the flesh-and-blood author already very adequately made by Wayne Booth" (Harris 1996, 32). But it is also more. For Foucault, the concept and practice of authorship itself become just another socially constructed function.

THEORY maintains Saussure's concept of *langue* and its concomitant reliance on signs consisting of signifiers and signifieds that form a fixed language-wide code. Among other things, this oversimplified code model of cognition has led to a profoundly mistaken THEORETICAL approach to reading. Frank Smith, psycholinguist and cognitive scientist, has studied actual human reading processes at length. In his book *Understanding Reading*, he makes it clear from the outset that reading is not mere decoding: "Reading is a matter of making sense of written language rather than decoding print to sound" (1994, 2). Agency, again, becomes the issue: "Reading is seen as a creative and constructive activity having four distinctive and fundamental characteristics—it is *purposeful*, *selective*, *anticipatory*, and based on *comprehension*, all matters where the reader must clearly exercise control" (3). THEORISTS consistently attribute control, or agency, to language itself and deny it to the subject. The encoding/decoding approach is irreconcilably at odds with contemporary cognitive linguistics, psycholinguistics, and pragmatics, and fundamentally flawed in its assumptions about perception, particularly visual perception.

From the start, structuralists, inspired by Saussure, assumed that absolute dichotomies or binary oppositions are fundamental. By all accounts, binarism is essential to THEORY. The poststructuralist repeatedly sets up a straw man binary dilemma (e.g., the one regarding Enlightenment thought: man/woman, logic/emotion, etc.) and then knocks it down. So, the deconstructionist first identifies an A/B hierarchy and then reverses it into B/A, as well as all other possible combinations: A but not B, B but not A, both A and B, and neither A nor B. The deconstructionist then claims to have reversed, problematized, and swept away the patriarchal Enlightenment hierarchy. And this process is repeated, over and over: the strategy guarantees victory.

An introduction to the science of fuzzy thinking would also be in order; see Kosko (1993) as discussed in Sect. 5 of Chapter 8. One of the main problems with writing and theorizing in literary studies (and the humanities and many social sciences in general) is that it reduces complex reality to simplistic binary alternatives. Many THEORISTS decry the reductionism of science while, ironically, reducing complex matters to nothing more than simple binaries and arbitrary codes. Although Saussure is credited with much of the theoretical grounding of THEORY, he would find much of what has been written in his name is radically at odds with what he actually thought. Saussure was enshrined and fetishized, while substantially betrayed as his theory was extended into realms that he would never have considered permissible (see Jackson 1991; Harris 2003).

In THEORETICAL discourse, psychology is generally reduced to psychoanalytic theory. What happened in linguistics paralleled what took place in psychology. THEORY enshrined Saussurean linguistics at about the same time that linguistics moved beyond the Saussurean concept of linguistics to Chomskyan and post-Chomskyan approaches. Similarly, while psychology was moving beyond Freudian psychology and psychoanalysis, THEORY enshrined the psychoanalytic approach alone, often elevating Lacan above Freud. The *Norton Anthology of Theory and Criticism* (1st ed. 2001; 3rd ed. 2018), edited by Vincent Leitch, for example, lists, in its otherwise helpful "Alternative Table of Contents," fourteen essays, most by major names in THEORY, under the rubric of "psychoanalysis," (xxiv), but has no such rubric for psychological, cognitive, or evolutionary approaches. In his 2011 edition of *Literary Theory: A Very Short Introduction*, Culler makes it too short of an introduction indeed, leaving out reference to cognitive literary studies altogether and mentioning only New Historicism/Cultural Materialism and Postcolonial Theory as the important movements of the past few decades.

At the turn of this century, cognitive literary specialist Craig Hamilton provided a perfect illustration of THEORETICAL scholarship's disciplinary insularity:

> Of course, memory is a major topic today in cognitive science research although many cultural critics show no awareness of that research. Disciplinary blindness like this leads to things like the March 1999 issue of *PMLA*, where 30,000 literary critics in North America and beyond were treated to three essays allegedly on memory and literature. Fortunately, one author did cite a psychologist to acknowledge that relevant research on memory had been done before and in a field outside of literary studies. Unfortunately, the psychologist cited was Freud, which naively assumes

that no reliable research on memory has been done since the early 1900s. (2002, 19–20)

Today, as important work is being done in the area of cultural and collective memory and trauma regarding the Holocaust, 9/11, the COVID-19 pandemic, and other collective tragedies, humanities scholars must consider how their analyses would be enhanced and made immeasurably more credible if informed by the memory research of cognitive psychologists. THEORY grounded in Freudian or Lacanian concepts of the psyche does not reflect today's understanding of the human brain and the activities of the human mind (see Chapter 8 for more on this).

We live in a media-dominated world of virtuality where both the post-post-Enlightenment subject and all discourse in the world are not merely constructs but illusions, simulations, falsehoods—in short, nothing but fictions, such that reality is hardly discernible: what Marie-Laure Ryan (1997) described as the "doctrine of panfictionality" has been taken to the extreme since she made this statement. The Trump era saw the U.S. administration's presentation of so-called alternative facts to further its conservative agenda. A theoretical discourse that essentially ignores or denies contemporary scientific thought is not tenable in this environment and must be countered by a model that takes scientific inquiry seriously. National Public Radio has highlighted in recent years people who are outraged by mistruths in the political arena yet tend to readily believe "alternative facts," i.e., falsehoods, when it comes to proven science; this kind of ignorance must be brought to light and remedied; see Shtulman 2017 and Brumfiel 2021. Certainly, we must be wary and retain a critical eye: all science is not good science, and science done poorly can reinforce gender stereotypes, racism, and sexism. Humanists are trained to detect such bias, but this only works if we are paying attention to the scientific literature (see Saini 2017).

2 Rewriting Saussure

The efforts of Roman Jakobson and Emile Benveniste, among others, to modify aspects of Saussurean theory amounted to little more than refinements; they did very little to change the basic underlying concepts of *langue* and *parole*, signifier and signified, coding and decoding (the conduit model), synchrony and diachrony, system of differences, the speech circuit, and the social reality of *langue*. The later, more dramatic, attempt by Derrida (1976, 27–73) to offer a revolutionary, subversive, deconstructive, critique of Saussure prominently features the binary two-step of reversing the logocentric primacy of speaking over writing, along with a plethora of other hierarchical reversals; in other words, it is little more than further tinkering, leaving all the basic concepts firmly in place (see Carroll 1995, 407). Saussurean theory is so problematic in all of its major tenets that the only thing to do with it is, as Chomsky and subsequent linguists have indeed done, to start over with a whole new set of premises and move beyond it. While Chomsky has at times paid homage to Saussure and his ideas (e.g., the notion of "Saussurean arbitrariness," see Chomsky 2000, 120), this should not be taken as a turn back to Saussurian principles.

Many of the intellectual and political goals that originally inspired much THEORY—freedom, equality and equity, understanding, tolerance, ethics, social justice, and opposition to sexism, racism, homophobia—are laudable in every way. But the means toward these ends taken by many THEORISTS over the past three decades or so have been counterproductive and greatly undermined these crucial goals. Not all practitioners of contemporary THEORY are equally guilty of the more extreme positions we have ascribed to it here. Some serious and conscientious THEORISTS qualify and nuance their work in important ways. Not all buy wholesale into linguistic determinism and social constructionism, or dishonestly misrepresent or deny human psychology and biology. But the theoretical assumptions and intellectual positions we have described here form an important and largely unacknowledged part of the contemporary theoretical context.

In summary: the great edifice of late twentieth-century and early twenty-first-century THEORY rests on a foundation that is weak in at least three important ways. The first of these weaknesses is linguistic theory. The newly ascendant semioticians and poststructuralists—budding THEORISTS—insisted back in the 1970s and 1980s that it will not do to have a naïve and inadequate concept of language. By the very same token, we now turn the tables on the THEORISTS and charge that to maintain a nineteenth-century Saussure-based theory of language into the twenty-first century is intellectually untenable.

The second area in which THEORY is grossly inadequate is in its understanding of psychology. Humans are not linguistically constructed subjects, nothing but pawns of social construction and power/knowledge relationships, who are hailed and interpellated by ideology. We are not mere speaking subjects who consist only of words, nor are we *its*, as the THEORETICAL subject is usually called, but *hes* and *shes* and gender-neutral *zes* and *theys*. The subject of THEORY is as passive and mechanical as was the subject of strong Skinnerian behaviorism, a thoroughly discredited branch of psychological theory. Human beings are embodied and contextualized cognitive agents with unique brains and minds.

And, finally, contemporary THEORY is also inadequate with regard to biology, for there is no biology in THEORY. It is rare indeed even to find reference to the body—except as inscribed by language or socially constructed. One can only agree with N. Katherine Hayles, who has argued that "one belief from the present likely to stupefy future generations is the postmodern orthodoxy that the body is primarily, if not entirely, a linguistic and discursive construction" (1999, 192; see 192–207). But for the THEORIST, especially under the influence of Lacan's "implacably anti-biological" stance (Sarup 1993, 7), it is imperative to debiologize the subject to make a case for ideological, linguistic, and social construction. By failing to acknowledge the reality of evolution and recent advances in neuroscience, THEORISTS are out of touch with much of the most exciting and important intellectual work being done in the world today.

It is curious how little attention the body of critique of THEORY cited throughout this book—some very uneven and problematic themselves—has received. The almost complete lack of reference to it, even in books and essays that do not use the hegemonic poststructuralist paradigm, is a surprising gap that we seek to remedy here. Before we offer alternatives to the flaws in the THEORETICAL concepts criticized in this chapter, we present some alternate views of what language is and how it functions, examining how and when linguistics and the Saussurean tradition definitively parted ways as Noam Chomsky revolutionized the study of language.

3 Generative Linguistics

One very important difference between Noam Chomsky (1928–) and Ferdinand de Saussure is their reception history. Saussure's *Course*, published in 1916, was his first and last word on the subject. Saussure's theory was received posthumously, maintained, and elaborated only slightly, first in specific European circles, and only decades later much more broadly. The Saussure whose *Course* is criticized today is the same

Saussure who was so revered by his disciples; he has been, in effect, frozen in time: an icon, an idol, a fossil. Meanwhile, Chomsky's theory has from the start, during his lifetime and with him as an active participant, been subject to heated, sometimes virulent, debate and revision.

The debate surrounding Noam Chomsky has not been confined to a few concentrated groups of the faithful but has been a worldwide phenomenon almost from the very beginning. Chomsky's 1956 paper (not published) at the MIT Symposium, his *Syntactic Structures* (1957), and his review of Skinner (1959) hit the scene as the work of an unknown thirty-year old. His work was greeted with shock and criticism, as well as with delight and praise. Over the years, Chomsky responded to both his admirers and his critics, wrote more, developed his ideas, constantly modified his position, and today sounds like a very different person. Chomsky's early career and the earliest articulations of his theory from around 1955 to 1970 still represent for many what Chomsky is all about. The approach to language presented in these early years revolutionized the study of linguistics and relegated Saussure to little more than a name in the history of the discipline.

Syntactic Structures (1957) is a small and unassuming book, barely 100 pages long, with a heavy style and a great deal of off-putting symbolic annotation. The aim is "to construct a formalized general theory of linguistic structure and explore the foundations of such a theory" (5). Chomsky defines "language" as "a set (finite or infinite) of sentences, each finite in length and constructed out of a finite set of elements" (13). A "grammar" of a language L is "a device that generates all of the grammatical sequences of L and none of the ungrammatical ones" (13). Anticipating the objection that it is not possible to study syntax in the complete absence of semantics, Chomsky responds in advance that it can and indeed must be done, affirming that his theory is "completely formal and non-semantic" (93).

Chomsky emphasizes scientific rigor and objectivity, decontextualization, abstract analysis, and formal logic: "this theory shall be a completely formal discipline" (103). It is a structural approach in the most literal sense (Newmeyer 1996, 26–28): "formal" is the text's key term. Unlike Saussure, however, whose emphasis is on word meaning, Chomsky concentrates on generativity and syntax, elements almost completely absent from Saussure's *Course*. He illustrates the independence of syntax from semantics with the single most famous sentence he ever wrote (sometimes simply called "the Chomsky sentence"):

i. Colorless green ideas sleep furiously.
ii. Furiously sleep ideas green colorless. (1957, 15)

Then he comments as follows:

> It is fair to assume that neither sentence (1) nor (2) (nor indeed any part of these sentences) has ever occurred in an English discourse. Hence, in any statistical model for grammaticalness, these sentences will be ruled out on identical grounds as equally "remote" from English. Yet (1), though nonsensical, is grammatical, while (2) is not. (16)

Syntax matters.

In *Syntactic Structures,* Chomsky introduces the idea of phrase structure and his famous system of sentence diagram. His interest is in how the deep structure of phrases, that is, their underlying grammatical relationships, can be transformed by procedures governed by a complex set of rules to produce an infinite number of sentences from a finite number of grammatical forms. The actual form of a verbal expression, then, is the structure that we see or hear on the surface, which is itself the result of the transformations from the deeper, underlying structure. The quintessential Chomskyan terms thus stem from this early work: deep structure, surface structure, phrase structure, rule-governance, transformational grammar, and generative linguistics. The theoretical, abstract, context-free structure of grammatical sentences as discussed by Chomsky places this work at the COGNITIVISM or disembodied end of the cognitive science spectrum (see Chapter 11). The speaker of a language in this approach is a kind of sequential computer, which also explains much of the enthusiastic reception Chomsky received in 1956 by those in computer science, mathematics, and cybernetics.

What sometimes passes unnoticed in the midst of Chomsky's elaboration of his rigorously scientific and formal system is his reference to language as "an instrument or a tool" (1957, 103). The implications of this statement are particularly important in recognizing the difference between a Saussurean, structuralist, or semiotic concept of *langue*—language as an independently existing social system of codes and signs—and the contextualized, cognitive concept of language as a tool, something that people use for practical purposes. Although almost a throwaway line in *Syntactic Structures*, this idea of language as a cognitive tool became an important theme in Chomsky's work and is one of the most important differences between the Saussurean and the Chomskyan approaches to language.

Chomsky's (1959) review of B. F. Skinner's *Verbal Behavior* (1957)—"perhaps the most devastating review ever written" (Smith 1999, 97)—is a scathing critique both of the book's approach to language learning and production and of behaviorism in general. The review is a landmark in the demise of the behaviorist enterprise, in part because of the audacity of a young and still relatively unknown scholar contemptuously dismissing the validity of a theoretical approach taken by one of the world's most respected scholars. But it is also the first time Chomsky explicitly suggests a biological basis for language (1959, 43). He also suggests children may accomplish their remarkable feat of acquiring a language rapidly and with little formal instruction by means of a biological structure that "may be largely innate, or developed by early learning processes about which very little is yet known" (57 n48). Chomsky cites in a note the (at that time still unpublished) work of Eric Lenneberg, a close friend who was studying the biology of language at the same time Chomsky was developing his transformational grammar, and the relationship between them was almost certainly one of mutual influence. It seems likely that Lenneberg is the primary source for what was later to become (erroneously) known as Chomsky's "innateness hypothesis."

The fundamental problem of language acquisition by children inspired Chomsky to propose a universal biological inheritance for language. How can it be, he asks, that children everywhere and at all times so effortlessly acquire their native language? Saussure's explanation is that mere exposure to a culture's language causes words to accumulate in identical mental dictionaries for the encoding, sending, and decoding of thoughts with perfect mutual comprehension. Clearly, this is inadequate. In spite of a lack of specific training, children somehow can speak and understand their native language in what seems like an effortless way simply by being in the culture. This is the problem that Chomsky dubbed "the argument from poverty of the stimulus" (1980, 34), a cornerstone in his concept of what language is and how children acquire it.

To explain this phenomenon, Chomsky argues that language acquisition is "a normal biological function of humans" (1975, 10), an innate function of the complex, modular human brain: "Language acquisition seems much like the growth of organs generally; it is something that happens to a child, not that the child does" (2000, 7). Language is a part of the biology of human beings, and its acquisition must take place within a critical period of childhood that is a part of normal and biologically programmed physical development (see the discussion of the emergence of language in Sect. 3 of Chapter 9). But language growth and development also necessarily involve the context of each individual's experiences; children do not learn language in isolation from familial, social, and cultural contexts. These factors trigger and make possible language acquisition. Chomsky insists that language is a species-unique biological characteristic; no other animal, including the great apes who are capable of learning rudimentary sign systems, has a generative capacity in any way comparable to that of human beings: "Ultimately, the study of language is a part of human biology" (Chomsky 1980, 226). The human brain everywhere is structured in such a way that it naturally generates a culturally specific correct language: global structure enables local instantiation. Language develops naturally in the brain in the same way that vision does; it is part of each person's biological heritage and always works correctly when there has been sufficient post-natal experience. This sort of structure is obviously shared by everyone because competent language acquisition is a worldwide and infallible phenomenon; therefore, such a structure must be universal within the human species: it is a "universal grammar" (UG).

Probably no proposal ever made by Chomsky has caused more confusion or evoked more negative reaction than his idea of UG. Recall that "grammar" for Chomsky is a "device" that can generate grammatical sequences: it is not a set of rules for usage. UG, then, is a tool used for generative or creative production of language. Chomsky employs the term "generative grammar" to distinguish his concept from "descriptive grammars," which describe or portray language already generated, and from "prescriptive grammars," which provide rules (in the ordinary sense) for everyday usage. This grammar device could in theory be either mechanical or biological. If mechanical, it would be a computer programmed to produce such grammatical sequences. If biological, it would be a brain structure that would allow any member of the species to acquire the ability to produce grammatical sequences in that person's native language. Chomsky has at various times referred to such a biological structure variously as a "language organ," a "language module" or "language faculty" in the brain, and a "language acquisition device" (LAD), to describe the innate brain-mind structure/function/capability that makes the acquisition of a language possible. His inability to identify and consistently use a term that would accurately describe the biological phenomenon and not create confusion with other uses of the same terms explain much of the problem that other linguists, and readers in general, have had with Chomsky's proposals. Possible confusion aside, Chomsky was correct from the start that the study of language should explicitly recognize there is something unique in the human brain that makes language possible in the human animal. Chomsky argues this faculty, or potential, is innate—not any language itself. Language is indeed best conceived of as a branch of (cognitive) psychology and also a matter of biology. Any approach to language that fails to take the biology and psychology of language into consideration is highly problematic.

Chomsky made a further distinction between what he called "competence," which is a person's general knowledge of a language (explicit and implicit, conscious and nonconscious), and "performance," which is what a person does with that knowledge; that is, individual instances of language production. This binary evokes a natural comparison with Saussure's *langue* and *parole*, and Chomsky has acknowledged the similarity. But the two concepts are not at all the same: Saussure's *langue* is an entity that exists in society and that is capable of generating by itself the process of signification, while Chomsky's *competence* is what an individual person is, in theory, capable of doing. Competence is not a private language, but essentially what has been called an idiolect, the (implicit) rules an individual has that guide her actual linguistic performance. *Langue* is external to and independent of the individual, while competence is internal to and dependent on the individual. *Langue* is also exactly the same for all members of a language community or culture, while *competence* is unique to each individual (which does not rule out a great deal of commonality among individual competences of a language community or culture, despite individual differences in intelligence, background, and training). To equate Chomsky's competence to Saussure's *langue* is to misunderstand (or distort) Chomsky's position, effectively gutting his concept of language as a biological phenomenon and reinstating the Saussurean structuralist language-as-independent-agent (Smith 1999, 37–38; Hogan 2000, 286–89).

While he believes that his emphasis on abstraction and the formal mechanisms of languages constitute the proper arena for linguistic study, Chomsky allows that in the future it may perhaps be possible to study more systematically that which he has relegated to "the waste-bin of 'pragmatics'" (1980, 112). Chomsky has suggested that pragmatics is even more difficult to study meaningfully than "empirical" (i.e., his brand of generative) linguistics (2000, 70). Ultimately, both the study of competence (UG) and that of performance (pragmatics) need to be carried out in such a way that they mutually complement and support each other. In other words, Chomsky recognizes the legitimacy of performance studies, but personally prefers (and finds much more interesting) the study of competence (Smith 1999, 151–54).

Chomsky's ideal speaker-listener, in spite of its generative ability, is in many ways hardly different from Saussure's passive individual used by *langue*. Both models permit the study of certain kinds of ideal functions, but Chomsky's model claims a far more complex and basic universal function. Furthermore, Chomsky freely admits that his is a theoretical construct, while Saussure claims to be describing what happens in real life. The abstract model can help us comprehend better both the species-wide phenomenon that is language, as well as the individual human realities of language use in context.

Chomsky has made modifications, both major and minor, in his stance on key issues repeatedly throughout his career, redefining his work and his terminology in light of reactions, criticism, and his further development of his own ideas. It is generally accepted that by the early 1980s Chomsky was headed in an important new direction and that this approximate time period marks the single most important division in his long career. His series of lectures delivered at the GLOW (Generative Linguists in the Old Worlds) conference, held in 1979 at the Scuola Normale Superiore in Pisa, Italy, was the first statement of an important reorientation of his theories. Chomsky refers to the new theoretical direction in his work as "a very radical departure from the long history of the study of language, much more so than early generative grammar" (1991, 23; see also 1995, 5–6). In the 1980s and 1990s, Chomsky no longer spoke or wrote about generative grammar, phrase structure, or rule systems, but about government and binding, principles and parameters, core and periphery, E-language and I-language, and minimalism. His most recent book, *What Kind of Creatures Are We?* (2015) synthesizes much of his own theoretical development over about half a century, while providing the political context that characterizes much of his later writing.

Chomsky has continued to stress the fact that a generative grammar is not a "theory" (as has often been maintained by his critics) but an aspect of individual biology and psychology, something that exists in specific form within each human being. Language, in contrast, is a concept derived "at a higher level of abstraction from actual neural mechanisms"

(Chomsky 1981, 4). Whereas "grammars have to have a real existence … there is nothing in the real world corresponding to language" (1982, 107). Theories of language that take a structural (Saussure), descriptive (Bloomfield), or behaviorist (Skinner) view of language conceive of it as "a collection of actions, or utterances, or linguistic forms (words, sentences) paired with meanings, or as a system of linguistic forms or events" (Chomsky 1986, 19). That is, language is conceived as something that exists in society, an aspect of culture, objective and external to any individual. Chomsky calls this commonly accepted idea an "E-language," a language that is externalized and extensional, a construct "understood independently of the properties of the mind/brain … a real object of study" (20).

In opposition to E-language, Chomsky proposes that language should be conceived as internalist, intensional, and individual—an "I-language." According to this approach, UG is "the theory of human I-languages, a system of conditions deriving from the human biological endowment that identifies the I-languages that are humanly accessible under normal conditions"; it is "a characterization of these innate, biologically determined principles, which constitute one component of the human mind—the language faculty" (Chomsky 1986, 23, 24). When we study I-language we study "a real object, a human being, whose I-language happens to be integrated into performance systems that yield such actions as articulation, expression of beliefs and desires, referring, describing and so on" (1997, 120).

At best, reference to a traditional E-language concept such as "English" or "Spanish" is a shorthand way of referring to the cognitive model we abstract out of a set of I-languages. Saussure's *langue* is, according to Chomsky, something non-existent and therefore by definition cannot be an object of scientific study. For Chomsky, the study of I-language precedes and enables or makes possible more pragmatic language study. UG is, "at its core, a computational system that is rich and narrowly constrained in structure and rigid in its essential operations" (1986, 43). Presumably the language faculty matures with the individual throughout life, though at some point an I-language achieves a fairly steady state.

Although Chomsky initially presented his new approach in terms of *government* and *binding* (hence the name GB theory), he later rejected these terms to characterize UG in favor of *principles* and *parameters* (P&P). Chomsky moves here toward a much simpler concept than the large and growing collection of transformations involved in EST (extended standard theory), though this new P&P formulation is basically consistent with EST in most details. Universal grammar consists of a limited number of essential principles (rather than rules, as earlier versions had it), which are common to all grammars and which indeed describe the limits of possible human language. One such principle of UG is "structure-dependence," which means that all grammars work by dealing with structural units (such as verb phrases or noun phrases) and relationships rather than with individual words in sequence.

Parameters by contrast are not universal but exist as essentially pre-wired systems that are set to one of a limited number of positions as determined by an individual's experience within a specific culture. Mark Baker's *The Atoms of Language* (2001) is perhaps the best and most readable book-length study of parameters. The "atoms" metaphor is explicitly intended to evoke a parallel with chemistry: "Just as atoms gave chemistry a way of resolving its foundational paradoxes, so parameters give linguistics a way of resolving its foundational paradoxes of similarity and difference. Parameters are the atoms of linguistic diversity" (45). Baker defines parameter as "simply a choice point in the general recipe for human language;" further, "I-languages are recipes, and parameters are the few basic steps in those recipes where differences among languages can be created" (57). For Baker, words are the atoms of a Saussure-like E-language, while parameters are the atoms of a Chomskyan I-language. A Saussurean theorist like Derrida who takes words to be the atoms of language is led to "the bewildering and ever-shifting postmodernist view of the world, in which nothing has a lasting or general meaning" (204).

In the mid-1990s Chomsky made another bold move by taking the conceptual simplicity (as well as the technical complexity) of P&P to an even greater extreme with *The Minimalist Program* (1995), which is not a complete rejection of P&P but a substantial modification to it, "a picture of language that differs considerably from even its immediate precursors" (10). This is the most recent version of Chomskyan linguistics, as complicated as ever, still in the process of development, not always clear in its elements or its implications, and not yet generally well understood. Just as P&P swept away many of the generative rules of EST (extended standard theory) in favor of a more limited number of principles and parameters, the minimalist program jettisons even more, including the very fundamentals of the original Chomskyan concept: D-structure, S-structure, and phrase structure. Chomsky's *New Horizons in the Study of Language and Mind* (2000) is a consistent, sustained reflection and commentary on the concept of I-language and the minimalist program. Even with all of these sometimes radical revisions, the program begun by Chomsky in the 1950s remains alive and vital and continues to provide the basis for much—though not all—cutting-edge linguistic theory in the twenty-first century.

Chomsky's linguistic theories have enormous potential to place literary studies into a new perspective, but it is important to see exactly what the implications of this work are and what influence they have had. For this reason, we intend to continue to evaluate his work and show where it stands today for literature scholars.

4 Chomsky Between Cognitivism and Contextualism

On the one hand, Chomsky makes language a function of an individual's biological inheritance, the species-specific, innate aspects of the brain. On the other hand, he insists on the universal, logical structure of a language. It is hard to reconcile these two in a single concept of language: the one is individualistic, subject to genetics, environment, culture, and personality, while the other is universal, abstract, logical, and mathematical. Chomsky has recognized this dilemma. He claims—not modestly—that the generative enterprise he began in the 1950s achieved its greatest successes as it led "to the discovery of a vast range of empirical phenomena in widely varied languages and to forms of explanation that much exceeded what could be contemplated not many years ago" (1995, 29), and he compares these results with the great achievements of the physical sciences since the seventeenth century. But, at the same time, he recognizes that "language is a biological system, and biological systems typically are 'messy,' intricate, the result of evolutionary 'tinkering,' and shaped by accidental circumstances and by physical conditions that hold of complex systems with varied functions and elements" (29). It seems Chomsky wants to have it both ways: language is abstract, logical, and mathematical, but it is also generated by individual biological organisms in specific contexts. Chomsky's work on formal grammar, mathematical or logical formulations, abstraction, and universalism is oriented toward COGNITIVISM. His insistence on a species-specific biological inheritance, psychology, and I-language, by contrast, is more CONTEXTUALIST. But although he may want to occupy both wings of the cognitive science spectrum at the same time, even Noam Chomsky can't be in two places at once.

In our opinion, literary scholars will not necessarily find a detailed knowledge of the Chomskyan program useful: the concept of I-language is directly applicable to literary study, but principles and parameters, core language, and other arcane, sterile, and decontextualized concepts are not likely to help elucidate the stuff of stories. Nor is competence more central to literary study and theory than performance. Quite the contrary, the devil is in the details of the lexicon, historical and cultural anomalies, and personal idiosyncrasy. Those of us who study literature need to be generally aware of theoretical linguistics, but perhaps not much more than to the level we have described here. What is of more interest, however, is the area that Chomsky acknowledges as legitimate but chooses not to study himself: *performance*. Every literary text, especially a novel, is a performance, in Bakhtinian terms, an utterance—"the novel is a secondary (complex) utterance" (1986, 62)—by a situated author, read and understood dialogically by equally situated readers. The study of literature should be grounded in an understanding of the language that constitutes it, language that must be understood individually, dialogically, and contextually.

Linguistics is a large field, and the majority of practicing linguists in the world do not carry out a research program exactly within the Chomskyan tradition. Work in historical linguistics, sociolinguistics, phonetics, and many other areas do not necessarily depend explicitly on the generative paradigm, although in many of these cases this is the paradigm that is assumed, even when this is not explicit in the research. But Chomsky has not worked in a vacuum; generative linguistics is not a one-person project. Two of the most prominent linguists in the world today are Ray Jackendoff and Steven Pinker, both of whom are firmly located in the Chomskyan tradition. A brief glance at some of their work will illustrate the current state of generative linguistics today.

Ray Jackendoff remained faithful to Chomsky during the bitter dispute over generative semantics in the late 1960s and early 1970s, and he is also in his own right the author of a series of books that have established him as one of the most versatile leaders in the field today. In addition to books on specific aspects of semantics and phrase structure, Jackendoff is one of the most prominent contemporary linguists to explore the relationships between language and cognition. In *Semantics and Cognition* (1983), he states, "to study semantics of natural language *is* to study cognitive psychology" (3). His *Patterns in the Mind* (1994) is an attempt to apply the principle of Chomsky's UG to the notion of reality altogether as a human construction. It also offers a good summary of generative linguistics with an emphasis on the biological basis of human language.

In *The Architecture of the Language Faculty* (1997), Jackendoff became one of the first linguists to elaborate on Chomsky's most recent model, the minimalist program (see also Lasnik 1999). He also proposed some alternatives to certain aspects of Chomsky's work and "a model of generative grammar that generalizes features of several alternative, non-Chomskyan generative frameworks," leading to "a more direct relationship between the theory of grammar and the theory of lexical and grammatical processing" (Jackendoff 1999, 393). In his massive *Foundations of Language* (2002), not a book for beginners, Jackendoff continues to refine his revision of Chomsky's theories by taking issue with the original "syntactocentric" orientation: "The alternative to be pursued here is that *language comprises a number of independent combinatorial systems, which are aligned with each other by means of a collection of interface systems*. Syntax is among the combinatorial systems, but far from the only one" (111). Stressing that everything about language is more complicated and involved than has been recognized, Jackendoff proposes to reintroduce semantics from its Chomskyan exile as one of the factors in this complex set of relationships. Jackendoff's extensive work is a fine example of how the approach to linguistics that Chomsky began in the 1950s has continued to grow and develop. Jackendoff's view of linguistic structure is subtle, compelling, and satisfying, but it is hardly the last word in the ongoing dialogue of what language is, where it came from, and how we make use of this cognitive tool—or, as Jackendoff prefers, this cognitive toolkit.

If Jackendoff is one of the most loyal and consistent of the contemporary Chomskyans, Steven Pinker is the most visible and controversial. In addition to his earlier work on language acquisition and visual cognition, Pinker is the author of the single most widely read and discussed book on linguistics in recent decades: *The Language Instinct* (1994). Pinker's enthusiasm, his light and breezy style, his popular and colorful examples, and his self-assured tone have made the book accessible and palatable to non-specialists—and offensive to certain kinds of purists. Its reception in the scholarly and popular press has been mixed, but its influence is undeniable. This is a version of Chomskyan linguistics that is accessible to anyone, but its view of language extends well beyond Chomsky's work alone. *The Language Instinct* is an excellent starting place for the lay reader who wants to learn what the modern discipline of linguistics is all about. Pinker's many critics are especially critical of his central metaphor, language as instinct, attributing to him the equation of human language with spiders building webs and salmon swimming upstream to spawn in a strong or literal sense, rather than in the more metaphoric way he intends the comparison.

The Language Instinct is a good brief introduction to the basics of Chomskyan linguistics, particularly Pinker's chapter on "How Language Works" (83–125). He renders Chomsky's sometimes impenetrable vocabulary quite accessible, explaining generativity, finite-state grammar, phrase structure grammar, recursion, arguments, X-bar theory, specifiers, heads, principles and parameters, case, inflection, deep and surface structure, trace, and more. Pinker grounds his work firmly in the camp of evolutionary psychology, and his final chapter (404–30) is a good summary of that position. Rejecting traditional positions of reductionism and causality, Pinker recognizes the complexity of human thought and the embodied nature of cognition. His subsequent book, *How the Mind Works* (1997) continues his work in these areas, introducing many of the biological and cognitive issues we discuss here in our book. In *Words and Rules* (1999) Pinker argues that language consists fundamentally of two elements: a mental lexicon and a mental grammar, that is, words and rules.

Words and Rules also illustrates modern cognitive concepts of categorization, presenting such corollary concepts as fuzzy borders, family memberships, and prototypes (1999, 270–87). *The Blank Slate* (2002) is a compelling argument against those who would deny the biological reality of the human animal (see Sect. 1 in Chapter 9). In *The Stuff of Thought: Language as a Window into Human Nature* (2007), Pinker summarizes and presents again many of the concepts discussed in his earlier works, using language as a means to explore aspects of human cognition. In his 2018 *Enlightenment Now*, he diverges from considerations of language altogether in favor of an optimistic take on humanity's progress.

5 Chomskyan Literary Theory

While Saussure's concept of language is the foundation upon which virtually all late-twentieth-century literary theory has been based, Chomsky's concept of language has provided the foundation for no literary theoretical structure whatsoever. There have been occasional, specific works of literary criticism that employ a Chomskyan concept of language, and some of them have potential theoretical implications, but so far no coherent body of literary theory, no list of well-known theorists, no school of literary thought traces its origins back to Chomskyan linguistics. Patrick Colm Hogan has recognized the implications of this fact: "It is arguable that little current literary theory would survive, were Chomsky's arguments even taken into consideration" (2000, 287–88). As Thomas Pavel has noted, "Contemporary linguistics has gradually … shifted its attention from semiosis—the arbitrary link between meaning and sound—to language universals, innate grammars, and the links to cognitive psychology. This shift has failed to make itself felt in literary theory" (1986, 116).[3] Any literary theory consistent with Chomskyan linguistics must be anchored in modern biology and psychology, given that for Chomsky and his theoretical successors, language is a biological and psychological function. The central chapters of this volume offer an introduction to these fields for the humanist, but before moving on to biology and psychology, we will turn to the more nuanced work of Mikhail Bakhtin and the contextualist approaches of modern pragmatics.

Notes

1. Anthropologists Edward Sapir and Benjamin Lee Whorf famously proposed in 1929 that language constrained an individual's ability to perceive the world. However, this linguistic relativism is not the basis of THEORY. The Sapir-Whorf approach works from the inside out, while poststructuralist linguistic determinism works from the outside in.
2. See Roy Harris (2003) for a detailed assessment of how Barthes and other poststructuralist thinkers misinterpreted Saussure.
3. See also Linda Flower (1994), whose work on "cognitive rhetoric" is grounded thoroughly in the Bakhtin-Vygotsky tradition and yet makes essential use of contemporary approaches to language coming from the study of cognitive science.

References

Baker, Mark C. 2001. *The Atoms of Language*. New York: Basic Books.

Bakhtin, Mikhail M. 1986. *Speech Genres and Other Late Essays*. Ed. Caryl Emerson and Michael Holquist. Trans. Vern W. McGee. Austin: U of Texas P.

———, and P. N. Medvedev. 1978 (1928). *The Formal Method in Literary Scholarship: A Critical Introduction to Sociological Poetics*. Trans. Albert J. Wehrle. Baltimore: Johns Hopkins UP.

Barthes, Roland. 1977. "The Death of the Author." In *Image, Music, Text*. Trans. Stephen Heath. New York: Noonday P. 142–48.

Best, Steven, and Douglas Kellner. 1991. *Postmodern Theory: Critical Interrogations*. New York: Guilford P.

Boyd, Brian. 2006. "Getting It All Wrong: Bioculture Critiques Cultural Critique." *The American Scholar* 75.4: 18–30.

Brockman, John. 1995. *The Third Culture: Beyond the Scientific Revolution*. New York: Simon and Schuster.

Brumfiel, Geoff. 2021. "For Some Anti-Vaccine Advocates, Misinformation is Part of a Business." *National Public Radio*, 12 May (Online).

Cameron, Deborah. 1992 (1985). *Feminism and Linguistic Theory*. New York: St. Martin's P.

Carroll, Joseph. 1995. *Evolution and Literary Theory*. Columbia: U of Missouri P.

Chandler, Daniel. 2002. *Semiotics: The Basics*. London: Routledge.

Chomsky, Noam. 1957. *Syntactic Structures*. The Hague: Mouton.

———. 1959. Review of *Verbal Behavior* by B. F. Skinner. *Language* 35: 26–58.

———. 1975. *The Logical Structure of Linguistic Theory*. New York: Plenum P.

———. 1980. *Rules and Representations*. New York: Columbia UP.

———. 1981. *Lectures on Government and Binding*. Dordrecht: Foris.

———. 1982. *The Generative Enterprise: A Discussion with Riny Huybregts and Henk van Riemsdijk*. Dordrecht: Foris.

———. 1986. *Knowledge of Language: Its Nature, Origin, and Use*. New York: Praeger.

———. 1991. "Linguistics and Adjacent Fields: A Personal View." In *The Chomskyan Turn*. Ed. Asa Kasher. Oxford: Blackwell. 3–25.

———. 1995. *The Minimalist Program*. Cambridge: MIT P.

———. 1997. "Language from an Internalist Perspective." In *The Future of the Cognitive Revolution*. Ed. David Martel Johnson and Christina E. Erneling. New York: Oxford UP. 118–35.

———. 2000. *New Horizons in the Study of Language and Mind*. Foreword Neil Smith. Cambridge: Cambridge UP.

———. 2015. *What Kind of Creatures Are We?* New York: Columbia UP.

Colapietro, Vincent Michael. 1989. *Peirce's Approach to the Self: A Semiotic Perspective on Human Subjectivity*. Albany: State U of New York P.

Culler, Jonathan. 1987. "Poststructuralist Criticism." *Style* 21: 167–80.

———. 2011 (1997). *Literary Theory: A Very Short Introduction*. Oxford: Oxford UP.

Cunningham, Valentine. 1994. *In the Reading Gaol: Postmodernity, Texts, and History*. Oxford: Blackwell.

Derrida, Jacques. 1976 (1967). *Of Grammatology*. Trans. Gayatri Chakravorty Spivak. Baltimore: Johns Hopkins UP.

———. 1988. *Limited Inc*. Evanston: Northwestern UP.

Elgin, Suzette Haden. 1999. *The Language Imperative*. Cambridge: Perseus Books.

Flower, Linda. 1994. *The Construction of Negotiated Meaning: A Social Cognitive Theory of Writing*. Carbondale: Southern Illinois UP.

Fokkema, Aleid. 1999. "The Author: Postmodernism's Stock Character." In *The Author as Character: Representing Historical Writers in Western Literature*. Ed. Paul Franssen and Ton Hoenselaars. Madison: Fairleigh Dickinson UP. 39–51.

Genette, Gérard. 1997 (1987). *Paratexts: Thresholds of Interpretation*. Trans. Jane E. Lewin. Cambridge: Cambridge UP.

Good, Graham. 2001. *Humanism Betrayed. Theory, Ideology, and Culture in the Contemporary University*. Montreal and Kingston: McGill-Queen's UP.

Guberman, Ross Mitchell. 1996. *Julia Kristeva Interviews*. New York: Columbia UP.

Hamilton, Craig. 2002. "Conceptual Integration in Christine de Pizan's *City of Ladies*." In *Cognitive Stylistics: Language and Cognition in Text Analysis*. Ed. Elena Semino and Jonathan Culpeper. Amsterdam: John Benjamins. 1–22.

Harris, Roy. 2003. *Saussure and His Interpreters*. Edinburgh: Edinburgh UP.

Harris, Wendell V. 1996. *Literary Meaning: Reclaiming the Study of Literature*. New York: New York UP.

Hayles, N. Katherine. 1999. *How We Became Posthuman: Virtual Bodies in Cybernetics, Literature, and Informatics*. Chicago: U of Chicago P.

Hilfer, Tony. 2003. *The New Hegemony in Literary Studies: Contradictions in Theory*. Evanston: Northwestern UP.

Hogan, Patrick Colm. 2000. *Philosophical Approaches to the Study of Literature*. Gainesville: U of Florida P.

Jackendoff, Ray. 1983. *Semantics and Cognition*. Cambridge: MIT P.

———. 1994. *Patterns in the Mind: Language and Human Nature*. New York: Basic Books.

———. 1997. *The Architecture of the Language Faculty*. Cambridge: MIT P.

———. 1999. "Parallel Constraint-Based Generative Theories of Language." *Trends in Cognitive Sciences* 3.10: 393–400.

———. 2002. *Foundations of Language*. Oxford: Oxford UP.

Jackson, Leonard. 1991. *The Poverty of Structuralism: Literature and Structuralist Theory*. London: Longman.

———. 2000. *Literature, Psychoanalysis and the New Sciences of Mind*. Harlow: Longman.

Kosko, Bart. 1993. *Fuzzy Thinking: The New Science of Fuzzy Logic*. New York: Hyperion.

Kristeva, Julia. 1980. *Desire in Language: A Semiotic Approach to Literature and Art*. Ed. Leon S. Roudiez. Trans. Thomas Gora, Alice Jardine, and Leon S. Roudiez. New York: Columbia UP.

———. 1986. *The Kristeva Reader*. Ed. Toril Moi. New York: Columbia UP.

Lanser, Susan Sniader. 1981. *The Narrative Act*. Princeton: Princeton UP.

Lasnik, Howard, 1999. *Minimalist Analysis*. Hoboken: Wiley-Blackwell.

Leitch, Vincent B., ed. 2018 (2001). *The Norton Anthology of Theory and Criticism*. New York: W. W. Norton.

Merrell, Floyd. 1992. *Sign, Textuality, World*. Bloomington: Indiana UP.

Morson, Gary Saul, and Caryl Emerson. 1990. *Mikhail Bakhtin: Creation of a Prosaics*. Stanford: Stanford UP.

Newmeyer, Frederick J. 1996. *Generative Linguistics: A Historical Perspective*. London: Routledge.

Pavel, Thomas G. 1986. *Fictional Worlds*. Cambridge: Harvard UP.

Pinker, Steven. 1994. *The Language Instinct: How the Mind Creates Language*. New York: William Morrow.

———. 1997. *How the Mind Works*. New York: W. W. Norton.

———. 1999. *Words and Rules: The Ingredients of Language*. New York: Basic Books.

———. 2002. *The Blank Slate: The Modern Denial of Human Nature*. New York: Viking.

———. 2007. *The Stuff of Thought: Language as a Window into Human Nature*. New York: Viking.

———. 2018. *Enlightenment Now: The Case for Reason, Science, Humanism, and Progress*. New York: Viking.

Ryan, Marie-Laure. 1997. "Postmodernism and the Doctrine of Panfictionality." *Narrative* 5: 165–87.

Saini, Angela. 2017. *Inferior: How Science Got Women Wrong and the New Research That's Rewriting the Story*. Boston: Beacon Press.

Sarup, Madan. 1993. *An Introductory Guide to Post-Structuralism and Postmodernism*. Athens: U of Georgia P.

Searle, John R. 1994. "Literary Theory and Its Discontents." *New Literary History* 25: 637–67.

Shtulman, Andrew. 2017. "In Public Understanding of Science, Alternative Facts are the Norm." *National Public Radio*, 29 May (Online).

Skinner, B.F. 1957. *Verbal Behavior*. New York: Appleton-Century Crofts.

———. 1971. *Beyond Freedom and Dignity*. New York: Alfred A. Knopf.

Smith, Frank. 1994. *Understanding Reading: A Psycholinguistic Analysis of Reading and Learning to Read*. Hillsdale: Lawrence Erlbaum Associates.

Smith, Neil. 1999. *Chomsky: Ideas and Ideals*. Cambridge: Cambridge UP.

Zbinden, Karine. 1999. "Traducing Bakhtin and Missing Heteroglossia." *Dialogism* 2: 41–59.

———. 2006. *Bakhtin Between East and West: Cross-Cultural Transmission.* Oxford: Legenda.

CHAPTER 4

Bakhtin and His Echoes

> *For Bakhtin, everything is "pragmatics," and semantic and syntactic codes are really "context in rigor mortis."*
>
> —Gary Saul Morson

Russian philosopher Mikhail M. Bakhtin (1895–1975) is the third figure after Saussure and Chomsky to consider seriously when exploring theories of language. The overarching idea unifying the work of Bakhtin is his insistence on situating all study of the human sciences, specifically including language and literature, within the context of human social realities. Bakhtin never considered himself a linguist and he does not provide a consistent and coherent linguistic theory. Throughout his writings, however, he deals specifically with language use in context. In this chapter, we discuss the role of Bakhtin and his friends and colleagues of the so-called Bakhtin Circle, outline Bakhtin's concept of metalinguistics, and review his signature concepts of dialogism and context. We also introduce linguistic approaches—cognitive, integrational, and pragmatic—that echo Bakhtinian thought in their emphasis on language usage in context.

H. Mancing and J. Marston William, *Restoring the Human Context to Literary and Performance Studies*, Cognitive Studies in Literature and Performance,
https://doi.org/10.1007/978-3-030-89078-0_4

1 The Bakhtin Circle

Bakhtin is one of the greatest anomalies of twentieth-century intellectual life. His life is the stuff of legend: he suffered great physical and economic hardship, and endured political exile and obscurity, yet he outlived all of his friends and collaborators. His interests ranged from moral philosophy and esthetics to language, literature, and cultural studies. Much of his writing may have been published under others' names or not at all. His complex, original, and at times difficult work is so attractive and protean that everyone wants to claim him as one of theirs: "He has been described as structuralist and poststructuralist, Marxist and post-Marxist, speech act theorist, sociolinguist, liberal, pluralist, mystic, vitalist, Christian, and materialist" (Morson and Emerson 1990, 4). He fits in nowhere, yet everywhere.

The history of linguistics would have been different if Bakhtin's work on language and linguistics had been generally accessible when it was written in the 1920s and 1930s. Bakhtin's devastating criticism of Saussure and his "metalinguistics" could have established pragmatics as a viable approach to language decades before it came to the fore in the 1960s and 1970s. Bakhtin's work might have even laid the groundwork for a literary theory to compete with Saussurean-based theory, but it was not broadly available until near the end of the twentieth century (and some is still not available). But as the Bakhtinian paradigm lay in obscurity, Saussurean-based approaches to literary theory during the development of semiotics and poststructuralism became firmly established as theoretical orthodoxy. Because of his work's chronologically staggered reception, Bakhtin is less a contemporary or immediate successor to Saussure and predecessor to Chomsky than he is an (anachronistic) participant in post-Chomskyan linguistic theory. His work can be integrated seamlessly into contemporary pragmatics and cognitive science.

It is difficult to determine just what Bakhtin wrote on his own, what he might have written in collaboration with his friends and colleagues, and finally what those friends might have written on their own but that some attribute to Bakhtin. But we can assume that many topics concerning moral philosophy, esthetics, psychology, literary theory, linguistics, and much more were discussed in detail during the long nights of strong tea, cigarettes, and heated conversation in the so-called "Bakhtin Circle" in Nevel, Vitebsk, and Leningrad in the years between 1918 and 1930 (which included not only Pavel N. Medvedev and Valentin N. Voloshinov, but also Bakhtin's wife Elena Aleksandrovna, and others, such as M. I. Kagan, I. I. Kanaev, L. V. Pumpinansky, V. Z. Rugevich, I. I. Sollertinsky, M. I. Tubyansky, M. V. Yudina, B. V. Zalessky, and B. M. Zubakin). Anything that came out of this group was to some extent collaborative. Since Bakhtin was apparently the strongest intellect and most powerful personality of that group (Perlina 1983, 35), everything in these works not in direct contradiction to what Bakhtin has written elsewhere (such as much overt Marxist and semiotic theory) can be attributed to him, at least in spirit if not necessarily verbatim.

The chronology of composition differs considerably from the chronology of reception of the Bakhtin's works. The first works known outside of Russia were the books on Rabelais and Dostoevsky, rather than his early writings of the 1920s. Some of these early—and seminal—works were the very last ones published, appearing only in the 1990s. The essays published in the *Speech Genres* (1986) collection were written between 1930 and 1974, so not all were "late essays" as advertised. For many scholars, the essays in the 1981 *Dialogic Imagination* (perhaps along with the Rabelais and Dostoevsky books) represent the core of Bakhtin's major work. This is a grave misunderstanding both of his production—which after all extended from 1919 through 1974—and of his stance on several crucial issues. For our purposes, the works from the 1920s are of central importance.

Although semioticians such as V. V. Ivanov (1976; see also Titunik 1976) would claim Bakhtin as one of their number, the totality of his work argues strongly against such an effort (Morson and Emerson 1990, 103). The single most specific and direct statement about semiotics from Bakhtin is in "From Notes Made in 1970–71": "Semiotics deals primarily with the transmission of ready-made communication using a ready-made code. But in live speech, strictly speaking, communication is first created in the process of transmission, and there is, in essence, no code" (1986, 147). There is neither a semiotic conceptual framework nor any semiotic vocabulary in Bakhtin's work on moral philosophy and esthetics, Dostoevsky, Rabelais and carnival, the theory of the novel, or dialogism. The entire spirit of the Bakhtin Circle's total corpus of writings is antithetical to semiotic thought, an "*anti-linguistics*, a systematic questioning and inverting of the basic premises and arguments of traditional linguistic theory" (Stewart 1986).

The first section of *Marxism and the Philosophy of Language* (Voloshinov 1986) contains a substantial semiotic framework, but even that is not sustained throughout the book: the first section of about 30 pages contains 267 instances of the words *sign*, *semiotic*, and their variants (nearly nine per page on average), while in the rest (about 100 pages) the same words appear about 40 times (less than one every three pages on average). No argument in the remainder of the book rests on semiotic concepts. Critics argue about whether the Marxism of the Voloshinov and Medvedev books represents an authentic conviction or a superficial appliqué, in spite of Bakhtin's emphatic denial of any adherence to Marxist thought (Bocharov 1994, 1016; Rzhevsky 1994, 433). While the sincerity of the Marxism is difficult to determine, the clear fact remains that the semiotic introduction to *Marxism and the Philosophy of Language* is the most easily detachable, irrelevant section of any of these books.

Notably, many readers of Bakhtin, especially those who attribute a Marxist stance to his work, blend the concepts of *dialogue* and *dialectic*. But Bakhtin distinguishes very clearly between them: "Dialogue and dialectics. Take a dialogue and remove the voices (the partitioning of voices), remove the intonations (emotional and individualizing ones), carve out abstract concepts and judgments from living words and responses, cram everything into one abstract consciousness—and that's how you get dialectics" (1986, 147). Overall, the evidence seems to support the conclusion that Bakhtin is not a Marxist.

2 Metalinguistics

Bakhtin emphasizes the social nature of discourse and the essential importance of context in a seminal early essay called "Discourse in Life and Discourse in Art" (1926, published under Voloshinov's name): "Discourse, taken broadly as a phenomenon of cultural intercourse, ceases to be a self-sufficient thing and cannot be understood independently of the social situation which engendered it" (Voloshinov 1983, 8). This approach stands in stark contrast to formalist and structuralist views that presuppose independence of context in discourse signification. Rather, Bakhtin insists, "[d]iscourse in life is obviously not self-sufficient. It arises from the non-verbal real-life situation and maintains a very intimate connection with it. Moreover, discourse is directly filled with that life and may not be detached from it without losing its sense" (10). Clark and Holquist call attention to the "great differences between the essentially mechanistic or mentalist impulse of Structuralism and the more organic predilection of Bakhtin" (1984, 7), and Karina Zbinden uses the term "sociality" for the emphasis Bakhtin places on the social nature of discourse (2006, 54).

Bakhtin illustrates the validity of his insistence on context with a classic example that is worth reproducing in its entirety because of its centrality to Bakhtin's conception of what language is and how it functions:

> A couple are sitting in a room. They are silent. One says, "Well!" The other says nothing in reply. For us who were not present in the room at the time of the exchange, this "conversation" is completely inexplicable. Taken in isolation the utterance "well" is void and quite meaningless. Nevertheless the couple's peculiar exchange, consisting of only one word, though one to be sure which is expressively inflected, is full of meaning and significance and quite complete.... However much we fiddle with the purely verbal part of the utterance, however finely we define the phonetic, morphological and semantic features of the word "well," we will not be a step closer to understanding the integral sense of the exchange.
>
> Let us suppose that we also know the intonation with which our word was articulated, and that it was indignantly reproachful, but softened with a touch of humour. Though this somewhat fills out for us the semantic void of the adverb "well," it does not reveal the meaning of the whole.
>
> So what are we missing? That "non-verbal context" in which the word "well" sounded intelligibly for the listener. This non-verbal context of the utterance is formed out of three factors: 1) a spatial purview common to the speakers (the unity of what is visible—the room, the window and so on), 2) the couple's common knowledge and understanding of the circumstances, and finally 3) their common evaluation of these circumstances.
>
> At the moment of the exchange both individuals glanced at the window and saw that it was snowing. Both knew that it was already May and long since time for spring, and finally, that they both were sick of the protracted winter. Both were waiting for spring and were annoyed by the late snowfall. The utterance depends directly on all this—on what was "visible to both" (the snowflakes beyond the window), what was "known to both" (the date was May) and what was "similarly evaluated" (boredom with winter, longing for spring); and all this was grasped in the actual meaning of the utterance, all this soaked into it yet remained verbally unmarked, unuttered. The snowflakes stay beyond the window, the date on a page of the calendar, the evaluation in the mind of the speaker, but all this is implied in the word "well".
>
> Now, once we have been introduced to what was "implied," that is, to the common spatial and semantic purview of the speakers, we understood

perfectly the integral meaning of the utterance "well," and its intonation too. (Voloshinov 1983, 10–11)

A real-life utterance, Bakhtin maintains, "always binds the participants of the situation together, as *co-participants*, who know, understand and evaluate the situation in the same way" (11). The non-verbal, contextual aspect of the situation is an essential part of the utterance. An utterance in real life, therefore, does not consist solely of the words (or signs), but instead is made up of two parts: "1) the verbally realized (or actualized) part, and 2) what is implied" (12). Bakhtin insists that "[t]he peculiarity of real-life utterances is that they are intertwined by a thousand threads into the non-verbal real-life context, and, when separated from it, almost entirely lose their meaning. Not to know their immediate real-life context means not to understand them" (12).

The real-life, social, contextual meaning of an utterance always involves evaluation, whose "pure expression is found in *intonation*. Intonation establishes an intimate connection between discourse and the non-verbal context. Living intonation, as it were, leads discourse beyond its verbal limits" (13; see also Bakhtin 1993, 32–33). Every utterance is concrete and once-occurrent; it expresses the subjectivity of the speaker, that individual's history, social reality, values, ideology, and point of view (see Bakhtin/Medvedev 1978, 119–26; Bakhtin 1990, 292). For Bakhtin, no utterance is ever separable from its speaker (or, for that matter, from its listener). Foucault could not be more profoundly mistaken when he rhetorically asks, "What matter who speaks?" For Bakhtin, *who speaks*—the individual—is precisely one of the things that matters most, along with *under what circumstances*—the context (1981, 340–41).

The stunning originality and modernity of Bakhtin's concept of language at this early point in time cannot be overstated. The word as a social event, the ideology of the speaker and listener, intonation, and, above all, the real-life context of an utterance—these elements are completely absent from traditional linguistic approaches, but they are at the heart of modern pragmatics. It would take nearly half a century before other linguists, namely H. P. Grice and other early pragmatists, would even begin to approach such a sophisticated concept of the social function of language in context.

In *Marxism and the Philosophy Language*, Bakhtin consistently rejects a concept of language as "a system of normatively identical forms" in favor of a focus on "the particular, concrete utterance" made by a speaker

(Voloshinov, 1986, 67). For the situated speaker, what matters does not lie in "the identity of the form but in that new and concrete meaning it acquires in the particular context" (68). The significant factor in linguistic form is its "specific variability" (69), and linguistic form "exists for the speaker only in the context of specific utterances, exists, consequently, only in a specific ideological context" (70).

All verbal intercourse involves two or more human beings in a historical and cultural context. The speaker's words are addressed not to a passive listener (or theoretical construct) but toward a flesh-and-blood addressee: "The *word is oriented toward an addressee*, toward *who* that addressee might be ... There can be no such thing as an abstract addressee" (85). The addressee always plays an active role as language is a "two-sided act": "A word is territory shared by both addresser and addressee, by the speaker and his interlocutor" (86).

Precisely in such cases, Bakhtin seems more at home in late twentieth-century and early twenty-first century pragmatics than anywhere else. His work is also in harmony with today's best approaches to embodied cognition, such as the position taken by philosopher Evan Thompson, who has done groundbreaking work in empathy studies, sounding like a modern Bakhtin when he writes of the "dialogical dynamic" that "emerges from and reciprocally shapes the nonlinear coupling of oneself and another in perception and action, emotion and imagination, and gesture and speech" (2007, 402).

Anticipating the philosophy of dialogism, Bakhtin at this early point defined dialogue in its broadest sense: "not only direct, face-to-face, vocalized verbal communication between persons, but also verbal communication of any type whatsoever" (Voloshinov 1986, 95). A book is thus just as dialogic as a face-to-face conversation: "the printed verbal performance engages, as it were, in ideological colloquy of large scale; it responds to something, objects to something, affirms something, anticipates possible responses and objections, seeks support, and so on" (95). Further, Bakhtin insists, "[*a*]*ny utterance*, no matter how weighty and complete in and of itself, *is only a moment in the continuous process of verbal communication*" (95). The act of understanding an utterance is never passive, as in Saussurean linguistics, but active and participative.

Bakhtin maintains that the truly meaningful unit of discourse in human life is the utterance. Linguistic approaches concentrating on language at the level of the word and sentence is insufficient for studying the utterance. To approach the "*social event of verbal interaction implemented in an utterance or utterances*" (94), Bakhtin coined a concept more inclusive than traditional linguistics: *metalinguistics* (sometimes translated as *translinguistics*). The term was introduced in the 1929 version of his book on Dostoevsky (Bakhtin 1984, 185) and included the phenomena of stylization, parody, *skaz*, and dialogue. The primary characteristic of these phenomena is that they are "directed both toward the referential object of speech, as in ordinary discourse, and toward *another's discourse*, toward *someone else's speech*" (185). Traditionally, linguistic inquiry was limited to simple and direct (monologic, in Bakhtin's terms) discourse and disregarded this more complex, but very common, type of discourse. Bakhtin did not develop this concept much beyond this brief introduction in the original version of the Dostoevsky book, but in the revised and expanded 1963 version, he elaborated further on the nature of metalinguistics.

Bakhtin's aim is to study the complex *discourse* of Dostoevsky's novels, or "language in its concrete living totality, and not language as the specific object of linguistics, something arrived at through a completely legitimate and necessary abstraction from various aspects of the concrete life of the word" (181). Therefore, his analysis is not linguistics, but metalinguistics, which is "the study of those aspects in the life of the word, not yet shaped into separate and specific disciplines, that exceed—and completely legitimately—the boundaries of linguistics" (181). Metalinguistics is essentially the study of language in context: it is pragmatics in the full, modern sense of that term—and Bakhtin and his colleagues were already thinking in these terms in the 1920s.

3 Dialogism

Dialogism is the term most frequently associated with Bakhtin and it is a theme throughout his writings. As opposed to the depersonalized, logical, and mechanical relationships among language units in structuralism, Bakhtin's theory personalizes everything: "But I hear *voices* in everything and dialogic relations among them" (Bakhtin 1986, 169). A Bakhtinian approach to literature, rather than focusing on structural relationships, signs, or gaps and traces, would seek to hear voices, to engage in dialogue—dialogue not with texts, but with authors, characters, and other readers. This is consistent with an approach to literature grounded in the cognitive sciences.

While dialogism is characteristic of novelistic discourse in particular, for Bakhtin all discourse is ultimately dialogic: the word is "born in a dialogue as a living rejoinder within it; the word is shaped in dialogic interaction with an alien word that is already in the object" (1981, 279). The double-voiced discourse that particularly characterizes the novel is also found throughout extranovelistic, real-life, discourse. "Dialogic relationships … are the subject of metalinguistics" (182); they exist among all words or utterances that are juxtaposed, and they exist among past and future uses of a word or phrase. Linguistics is incapable of dealing with dialogic relationships (182–85). Bakhtin often writes of a word, text, or utterance as a "link in the chain of speech communication."

Bakhtin anticipates Chomsky in specifically rejecting a Saussurean E-language and proposing the idea of an I-language:

> If we claim that language as a system of incontestable and immutable norms exists objectively, we commit a gross error. But if we claim that language, with respect to the individual consciousness, is a system of immutable norms, that such is the mode of existence of language for each member of any given language community, then what we are expressing in these terms is a completely objective relationship. (Voloshinov 1986, 66–67)

Even the most monologic utterance necessarily has dialogic overtones. Bakhtin, drawing on the tradition of Wilhelm Dilthey, distinguishes between the natural sciences (*Naturwissenschaften*), in which a speaking subject deals with a mute object, and the human sciences (*Geisteswissenschaften*), which always involve "a special kind of dialogue: the complex interrelations between the *text* (the object of study and reflection) and the created, framing *context* (questioning, refuting, and so forth) in which the scholar's cognizing and evaluating thought takes place" (Bakhtin 1986, 106–07). In deconstruction, everything is text; there is no context. In Bakhtin, by contrast, everything depends on the contextualized text, on the relationships between text and context.

Bakhtin never sees meaning as an infinite sequence; he speaks of a never-ending chain of utterances, but he never implies in any way that individual contextualized meaning, or understanding, cannot be achieved. Zbinden has written carefully on the way structuralists (especially Todorov and Kristeva) appropriated and distorted Bakhtin's writings: "It might indeed be the case that it was the structuralists who turned him into a fashionable semiologist who had already foreseen the advent of deconstruction" (2006, 5). Understanding itself is necessarily dialogic because it involves utterances (Voloshinov 1986, 102; Bakhtin 1986, 111, 125–26). Dialogue is present everywhere; it even "moves into the deepest molecular and, ultimately, subatomic levels" (Bakhtin 1981, 300).

Just as understanding an utterance is an active dialogic process by the listener that vitally involves the speaker, so too is the speaker's utterance creation an active dialogic process that vitally involves the listener. The listener dialogically helps to shape the speaker's utterance through what Bakhtin calls "addressivity." *Addressivity* was introduced in *Marxism and the Philosophy of Language* (Voloshinov 1986, 85) and is an important element throughout the Dostoevsky book (Bakhtin 1984), but its fullest development comes in the much later essay on speech genres:

> The role of the *others* for whom the utterance is constructed is extremely great. The role of these others, for whom my thought becomes actual thought for the first time (and thus also for my own self as well) is not that of passive listeners, but of active participants in speech communication [sic]. From the very beginning, the speaker expects a response from them, an active responsive, understanding. The entire utterance is constructed, as it were, in anticipation of encountering this response.
>
> An essential (constitutive) marker of the utterance is its quality of being directed to someone, its *addressivity*. (1986, 94–95)

The significant, formative role of the addressee as "an immediate participant-interlocutor" (95) in all linguistic communication, both textual and verbal, is one of the most original and potentially useful of Bakhtin's concepts. Even inner speech (thought) involves such an addressee—one's dialogic self. The addressee need not be present physically or even exist as a flesh-and-blood person; it can be "an indefinite, unconcretized *other*" (95). In a later essay Bakhtin also introduces the concept of the *superaddressee*: a third party "whose absolutely just responsive understanding is presumed, either in some metaphysical distance or in distant historical time" (126).

Bakhtin's last written words refer to the universality of dialogic relationships and dialogic contexts:

> There is neither a first nor a last word and there are no limits to the dialogic context (it extends into the boundless past and the boundless future). Even *past* meanings, that is, those born in the dialogue of past centuries, can never be stable (finalized, ended once and for all)—they will

> always change (be renewed) in the process of subsequent, future development of the dialogue. At any moment in the development of the dialogue there are immense, boundless masses of forgotten contextual meanings, but at certain moments of the dialogue's subsequent development along the way they are recalled and invigorated in renewed form (in a new context). Nothing is absolutely dead: every meaning will have its homecoming festival. (170)

Dialogism is thus the contextualizing frame within which language is understood (1981, 340).

The concept of *context* informs everything Bakhtin wrote probably as much as dialogism. Significantly, in 1964, shortly after Bakhtin's discovery by younger scholars, Viach Ivanov and other semioticians began an annual journal titled *Sign Systems Studies* (*SSS*), but Bakhtin never submitted anything to *SSS* for publication. In 1972, other Bakhtin followers began the journal *Kontekst*, and that is where Bakhtin chose to publish three late essays (see Reid 1990, 38). From his very earliest works in the 1920s on esthetics (1990) and moral philosophy (1993), Bakhtin emphasizes context. In his discussion of the architectonics of answerability, for example, Bakhtin writes:

> The emotional-volitional tone and an actual valuation do not relate at all to content as such in its isolation, but relate to it in its correlation with me within the once-occurrent event of Being encompassing us. An emotional-volitional affirmation acquires its tone not in the context of culture; all of culture as a whole is integrated in the unitary and once-occurrent context of life in which I participate. Both culture as a whole and every particular thought, every particular product of a living act or deed, are integrated in the once-occurrent, individual context of actual thinking *qua* event. The emotional-volitional tone opens up the self-seclusion and self-sufficiency of the possible content of a thought, makes it a participant in unitary and once-occurrent Being-as-event. Any universally valid value becomes *actually* valid only in an individual context. (1993, 35–36)

Bakhtin is never interested in "universally valid" value, but in unique individuals in context. Neither Saussure nor Chomsky raised such a concern.

For Bakhtin, life itself has meaning only in context: we are answerable in context. Meaning and truth are individual (*pravda*) and not universal (*istina*) (37). In discussing the context of artistic creation, Bakhtin similarly asserts that "the author's act of creation is accomplished totally within the purely literary value-context (without in any respect exceeding its bounds) and is rendered meaningful in all of its constituents solely by that context: It is born (with respect to value) in this context, is consummated in this context, and dies in this context" (1990, 196). Meaning does not inhere in an utterance, nor is meaning transmitted in identical form from one person to another. Meaning is not an abstract process; all meaning is meaning *to* someone; all understanding is understanding *by* someone. "The only way that words can mean is to be understood" (Clark and Holquist 1984, 213). Meaning *emerges* in dialogic interaction:

> In the actual life of speech, every concrete act of understanding is active: it assimilates the world to be understood into its own conceptual system filled with specific objects and emotional expressions, and is indissolubly merged with the response, with a motivated agreement or disagreement. *To some extent, primacy belongs to the response, as the activating principle*: it creates the ground for understanding, it prepares the ground for an active and engaged understanding. Understanding comes to fruition only in the response. Understanding and response are dialectically merged and mutually condition each other; one is impossible without the other. (Bakhtin 1981, 282; emphasis added)

For Bakhtin, understanding is as an active process (Voloshinov 1986, 73), never passive as in Saussurean linguistics. Understanding is furthermore "creative": "*Creative understanding* does not renounce itself, its own place in time, its own culture; and it forgets nothing. It is immensely important for the person who understands to be *located outside* the object of his or her creative understanding—in time, in space, in culture" (Bakhtin 1986, 7).

The phrase "contextual meaning" occurs for the first time in *Marxism and the Philosophy of Language* (Voloshinov 1986, 77). The notion of meaning in context appears throughout Bakhtin's work, especially in some of his last writings. In "The Problem of Speech Genres," for example, written in 1952–1953, Bakhtin uses context to explain the distinction between sentence versus utterance:

> The context of a sentence is the speech of one speaking subject (speaker). The sentence itself is not correlated directly or personally with the extraverbal context of reality (situation, setting, prehistory) or with the

> utterances of other speakers; this takes place only indirectly, through its entire surrounding context, that is, through the utterance as a whole. (1986, 73–74)

In his essay on "The Problem of the Text ...," from 1959–1961, Bakhtin defines "contextual meaning" as follows: "integrated meaning that relates to value—to truth, beauty, and so forth—and requires a *responsive* understanding, one that includes evaluation" (125). In the fragmented style that characterizes much of his later writings, Bakhtin meditates again in "From Notes Made in 1970–71" on the concept of contextual meaning as "potentially infinite, but it can only be actualized when accompanied by another (other's) meaning, if only by a question in the inner speech of the one who understands" (145–46). In this essay, Bakhtin also differentiates pragmatic, dialogic context from structuralist, semiotic code: "A context is potentially unfinalized; a code must be finalized. A code is only a technical means of transmitting information, but it also has cognitive, creative significance. A code is a deliberately established, killed context" (147). Such statements make it impossible to consider Bakhtin a semiotician in any way.

Bakhtin's final essay, "Toward a Methodology for the Human Sciences," written shortly before his death in 1975, is primarily a study of the significance of context and contextual meaning. The words "context" and "contextual" appear some 51 times in the 12 pages of the brief essay, suggesting that Bakhtin wanted his final message to emphasize the centrality of context in all communication (1986, 159–70).

In real-life situations, neither a Saussurean concept of *langue* nor a Chomskyan mastery of syntactic rules, nor a combination of the two is sufficient to facilitate understanding of actual utterances. Bakhtin went two steps beyond Saussure's words and one beyond Chomsky's rules by adding that third crucial element of context, which the two previous linguists neglected. Context, shared knowledge and experience, and an understanding of intentions are all necessary to comprehend and engage dialogically with another human being. Consider the following apt illustration of the Saussure-Chomsky-Bakhtin continuum in language from a textbook on evolutionary psychology: "The great bulk of the literature on language has concentrated on the structural content of language (grammar) [Chomsky] and the meanings of words [Saussure]. However, much of what we say is given meaning by the nonverbal behaviour that we wrap around our utterances [Bakhtin]" (Barrett et al. 2002, 345).

We have concentrated here on Bakhtin's signature concepts of dialogism and language use in context, especially as these ideas contrast with the approaches of Saussure and Chomsky. But Bakhtin's extensive writings include numerous other explorations of language-related phenomena, for example, the distinctions he draws between monologic and dialogic discourse (especially in Bakhtin 1981, 1984, 1986; Voloshinov 1986), and between centripetal (unifying, centralizing) linguistic/ideological forces and centrifugal (disunifying, decentralizing) ones, as well as heteroglossia (Bakhtin's term for the multiplicity of languages, cultures, and consciousnesses) and other related concepts on language variation (particularly prominent in Bakhtin 1981, 1984, 1986). Bakhtin is far more than a proto-pragmatist, but these other phenomena fall outside the purview of this book and thus we can only give them this brief mention here. Suffice it to say that Bakhtin's ideas on language—especially insofar as they are confirmed and expanded by newer findings from biology, psychology and evolutionary studies—provide the grounds for a new, contextualized, approach to literary theory.

The following passage by Bakhtin could summarize a lifetime's concern with meaning in context: "There is neither a first nor a last word and there are no limits to the dialogic context (it extends into the boundless past and the boundless future)" (1986, 170). A constant emphasis on context sets Bakhtin apart from all other philosophers and linguists of his time, and it places him at the forefront of the modern pragmatic study of language. Bakhtin reformulated and recontextualized his major concepts throughout his life, a fact he commented on in "From Notes Made in 1970–71" when he acknowledged his "love for variations and for a diversity of terms for a single phenomenon" (155). Still, his work is surprisingly consistent throughout the more than half a century in which he wrote, and no concept more thoroughly permeates his writings from the earliest days to his last words than *context,* which is also the overarching concept of this book.

The next section deals with developments in linguistics that have emerged both directly and indirectly out of the pioneering work of Bakhtin, a pragmatist *avant la lettre*. The common theme running through these studies is that language in real life is indeed language in context.

4 Cognitive Linguistics, Integrational Linguistics, and Pragmatics

Two strands of linguistics claim explicitly to be grounded in human cognition. The first is the primarily European tradition of social-cognitive and sociocultural studies that grew directly out of the work of Bakhtin and his contemporary, Lev Vygotsky (1896–1934), a pioneering developmental psychologist whose research (see 1962, 1978) was marginalized during the Soviet era. According to James Wertsch, three main themes in Vygotsky's work are "1) a reliance on genetic, or developmental, analysis, 2) the claim that higher mental functioning in the individual derives from social life; and 3) the claim that human action, on both the social and individual planes, is mediated by tools and signs" (1991, 19). Vygotsky emphasizes speech, both vocal utterances and inner speech, as well as the role of teachers (parents, monitors, and friends) in leading learners (especially young children) into a "zone of proximal development" by way of "scaffolding" or modeling with level-appropriate guidance. As Vygotsky's work became more widely known outside the Soviet Union from about the 1970s onward, it was clear that it dovetailed nicely with the approach to language that Bakhtin and his friends and collaborators had proposed, as well as with the developing field of cognitive psychology. The Vygotskyan legacy has been most notable in psycholinguistics, the Bakhtinian legacy mostly in pragmatics; both have influenced sociolinguistics and have been increasingly incorporated into cognitive linguistics (which is very much a hybrid field, as indeed are all of the approaches to language described in this chapter).

The social-cognitive approach to language "reflects an insistence on the necessity to study language use, a conception of the world as multidimensional and always only partially understood, and Man as a social being in search of meaning with individual minds embedded in a cultural collectivity" (Wold 1992, 1). Foremost among those who have worked in this tradition is the Norwegian linguist Ragnar Rommetveit. Beginning in the 1960s, Rommetveit published a series of studies that place contextualized utterances at the heart of communication (see e.g., 1992). Rommetveit then increasingly incorporated his work into various cognitive paradigms (see Wertsch 2003). Bakhtinian and Vygotskyan concepts—context, dialogue, perspective, intersubjectivity, mediation—are central to Rommetveit and others working in the social-cognitive movement. This work may not be well known outside a certain circle of writers, but it is thoroughly consistent with the cognitive, integrational, and pragmatic approaches to linguistics discussed in the following sections.

The second approach grounded explicitly in cognitive studies could be called the "Lakoff School" because it tends to take shape around George Lakoff's research and prominently involves his colleagues' and his students' work. Central to the concerns of these linguists is the essential metaphoricity of human thought. This approach has a precise beginning in 1979 when Michael Reddy published an essay on the "conduit metaphor" in an important collection of very disparate studies of metaphor assembled by Andrew Ortony (see Reddy 1993). Reddy notes that our discourse about language is frequently characterized by a set of metaphors, suggesting that we somehow package language and send it to others via some sort of conduit. This is an accurate description of the telementation process at the heart of the Saussurean concept of how language functions. Reddy asks why we use this metaphoric way of describing how we communicate, for surely it cannot be an even remotely accurate description of the process. When we realize both how widespread and how absurd the metaphor is, it becomes clearer how easily it can hide in plain sight and how weak the linguistic foundation of theory truly is.

Reddy's essay first appeared in 1979. In January of that year, Lakoff met philosopher Mark Johnson and they realized both the potential in Reddy's work and how their own interests, from their different disciplines, coincided in the question of metaphorical meaning.[1] They began to collaborate on what was originally to be a brief paper on the subject, and within six (obviously very intense) months they had completed a book, the first to outline and demonstrate in detail the cognitive theory of metaphor, that was published the next year: *Metaphors We Live By* (1980, ix–x).

For Lakoff and Johnson, metaphor is not, as has traditionally been assumed, "a characteristic of language alone, a matter of words rather than thought or action" but, rather, it is "pervasive in everyday life, not just in language but in thought and action. Our ordinary conceptual system … is fundamentally metaphorical in nature" (3). This modest and easily readable book, in which Lakoff and Johnson work out the implications of this fundamental truth about cognition and language, is a landmark in cognitive science. The authors define the "essence of metaphor" as "*understanding and experiencing one thing in terms of another*" (5). By means of a series of examples of basic metaphors, such as ARGUMENT IS

WAR, TIME IS MONEY, HAPPY IS UP/ SAD IS DOWN, and LIFE IS A STORY (the practice of using small caps to identify metaphors is the authors'), they illustrate how we understand metaphoric language, mapping from a source domain (e.g., WAR) onto a target domain (e.g., ARGUMENT): your claims are *indefensible*, he *attacked every weak point* in my argument, his criticisms were *right on target*, and so forth (4). Metaphor is a standard means by which we communicate, not a mere rhetorical embellishment. Human thought itself is largely metaphorical, not literal. Importantly, Lakoff and Johnson stress that "*no metaphor can ever be comprehended or even adequately represented independently of its experiential basis*" (19). By "experiential" they mean the knowledge gained through embodied, contextualized experiences in the world. This work is crucial to the contextualist approach to cognition that is in turn vital to literary and performance studies.

Bakhtin insisted that meaning is always "meaning to" someone, that understanding is always "understanding by" someone, and that this "someone" is always situated in a unique social, historical, political, gendered, and personal context. Every act of reading, every literary experience, and every corresponding act of understanding what is read is unique to the person, the place, and the time. Although Norman Holland, in his otherwise excellent applications of Lakoff and Johnson's metaphor theory to concepts of literacy (1988, 1992), unfortunately does not cite the work of Louise Rosenblatt, she was the first to express this line of thinking—she was also the first to use the term "transaction"—in the 1930s, and it is found throughout her later work: "A specific reader and a specific text at a specific time and place: change any of these, and there occurs a different circuit, a different event—a different poem" (Rosenblatt 1994, 14; see also Rosenblatt 1995). Rosenblatt was, in effect—rather like Bakhtin—a cognitive scientist *avant la lettre*.

The cognitive theory of metaphor has now been in existence for over four decades (a long time by the standards of the field) and has already substantially influenced the study of cognition, language, and literature. Mark Turner has done the most to incorporate the theory into a significant new understanding of literary texts, writing a series of books (1987, 1989 with Lakoff, 1991, 1996, 2002 with Fauconnier, 2014) informed by metaphor theory and work in cognitive science in general. Turner's work primarily focuses on English poetry, but whether his interest is directed toward themes of kinship or the nature of parable, he always focuses on the mind's metaphoric nature. He provides insights of a sort not normally found in literary theory and criticism, which can be hampered by a traditionally naive concept of human cognition and understanding. Turner has written with the pioneering goal of reorienting literary study "in the age of cognitive science" (1991) and of embodied cognition (1996). He explicitly grounds his work in a research paradigm radically different from the one derived from Saussurean linguistic theories. The books by Turner are exemplary of work grounded in cognitive science that can inspire and inform the study of literature.

Turner carefully links literature to linguistics via the study of human cognition. He provides a concise description of cognitive linguistics and its implications:

> Cognitive linguistics is concerned with human concepts as the basis of meaning, rather than with truth-conditions as the basis of meaning; with the role of conventional imagery in cognition and language; with figuration in thought and speech; and with grammar as symbolic phenomenon. Much of the empirical research in cognitive linguistics has led to a rejection of those dichotomies that have in practice separated the study of language from the study of literature, namely, the putative autonomy of syntax from meaning, the putative autonomy of the language system from other cognitive systems, the putative separation of the grammar and the lexicon, the putative separation between linguistic (semantic) meaning and extralinguistic (pragmatic) meaning, and the putative separation between literal language and figurative language. (1991, 20–21)

Cognitive linguistics can be considered a subfield of psycholinguistics—a discipline defined by Gerry Altmann as "the study of how the mind turns language into meaning, and back again" (1997, vi)—and it is also related to sociolinguistics and pragmatics (cognitive studies does indeed make disciplinary boundaries fuzzy). Cognitive linguistics blends the study of fundamental cognitive processes with the study of language proper, so that some topics in this chapter—metaphor, figurative language—are inseparable from topics central to a discussion on mind—schema theory, categorization, conceptual blending. It is sometimes impossible to distinguish what is linguistic from what is cognitive, hence the accuracy of the term cognitive linguistics (see Langacker 1987 for an extensive approach to cognition-based linguistics, an integral part of CONTEXTUALISM).

The work of cognitive psychologist Raymond W. Gibbs, Jr. is closely related to the Lakoff School, as seen in his book *The Poetics of Mind* (1994), in which he closely examines the figurative nature of ordinary human thought processes. Gibbs's position is clear from the start: "An old but still prevailing view among students of mind holds that thought and language are inherently literal. ... This book advances the idea that the traditional view of mind is mistaken, because human cognition is fundamentally shaped by various poetic or figurative processes" (1–2). Gibbs's persistent, overwhelming argument of the case against the commonly accepted idea of literal meaning greatly compromises any sense this term might have. Careful readers of Gibbs's book will never conceive of supposed literal language in the same way. Consistent with Lakoff and Johnson, Gibbs makes clear the case for the presence of figurative, rather than literal, thought and language at the heart of human cognitive and linguistic enterprises.

Roy Harris coined the term "integrational linguistics" and first wrote of its potential in his book *The Language Myth* (1981), a work that starts from the (Chomskyan) premise that language "exists in the individual as a form of neurophysiological programming, associated with the control of certain specific motor activities and with certain centres in the brain" (4). Harris writes to dispel a common notion, a "myth" about what language is and how it functions, based on two fallacies. The first is the fallacy of *telementation*, according to which "linguistic knowledge is essentially a matter of knowing which words stand for which ideas. For words, according to this view, are symbols devised by man for transferring thoughts from one mind to another. Speech is a form of telementation" (9). The second is the fallacy of *determinacy* (or fixed code), which "provides the explanation of how the telementation process works, and indeed of how telementation is possible" (10). Citing Reddy's conduit metaphor, Harris describes the process by which one person has a concept, encodes it in speech or writing, sends it to a second person, who then decodes it and thus has the same concept as the first person. The language myth, in other words, is that Saussure's concept of language describes what language is and how it functions.

According to Harris, one alternative to the mythical view of language is the approach taken by Chomsky in "the orthodox tradition of modern linguistic theory" (32). The major fault Harris finds with Chomskyan linguistics is that it makes no provision for real human use in context (performance), concentrating exclusively on underlying, theoretical, abstract principles (competence) that make performance possible. He suggests an integrational approach to language that begins with the recognition that human realities take place in time, that events occur in succession in time: "The contextualisation provided by succession in time ensures that every linguistic act is integrated into the individual's experience as a new event, which has never occurred before and cannot occur again" (155). Harris uses the term "cotemporality" to stress the idea that "[l]inguistic acts could accordingly be said to be cotemporal in our experience with non-linguistic events and circumstances of all kinds" (157). For Harris, language use should be conceived of as integrated into everything else in our lives, and language should be conceived not as an independent social system nor as an abstract formal competence, but instead as "a process of making communicational sense of verbal behavior" (165). Harris's pragmatic idea of integrating actual language use into the overall context of human life struck a responsive chord with scholars such as Deborah Cameron and Michael Toolan.

In *Feminism and Linguistic Theory* (1992), Cameron sketches an alternative to the linguistic determinism that has characterized much poststructuralist thought with an empowering approach to language:

> Firstly, language is *radically* contextual. It is not just a matter of context affecting the system, the system has no existence outside a context. Thus language cannot be abstracted from time and space, or from the extralinguistic dimensions of the situation in which it is embedded. … Secondly, language-using is a creative process. There is virtually no limit to the novel situations humans may encounter, and therefore to the communicational demands that may be placed upon them. (192)

For Cameron, the concept of languages of power—"those which in a given situation maximise the speaker's personal and social control" (200)—are meaningful for women who have typically been "denied access to them, or defined negatively in relation to them" (200). This concept of language use in an integrated context of the political and social reality of women's lives can be a tool far more powerful than those within the Saussurean-Lacanian tradition.

Toolan's *Total Speech* (1996) is a comprehensive presentation of the integrational approach to language. For Toolan, the "key tenet" of Harris's term *integrational* is "the contextual embeddedness of language" (3). Both "language" and "context," Toolan argues, are often considered foundational categories, but in fact "they are (and not so much 'merely' as 'importantly') contingent ones" (4). A traditional approach to the text-context binary is "to separate text from context, that is, raise up the 'linguistic' and set aside the 'nonlinguistic,'" but, he insists, "ultimately there is no absolute separability of text from context; text (language) is never autonomous, and context is never permanent or stably linked to but distinct from text" (5).

Harris, Cameron, and Toolan all acknowledge that the integrationist approach to language inherently recognizes that linguistic communication is not infallible, that there is no fail-safe process of telementation, that misunderstandings and misinterpretations can and do frequently take place (see e.g., Toolan 67). Ultimately, integrational linguistics is another name for pragmatics. Pragmatics is the study of language in context; it has been defined as "the study of the relations between language and context that are basic to an account of language understanding" (Levinson 1983, 21). Pragmatics' rich tradition has influenced literary theory and criticism significantly, including both the speech-act theory of Austin and Searle, and the program of "conversational implicature" of J. P. Grice from the 1960s and 1970s. Pragmatics serves as a theoretical and conceptual overview for both the Lakoff-Johnson cognitive theory of metaphor and Harris's integrational linguistics.

An important earlier contribution to pragmatics is *Relevance* by Dan Sperber and Deidre Wilson (1986). The authors present "a new approach to the study of human communication," one "grounded in a general view of human cognition" (vii). Coming from different disciplinary orientations (Sperber from anthropology, Wilson from linguistics), their interests met on the common ground of pragmatics, "the study of contextual factors in verbal communication" (vii). Like Lakoff and Johnson, their original plan was to collaborate on an essay, but as work progressed they realized nothing less than a book would suffice for their purposes (vii–viii).

Sperber and Wilson acknowledge that linguistic communication can be described in a limited way by the coding and decoding of the traditional semiotics model, but it also involves the *inference* of the pragmatics model. While the code model is explanatory, "it is descriptively inadequate: comprehension involves more than the decoding of a linguistic signal" (6).

While code theories cannot demonstrate how human communication functions, the inferential process can. Derived primarily from the research tradition of H. P. Grice, the inferential process "starts from a set of premises and results in a set of conclusions which follow logically from, or are at least warranted by, the premises" (12–13). The understanding of an utterance rests on a set of inferential premises, and these inferential premises are the context, which the authors define as "a psychological construct, a subset of the hearer's assumptions about the world" (15). Significant are only the hearer's assumptions about the world, not the way the world actually is. Context is therefore not limited to an utterance's setting, time, and place, but explicitly includes the individual hearer's beliefs, values, feelings, experiences, and many other subjective factors.

Sperber and Wilson's central idea is that "there is a single property—relevance—which makes information worth processing for a human being" (46). Communication is an attempt to alter others' cognitive environment. *Ostensive behavior* (or *ostention*) is important in attempting to guarantee relevance in communication. The authors define this process as "behaviour which makes manifest an intention to make something manifest" (49), such as showing something (often via a gesture such as pointing) to another person. Ostensive-inferential communication takes place when "the communicator produces a stimulus which makes it mutually manifest to communicator and audience that the communicator intends, by means of this stimulus, to make manifest or more manifest to the audience a set of assumptions" (63).

Sperber and Wilson's basic-level, pragmatic, intelligent presentation of inferential processes within the context of modern cognitive science makes it invaluable. Relevance theory is still considered "an important contribution to our understanding of the pragmatics of communication" (Wearing 2015), and its implications for literary theory are clear. Literary style and technique can be a form of ostensive efforts by an author to make available information that might be relevant to a reader. Dialogue within a text can be understood in the same terms.

Ian MacKenzie's *Paradigms of Reading: Relevance Theory and Deconstruction* (2002) is an exemplary approach to literature in terms of Sperber and Wilson's relevance theory rather than deconstruction. MacKenzie begins by affirming that both pragmatics and deconstruction acknowledge that no text can be understood simply in terms of the words on the page, but he points out that the two approaches immediately part ways from there. Taking Paul de Man as his prime exemplar of deconstruction, MacKenzie describes the position that because words (signs) can signify a multiplicity of things, there is no way to ascribe any specific meaning to any utterance or text. MacKenzie proposes that "an effective pragmatic remedy to de Man's doom-laden account of language" lies in the emphasis that pragmatics (and relevance theory) places on a situated agent's ability to infer relevant meaning in context (3). The contrast between these two positions, deconstruction and pragmatism, has substantial implications for literary theory and criticism. Deconstructionists in general pay no attention to any non-Saussurean linguistic theory, a strategy that enables them to "*choose* to disregard the constraints of intentionality, meaningfulness and context, and prefer to look for all the discordant signifying elements that pragmatists conventionally disregard" (13).

Echoing Holland, MacKenzie charges that theory conceives of language as a powerful social force with which texts control their passive subject readers so that language is the agent and human beings are objects. He recognizes "[t]he fact that words do not coincide with the world does not necessarily entail that language is false, or that no utterance can come into being without a non-referential moment, or that all utterances are at one level 'fictions.' It merely reveals that people are able to do things with language" (151). Language is a cognitive tool that embodied agents use to do things; it is not an active, animated power that does things to people. MacKenzie thus rejects Saussure's "false model" of active texts that has shaped literary theory as well (154).

As MacKenzie's title succinctly states, relevance theory (pragmatics, Bakhtinian metalinguistics, cognitive linguistics, etc.) and deconstruction (and all other versions of theory) do indeed inhabit radically different and profoundly incompatible paradigms. Relevance theory "demonstrates the untenability of all critical approaches to literature grounded on the notion of the passive reception or understanding of an author's meaning or message. Language cannot 'posit' or 'signify' on its own, nor operate mechanically" (197). MacKenzie's invitation to think and write about literature based on a pragmatic paradigm rather than on one of signification represents a radical and logical reorientation for literary theory.

5 Using Language

Herbert H. Clark begins his book *Using Language* (1996) with this assertion: "Language is used for doing things" (3). Clark's primary thesis is that "[l]anguage use is really a form of *joint action*," which he defines as an action "that is carried out by an ensemble of people acting in coordination with each other" (3). Moving beyond the speaker-listener model for linguistic communication, Clark conceives of language use as "the joint action that emerges when speakers and listeners—or writers and readers—perform their individual actions in coordination, as ensembles" (3). Thus, people using language are engaged simultaneously in individual and social activities; they "work together as participants in the social units" (3) as well as carry out their individual activities. Clark's concept of "common ground" is very similar to Sperber and Wilson's cognitive environment. Individuals involved in the joint activity of language use can employ their basic inferential processes to understand each other: "Common ground is the foundation for all joint actions, and that makes it essential to the creation of speaker's meaning and addressee's understanding as well" (14).[2]

Clark's distinctions among types of participants in language use go far beyond the traditional speaker-addressee binary. *Side participants*, for example, are "taking part in the conversation but not currently being addressed," while *overhearers* "have no rights or responsibilities" in the conversation (14). Overhearers can be further divided into *bystanders*, who "are openly present but not part of the conversation," and *eavesdroppers*, who "are those who listen in without the speaker's awareness," as well as some positions in between these categories (14). These distinctions are important in understanding how people read narrative. For example, Richard J. Gerrig proposes that "[a]uthors and readers most often behave as if readers are side-participants; in that role, authors intend readers to be genuinely informed by narrative utterances" (1993, 110). Such a concept of the author-reader relationship has substantial implications for reader-response theory in general, as well as for our evaluation of literary characters, our understanding of plot, and much more.

Michael Tomasello's *Constructing a Language* (2003) presents a running critique of Chomsky and the idea of an innate UG. In its place, Tomasello proposes an approach to language acquisition based on what is often called "cognitive-functional linguistics," but which he prefers to call "usage-based linguistics" by which "the grammatical dimension of language is a product of a set of historical and ontogenetic processes referred to collectively as *grammaticalization*" (5). While Tomasello resoundingly rejects a Chomskyan UG approach, he still recognizes our species-specific biological capacity to learn a language. He prefers to describe language acquisition in terms of context-based learning experiences rather than as abstract rule construction or parameter setting. Tomasello, along with other cognitive scientists in the twenty-first century, has continued to present research evidence against UG. Not long ago, he teamed up with developmental psychologist Paul Ibbotson to demonstrate how Chomsky's theories don't hold up when applied, for instance, to a child's language learning. It's important to understand, the authors argue, that non-linguistic capabilities such as categorization rather than innate grammar structures are foundational for language acquisition because "the study of language plays a central role in diverse disciplines—from poetry to artificial intelligence to linguistics itself; misguided methods lead to questionable results. Further, language is used by humans in ways no animal can match; if you understand what language is, you comprehend a little bit more about human nature" (Ibbotson 2016, 70). Humanists need a basic understanding of contemporary linguistics to do *our* jobs properly.

Tomasello, in his 2003 book, invoked a Bakhtinian emphasis on the utterance as the unit of language rather than the Saussurean word or Chomskyan sentence:

> Utterances are the primary reality of language from a communicative point of view because they are the most direct embodiment of a speaker's communicative intentions. And so it is [Bakhtinian] utterances—not [Saussurean] words or [Chomskyan] abstract categories—that children are initially focused on learning. (325–26)

This returns us to the Bakhtinian origins of contextualized and embodied language. Nearly a century ago, Bakhtin knew that human beings understand intentional utterances in specific contexts. After the long, dry era of Saussurean semiotics, structuralism, and poststructuralism, in the age of cognitive science, we once again recognize this basic truth.

Together, the metalinguistic, cognitive, integrational, and pragmatic approaches to language offer a basis for understanding literary texts in a way consistent with a contextualized cognitive science. Literary theory must have at its core a coherent theory of language, for the material substance of literature is language. Having described how the concepts of language in contemporary theory are fatally flawed and having outlined alternatives from contemporary linguistics, we turn in the book's remaining two parts to the biology, psychology, and contexts of embodied cognitive science.

Notes

1. Lakoff's contribution to the second edition of the Ortony anthology is an overview of the development of (cognitive) metaphor theory since Reddy's original essay fourteen years earlier and stressed Reddy's pioneering status in the field (1993, 203–4).
2. See William (2012) for an application of the common ground concept to multicultural poetry.

References

Altmann, Gerry T. M. 1997. *The Ascent of Babel: An Exploration of Language, Mind, and Understanding*. Oxford: Oxford UP.

Bakhtin, Mikhail M. 1981. *The Dialogic Imagination: Four Essays*. Ed. Michael Holquist. Trans. Caryl Emerson and Michael Holquist. Austin: U of Texas P.

———. 1984. *Problems of Dostoevsky's Poetics*. Ed. and Trans. Caryl Emerson. Minneapolis: U of Minnesota P.

———. 1986. *Speech Genres and Other Late Essays*. Ed. Caryl Emerson and Michael Holquist. Trans. Vern W. McGee. Austin: U of Texas P.

———. 1990. *Art and Answerability: Early Philosophical Essays*. Ed. Michael Holquist and Vadim Liapunov. Trans. and notes Vadim Liapunov. Suppl. trans. Kenneth Brostrom. Austin: U of Texas P.

———. 1993. *Toward a Philosophy of the Act*. Ed. Michael Holquist and Vadim Liapunov. Trans. and notes Vadim Liapunov. Austin: U of Texas P.

———, and P. N. Medvedev. 1978 (1928). *The Formal Method in Literary Scholarship: A Critical Introduction to Sociological Poetics*. Trans. Albert J. Wehrle. Baltimore: Johns Hopkins UP.

Barrett, Louise, Robin Dunbar, and John Lycett. 2002. *Human Evolutionary Psychology*. Princeton: Princeton UP.

Bocharov, Sergey. 1994. "Conversations with Bakhtin." *PMLA* 109: 1009–24.

Cameron, Deborah. 1992 (1985). *Feminism and Linguistic Theory*. New York: St. Martin's P.

Clark, Herbert H. 1996. *Using Language*. Cambridge UP.

Clark, Katerina, and Michael Holquist. 1984. *Mikhail Bakhtin*. Cambridge: Belknap P or Harvard UP.

Fauconnier, Gilles, and Mark Turner. 2002. *The Way We Think: Conceptual Blending and the Mind's Hidden Complexities*. New York: Basic Books.

Gerrig, Richard J. 1993. *Experiencing Narrative Worlds: On the Psychological Activities of Reading*. New Haven: Yale UP.

Gibbs, Raymond W., Jr. 1994. *The Poetics of Mind: Figurative Thought, Language, and Understanding*. Cambridge: Cambridge UP.

Harris, Roy. 1981. *The Language Myth*. New York: St. Martin's P.

Holland, Norman N. 1988. *The Brain of Robert Frost: A Cognitive Approach to Literature*. New York: Routledge.

———. 1992. *The Critical I*. New York: Columbia UP.

Ibbotson, Paul, and Michael Tomasello. 2016. "Language in a New Key." *Scientific American* 315.5: 70–75.

Ivanov, Viach. Vs. 1976. "The Significance of M. M. Bakhtin's Ideas on Sign, Utterance, and Dialogue for Modern Semiotics." In *Semiotics and Structuralism: Readings from the Soviet Union*. Ed. Henryk Baran. White Plains: International Arts and Sciences P. 310–67.

Lakoff, George, and Mark Johnson. 1980. *Metaphors We Live By*. Chicago: U of Chicago P.

———. 1993. "The Contemporary Theory of Metaphor." In Ortony, 202–51.

———, and Mark Turner. 1989. *More than Cool Reason: A Field Guide to Poetic Metaphor*. Chicago: U of Chicago P.

Langacker, Ronald W. 1987. *Foundations of Cognitive Grammar, Vol. I: Theoretical Prerequisites*. Stanford: Stanford UP.

Levinson, Stephen C. 1983. *Pragmatics*. Cambridge: Cambridge UP.

MacKenzie, Ian. 2002. *Paradigms of Reading: Relevance Theory and Deconstruction*. Houndsmills: Palgrave Macmillan.

Morson, Gary Saul, and Caryl Emerson. 1990. *Mikhail Bakhtin: Creation of a Prosaics*. Stanford: Stanford UP.

Ortony, Andrew, ed. 1993 (1979). *Metaphor and Thought*. Cambridge: Cambridge UP.

Perlina, Nina. 1983. "Bakhtin-Medvedev-Voloshinov: An Apple of Discourse." *University of Ottawa Quarterly* 1: 35–47.

Reddy, Michael J. 1993 (1979). "The Conduit Metaphor: A Case of Frame Conflict in Our Language about Language." In Ortony, 164–201.

Reid, Allan. 1990. *Literature as Communication and Cognition in Bakhtin and Lotman*. New York: Garland.

Rommetveit, Ragnar. 1992. "Outlines of a Dialogically Based Social-Cognitive Approach to Human Cognition and Communication." In Wold, 19–44.

Rosenblatt, Louise M. 1994 (1978). *The Reader, the Text, the Poem: The Transactional Theory of the Literary Work*. Carbondale: Southern Illinois UP.

———. 1995 (1938). *Literature as Exploration*. New York: Modern Language Association.

Rzhevsky, Nicholas. 1994. "Kozhinov on Bakhtin." *New Literary History* 25: 429–44.

Sperber, Dan, and Deirdre Wilson. 1995 (1986). *Relevance: Communication and Cognition*. Cambridge: Harvard UP.

Stewart, Susan. 1986. "Shouts on the Street: Bakhtin's Anti-Linguistics." In *Bakhtin: Essays and Dialogues on His Work*. Ed. Gary Saul Morson. Chicago: U of Chicago P. 41–57.

Thompson, Evan. 2007. *Mind in Life: Biology, Phenomenology, and the Sciences of Mind*. Cambridge: Belknap P of Harvard UP.

Titunik, I. R. 1976. "M. M. Baxtin (The Baxtin School) and Soviet Semiotics." *Dispositio* 1.3: 327–38.

Tomasello, Michael. 2003. *Constructing a Language: A Usage-Based Theory of Language Acquisition*. Cambridge: Harvard UP.

Toolan, Michael. 1996. *Total Speech: An Integrational Linguistic Approach to Language*. Durham: Duke UP.

Turner, Mark. 1987. *Death is the Mother of Beauty: Mind, Metaphor, Criticism*. Chicago: U of Chicago P.

———. 1991. *Reading Minds: The Study of English in the Age of Cognitive Science*. Princeton: Princeton UP.

———. 1996. *The Literary Mind*. New York: Oxford UP.

———. 2014. *The Origin of Ideas. Blending, Creativity, and the Human Spark*. New York: Oxford UP.

Voloshinov, V. 1983 (1926). "Discourse in Life and Discourse in Poetry: Questions of Sociological Poetics." In *Bakhtin School Papers*. Ed. Ann Shukman. Oxford: RPT Publications. 5–29.

———. 1986 (1929). *Marxism and the Philosophy of Language*. Trans. Ladislav Matejika and I. R. Titunik. Cambridge: Harvard UP.

Vygotsky, L. S. 1962. *Thought and Language*. Ed. and trans. by Eugenia Hanfmann and Gertrude Vakar. Cambridge: MIT P.

———. 1978. *Mind in Society: The Development of Higher Psychological Processes*. Ed. Michael Cole, et al. Cambridge: Harvard UP.

Wearing, Catherine J. 2015. "Relevance Theory: Pragmatics and Cognition." *WIREs Cogn Sci* 6: 87–95.

Wertsch, James V. 1991. *Voices of the Mind: A Sociocultural Approach to Mediated Action*. Cambridge: Harvard UP.

———, ed. 2003. *Ragnar Rommetveit: His Work and Influence*. Mahwah, NJ: Lawrence Erlbaum.

William, Jennifer Marston. 2012. "Cognitive Poetics and Common Ground in a Multicultural Context: The Poetry of Zehra Çirak." *The German Quarterly 85.2* (Spring 2012): 173–92.

Wold, Astri Heen. 1992. "Introduction." In Wold, 1–18.

———, ed. 1992. *The Dialogical Alternative: Towards a Theory of Language and Mind*. Oslo: Scandinavian UP.

Zbinden, Karine. 2006. *Bakhtin between East and West: Cross-Cultural Transmission*. Oxford: Legenda.

PART II

Biology, Language, and the Brain

The biological science of context, ecology, furnishes us with a rich model for the understanding of context in human culture.

—Ben Ami Scharfstein

In biology, everything is context-dependent.

—Robin Dunbar

CHAPTER 5

Evolution and Language

If God is in the details, then evolution is in the context.

—Robert Foley

Human beings are animals.

This biological reality of our species is undeniable; it is a necessary assumption for any approach to all things human, including the sciences, the social sciences, the arts—and the humanities. The theory of evolution by natural selection as first articulated by Charles Darwin (1809–1882) is the one great theory of the nineteenth century that has survived and continues to flourish at the beginning of the twenty-first and that stands today as the firm foundation upon which all biological and psychological research and theory is based. This chapter provides a baseline set of assumptions and a historical-conceptual outline of evolution. Here, we review the status of the modern synthesis of the theory of evolution by natural selection that integrates newer work in genetics into the original Darwinian theory; review Darwin's second major theory, that of sexual selection, and with it the role of women in evolution; and provide a brief outline of the evolutionary history of the human species. We close the chapter by examining language's origins and its presumed coevolution with the brain.

H. Mancing and J. Marston William, *Restoring the Human Context to Literary and Performance Studies*, Cognitive Studies in Literature and Performance,
https://doi.org/10.1007/978-3-030-89078-0_5

1 Neo-Darwinism

A foundational assumption of THEORY that language *precedes* and even *makes* the body is incompatible with the basic fact of human biology. As Kieran Egan observes: "We had, as a species, and have, as individuals, bodies before language. Language emerges from the body in the process of evolutionary and individual development, and it bears the ineluctable stamp of the body" (1997, 5). Underlying all our premises about the foundations for literary and performance studies is this fact about humanity: We are animals. And like all other animals we have a unique evolutionary history. Yet it is precisely our biological reality that most THEORY ignores, perhaps not always by denying it outright, but by rejecting it inferentially by formulating theoretical approaches inconsistent with it in important ways. Linguistic determinism along with the notion of a debiologized psyche, the concept of infinite textuality, and other essential characteristics of THEORY are all incompatible with an explicit recognition of the biological reality of the human animal. Barash and Barash (2000, 317–20) use the term "Brahean thinking" (for the astronomer Tycho Brahe) to describe the process by which people are induced

> to accept, grudgingly if need be, awkward and discomfiting facts (what the facts demand), while at the same time retaining the core of their beliefs (what their bias demands). In Tycho Brahe's case, the core belief was that the solar system if not the whole universe was geocentric, not heliocentric. In the case of many people considering evolution, the core belief is that at some crucial, fundamental level, human beings are not really animals after all. (318)

To this, Robert Jay Russell adds, sadly, "it is an extreme irony of evolution that the only creatures capable of uncovering their own evolutionary history choose so often to cover it up" (1993, 85). Steven Pinker observes that "people who study the mind would rather not have to think about how it evolved because it would make a hash of cherished theories" (1997, 165). Our concepts of person, consciousness, self, mind, thought, language, society, culture, and more all appear radically different when we remind ourselves of our biological bases. The embodied reality of the human species places literary theory in a new light.

Embodied cognition is crucial to contextualized approaches to literature and performance. Paul Ehrlich's definition meshes with how we are using the term: "the brain is thoroughly integrated with the rest of the body, not only in its reliance on the body for physical and physiological support but also through its control functions and, especially, through feelings and emotions. This integration is sometimes described as our conceptual life being 'embodied'" (2000, 109). More concise is Esther Thelen's definition: "To say that cognition is *embodied* means that it arises from bodily interactions with the world and is continually meshed with them" (2000, 5). Evan Thompson expands upon this idea of organism-environment relationships: "The roots of mental life lie not simply in the brain, but ramify through the body and environment. Our mental lives involve our body and the world beyond the surface membrane of our organism, and therefore cannot be reduced simply to brain processes inside the head" (2007, ix).

Among the great nineteenth-century thinkers, Karl Marx was the one who most admired Darwin. He read Darwin's *Origin of Species* in 1860 and was greatly impressed by it, apparently believing that he and Darwin were "engaged in a similar enterprise" (Dissanayake 1992, 21). Marx wrote that "Darwin's book is very important and serves me as a basis in natural science for the class struggle in history" (qtd. in Lopreato and Crippen 1999, 9). Later, he sent Darwin a copy of the first volume of his *Das Kapital*, with the inscription from "a sincere admirer," but Darwin never read it—the pages of the book remained uncut. Still later, Marx offered to dedicate the second volume to Darwin, who "gently declined" (Gould 1977, 26). Marx would probably be sorely disappointed to find that many of today's Marxists maintain that the social (Marx) and the biological (Darwin) are irreconcilable opposites.

Overall, Marx and Freud invented supposedly universal (grand) *narratives* which have turned out to be fairy tales; Saussure built a great *structure* which has collapsed; Nietzsche killed God, leaned toward nihilism, and wrote about the will to power, paving the way for Foucault and strong social constructionism. But Darwin described a continually ongoing *process*, and that process is as valid today as it was then. Modern science has elaborated on this process, filling in the way it works, making it the firm foundation of all modern biological and human science. Philosopher Daniel C. Dennett (1995, 21) has gone so far as to place Darwin's theory of evolution above all other major scientific discoveries, including those of Newton and Einstein. The study of biological organisms starts by definition from a Darwinian base; virtually nothing about any animal can be fully understood in the absence of suppositions about the evolutionary history of that animal. The theory of evolution by natural selection stands on a par with the theories of relativity and quantum physics as one of the cornerstones of modern knowledge. As philosopher John Searle notes, evolution joins the atomic theory of matter as one of the concepts "so fundamental and so well established as to be no longer optional for reasonably well-educated citizens of the present era; indeed they are in large part constitutive of the modern world view" (1992, 86). Biologist Ernst Mayr (2001, 264, 275) has suggested that we should stop referring to evolution as a theory and recognize it explicitly as a *fact*. Facts are defined by Mayr as "empirical propositions (theories) that have been repeatedly confirmed and never refuted" (1997, 61). In this sense, evolution is, beyond question, a *fact*. Varki and Brower also state it in blunt terms: "biological evolution by natural selection is now established *as an incontrovertible fact*. The only reason we refer to evolution as a 'theory' is that biologists are perhaps more reticent then physicists, who might have declared a 'law' of evolution long ago" (2013, 27). The frequent assertion that "evolution is only a theory" is misleading: "Theories embody the highest level of certainty for comprehensive ideas in science. Thus, when someone claims that evolution is 'only a theory,' it's roughly equivalent to saying that the proposition that Earth circles the sun rather than vice versa is 'only a theory.' Evolution is, in fact, a very useful theory" (Ehrlich 2000, 74). But if *the* (general, Darwinian) theory of evolution by natural selection is not in doubt, many aspects of it are open to intense discussion (Barash and Barash 2000, 299).

The title of Charles Darwin's famous book is *The Origin of Species by Means of Natural Selection* (1998, 1859). His key insight was the realization that living things can change from one generation to the next and that of all the changes (randomly, blindly) introduced into any species, those that confer the greatest adaptive advantage are those most likely to get passed on to the next generation. He called this process *natural selection*, stressing the fact that it involves no supernatural intervention or overall purpose or goal, and this is the central idea that remains at the heart of all evolutionary thought. Over great periods of time, circumstance, particularly geographical separation, meant that one population of a species could grow reproductively incompatible with others of the original species and thus begin a new species. Often, as a result of chance modifications, changes in ecology (such as climate), disease, or competition with other species, a species would go extinct. In Darwin's words: "This preservation of favourable individual differences and variations, and the destruction of those which are injurious, I have called Natural Selection, or the Survival of the Fittest" (108). *Fitness*, as Darwin uses the term, refers without any moral or ethical connotation to the ability of an organism to survive and reproduce in a given environmental context.

Darwin also referred to the process of natural selection as a "struggle for existence" (117), a phrase that led to a simplistic and mistaken view of the process of evolution as an ongoing life-or-death fight as in Alfred Lord Tennyson's famous line "Nature, red in tooth and claw" from his poem *In Memoriam, A.H.H.* The poem was published in 1850, nine years *before* Darwin's book, and is therefore not an articulation of a specifically Darwinian concept but a reflection of the prevailing spirit or mind-set of the time (Cronin 1991, 273). Natural selection is instead, however, the process by which individuals, and thus species, are born, are modified, and die: It is the engine of life and the basis for all research and theorizing in biology. In the classic and often cited words of Theodosius Dobzhansky, "Nothing in biology makes sense except in the light of evolution" (1973).

The twentieth-century research that culminated in the discovery of DNA (deoxyribonucleic acid) in 1953 by James Watson and Francis Crick, which provided proof that all life was related, solidified even further the conceptual basis of Darwinism. Today's widely accepted neo-Darwinism, or the "modern synthesis," incorporates work in genetics and DNA that was unknown in Darwin's time but that is fully integrated into the original theory, strengthening and extending it. The Central Dogma of molecular biology, as formulated by Francis Crick and others, can be outlined in simple, linear terms: DNA is the molecule that, via RNA, transmits genetic information that creates the enzymes that construct biological traits. It is a one-way process; once genetic information gets into protein, it cannot get back out.

In the second half of the twentieth century, the study of evolution became ever more sophisticated, as refinements to and extensions of the theory brought the field increased subtlety and complexity. Some of the most significant advances in the last half-century have been William D. Hamilton's work on kin selection and inclusive fitness (1964), John Bowlby's theory of attachment (1969), Robert L. Trivers's theories of reciprocal altruism (1971), parental investment (1972), and parent-offspring competition (1974), Donald T. Campbell's idea of evolutionary epistemology (1974), John Maynard Smith's evolutionary game theory (1974), Richard Alexander's (1974) and Nicholas K. Humphrey's (1976) emphasis on the social function in evolution; Edward O. Wilson's sociobiology (1975), Amotz Zahavi's handicap principle (1975), Richard Dawkins's "selfish gene" theory (1976), and developmental systems theory (or evolutionary developmental biology, also called "evo devo") (Carroll 2005). Perhaps most important is the large, growing, influential, and controversial field of evolutionary psychology (see Chapter 13).

Language, in a Chomskyan sense, is a perfect example of Henry Plotkin's *secondary heuristic* at work, an understanding of human intelligence and its manifestations like memory and learning, reasoning and thinking, and culture and the ability to share knowledge.[1] The human species has the unique ability to "use a relatively limited—albeit quite large—set of symbols to generate a virtually infinite number of meaningful combinations to form utterances, each of which has meaning" (Plotkin 1993, 199). As a capacity that is innate to our species, language "fits well with the picture of the evolution of intelligence" (205) that has been outlined by Plotkin. Language becomes the most important tool that makes possible what we call culture: "learning about what others have learned, created or invented" (213). If a good case can be made for the evolutionary origins and biological function of language, then the social and human sciences, specifically including literary theory, will have to take such a concept of language into consideration.

The question of discussing evolution brings up a vexing issue, a problem that Darwin recognized: How do you tell a story without anthropomorphism and agency? What metaphors must you employ to make the process of evolution by natural selection understandable? In an original and insightful essay, H. Porter Abbott (2003) analyzes the difficulty of telling Darwin's story of evolution by natural selection. Early in *The Origin of Species* Darwin acknowledges the non-literal use of the term "struggle for existence" by saying, "I should premise that I use this term in a large and metaphorical sense including dependence of one being on another, and including (which is more important) not only the life of the individual, but success in leaving progeny" (1998, 90). In the third edition of *The Origin of Species* (published in 1861), Darwin wrote

> It has been said that I speak of natural selection as an active power or Deity; but who objects to an author speaking of the attraction of gravity as ruling the movement of the planets? Everyone knows what is meant and is implied by such metaphorical expressions; and they are almost necessary for brevity. So again it is difficult to avoid personifying the word Nature;

> but I mean by Nature, only the aggregate action and product of many natural laws, and by laws the sequence of events as ascertained by us. With a little familiarity such superficial objections will be forgotten. (109)

Human beings often attempt to make complicated matters more comprehensible by talking and writing of them as narratives. Narrative is a universal human cognitive function, and it is the most important way in which we know ourselves and our world. Here we consider, along with Abbott, the implications of presenting evolution by natural selection as a narrative.

"The special problem Darwin faced," Abbot writes, "and that Darwinians still face, is the combination of his idea's immense, wide-ranging importance and the fact that there is no unencumbered way of packaging it in narrative form without serious distortion" (2003, 144). Just as Darwin believed that he could present his work by recourse to figurative language, so have subsequent Darwinians striven to employ metaphoric language to make their points. The best example, perhaps, is Richard Dawkins's presentation of genetics in *The Selfish Gene* (1989, first published in 1976). Dawkins not only animated the stubborn "replicator" known as the gene, giving it agency and purpose, but even described its personality: "selfish." In what is perhaps his most frequently quoted statement, Dawkins describes these "engines of self-preservation" as follows:

> Now they swarm in huge colonies, safe inside gigantic lumbering robots, sealed off from the outside world, communicating with it by tortuous indirect routes, manipulating it by remote control. They are in you and in me; they created us, body and mind; and their preservation is the ultimate rationale for our existence. They have come a long way, those replicators. Now they go by the name of genes, and we are their survival machines. (19–20)

This does sound as if these selfish genes completely rule us; it sounds, in short, like genetic determinism. Here is where Dawkins's enthusiasm for metaphor lets him get carried away by his own rhetoric. His critics take this passage out of context when they fail to mention that Dawkins also explains how genes operate in collaboration with the environment:

> However independent and free genes may be in their journey through the generations, they are very much *not* free and independent agents in

> their control of embryonic development. They collaborate and interact in inextricably complex ways, both with each other, and with their external environment. Expressions like "gene for long legs" or "gene for altruistic behavior" are convenient figures of speech, but it is important to understand what they mean. There is no gene which single-handedly builds a leg, long or short. Building a leg is a multi-gene cooperative enterprise. Influences from the external environment too are indispensable: after all, legs are actually made of food! But there may well be a single gene which, *other things being equal*, tends to make legs longer than they would have been under the influence of the gene's allele. (36–37)

Later in his book, he makes it perfectly clear that he is in no way a genetic determinist, proposing that we "begin by throwing out the gene as the sole basis of our ideas on evolution" (191). Dawkins asserts that this sort of presentation is not meant to be taken literally, referring to such personification as "the license of talking about genes as if they had conscious aims," and then insisting that it is always possible to "[reassure] ourselves that we could translate our sloppy language back into respectable terms if we wanted to" (88). But writing about the matter in terms of narrative without agency and anthropomorphism is impossible, as is avoiding "sloppy language." And Dawkins's "story" then became the target of unrelenting criticism by those who misunderstood or willingly misrepresented what he was attempting to say. Thus, Dawkins became the poster-boy for crude biological determinism, in a sadly typical demonstration of how ignorance and dishonesty often work together (see Chapter 13).

2 Sexual Selection and the Role of Women in Evolution

In *The Origin of Species*, Darwin introduced a second theory that complements the more rigorous natural selection, the theory of "sexual selection," which he describes as a "struggle between the individuals of one sex, generally the males, for the possession of the other sex" (1998, 117). The idea was mentioned only briefly in *The Origin of Species*, but Darwin then went on to develop the idea at length in the two-volume *The Descent of Man, and Selection in Relation to Sex* (1871).

The competition for reproduction is an essential subset of the competition for survival, but it is among the most important factors in shaping the biology and psychology of any species, including human beings. The essence of sexual selection is that males compete with each other and display for females, who then choose their mates. While the male activity is often visible—stags fighting with their large racks of horns, elephant seals patrolling the beach to be sure no other male comes near their harems of females, peacocks displaying their enormous tails, bower birds building elaborate nests—actual mate *choice* is made by the female. As Premack and Premack put it, "A simple rule appears to govern choice of mate: The gender that feeds and cares for the young has the privilege of selecting the mate" (2003, 65). This is borne out by the fact that for those few (egg-laying) species in which the male provides most of the care and feeding, the male also makes the mate selection. Some further important concepts in the process that drives sexual selection include *feedback* (when a process, once begun, feeds back into itself and exaggerates the development of that process), *runaway* (a process that gets out of control and is taken to an extreme), and *handicap* (the notion that many animals have developed systems of signals that seem to put themselves at a disadvantage—large horns, bright coloration, daring actions—as reliable indications of fitness). One of the best books on sexual selection is Helena Cronin's *The Ant and the Peacock* (1991), which presents the history of controversies in the study of evolution, especially sexual selection, beginning with conflicts between Darwin and codiscoverer Alfred Russell Wallace; see also the books by Small (1993) and Andersson (1994), and Vandermassen's (2004) short essay on sexual selection and feminism.

Because of its threat to the privileged status of humans in the order of things, Darwin's core theory of evolution was already controversial, and the subject of sexual selection therefore received relatively little attention for about a century. During this time, male dominance and choice were widely assumed (especially by the males who were doing the studies), and females were considered generally passive, or coy, little more than vessels for the continuation of the male line. What little attention it did receive was mostly negative. Alfred Russell Wallace, codiscoverer of evolutionary theory, fought against the theory of sexual selection from the beginning. George Levine is undoubtedly correct when he writes that Darwin's theory of sexual selection "had very little immediate influence and largely because the scientific community found it impossible to credit the idea that the female could have had much to do with evolutionary development" (2006, 189). But the active role of female choice was an integral part of Darwin's idea from the very beginning.

Two positive developments in the 1960s helped create a new climate for the concept of sexual selection: the rise of feminism (Jolly 1999, 151) and the decline of behaviorism. The mechanistic, deterministic, and linear cause-and-effect mind-set of behaviorism cannot explain such nuanced concepts as sexual strategies, ideas about fitness, and active choices. The "revival" of sexual selection, according to Geoffrey Miller, "has been swift, dramatic, and unique. It may be the only major scientific theory to have become accepted after a century of condemnation, neglect, and misinterpretation" (2000, 65). David C. Geary's heavily-documented *Male, Female* (1998) is largely an extended argument that "the ultimate source of most human sex differences appears to be sexual selection and not the sexual division of labor or other economic or cultural patterns that are—from an evolutionary perspective—relatively recent phenomena" (94).

The new freedoms and self-confidence made possible by social advances toward sexual equality led, among other things, to more women choosing to study biology, especially primatology. These biologists were less likely to accept the active male/passive female dichotomy and were more able to take a fresh look at subjects such as parenting, male-female relationships, and female choice. So while some feminist scholars have rejected outright the possibility of any biological influence, positing gender, sexuality, and patriarchy as purely social constructions, "biodenial" (a term coined by Patai and Koertge 1994) does not characterize the many feminist contributions to evolutionary theory. Some of the most prominent of these women evolutionary scholars are the anthropologists Sarah Blaffer Hrdy and Alison Jolly, evolutionary psychologist Anne Campbell, and philosopher Griet Vandermassen. In addition, the physician Leonard Shlain has substantially contributed to our understanding of the tremendous role that the female has played in evolution.

Sarah Blaffer Hrdy established her reputation as a primatologist with her studies of langur monkeys in India (1977). Her analysis of the social role of infanticide in langurs and the different reproductive strategies used by males and females was a breakthrough in understanding why adults kill infants (and caused no small amount of outrage when she first presented the idea that the practice was adaptive for males). She turned to the role of women in evolution in *The Woman that Never Evolved* (1981; see the 2nd ed., 1999a), a revolutionary work as the first major book of evolutionary theory to take the role of women and of female primates in general into detailed consideration. Her arguments concerning the complexity of primate sexuality, especially female primate sexuality, were stunningly original (if long overdue) in the 1980s. She argues that female primates, including human women, are not passive beings who await mating by the victor of active male competition, but instead that primate females are actively competitive and sexually assertive. Among other things, she points out that the sexual practices of many primates—with the glaring exception of *Homo sapiens*—have been observed comprehensively over long periods of time, but that the only way to gather information about human sexuality is by means of questionnaires and interviews: "In sum, we have no information about the sexual lives of women comparable in its validity to what is known about savanna baboons" (1999a, 163). After reviewing all the atrocious things that men (along with their female accomplices) have done through the centuries to restrict female sexuality—female infanticide, claustration, foot-binding, clitoridectomy (euphemistically also called "female circumcision"), infibulation (labial fusion), the suttee (immolation of widows), and much more—she remarks that "women in so many human societies occupy a position that is far worse than that of females in all but a few species of non-human primates" (184–85). Her compelling conclusion is that, among the primates, it is the human female who "never evolved."

Hrdy continued and extended her work in *Mother Nature: A History of Mothers, Infants, and Natural Selection* (1999b), a book that should be read particularly by social constructionist humanists. In this long and complex study, Hrdy demonstrates convincingly that women and motherhood played a role in evolution that was not appreciated in most of the earlier studies carried out by men. She begins by facing squarely an issue of deep concern to many, that of attachment theory, which proposed that "human infants have an innate need for a primary attachment figure in the first years of life, a role that mothers are uniquely qualified to fill, and that human babies deprived of such attachments suffer irreparable damage"; Hrdy argues that, faced with the demand that they be primary caregivers and bound to their infants in order not to harm them, many feminists began to actively deny the role of biology in human relationships and to assert "that maternal love was a socially constructed sentiment without any biological basis, a 'gift given'" (24–25). Hrdy's goal is to place "human mothers and infants in a broader comparative and evolutionary framework" to "offer a new slant on what babies need from their mothers, what mothers need from others in order to provide it, and explain why" (117). She makes crucial use of a term coined by E. O.

Wilson: *alloparent* (*allo* from the Greek "other than"), applying it specifically to women: "*allomothers*—meaning caretakers other than the mother ... who help care for or provision young" (91). Hrdy extends and updates her study of alloparents in her book, *Mothers and Others* (2009), in which she argues (quite convincingly) that alloparenting is not a recent evolutionary development (see, e.g., 67, 109, 280–84). Hrdy's compelling, detailed inquiry into the relationships among evolutionary history, biological reality, Bowlby's attachment theory, and legitimate modern feminist concerns is a thorough and original combination of biology and social theory.

In her major work, *Lucy's Legacy* (1999), primatologist Alison Jolly works with the premise that "[b]iology offers an increasingly coherent view of human nature and humanity's place in nature" (2). Jolly, too, rejects the male-centric and competitive view that once dominated the field and specifically examines the role of women in human evolution. She contrasts human sexual practices (such as pair bonding, concealed ovulation, concealed sex, menstruation, and menopause; see especially 180–98) and the extreme dependency of infant humans, with their counterparts among other primates, especially chimpanzees. Her focus is on social intelligence rather than on tool-using or hunting intelligence.

In *A Mind of Her Own*, evolutionary psychologist Anne Campbell begins by directly confronting the social constructionist view, arguing that feminism's "laudable quest for social equality" has led to a position that she describes as follows:

> The only acceptable account of sex differences was one which explicitly acknowledged the socially constructed, arbitrary and malleable nature of sex differences. Women's studies became steeped in a politically-driven rejection of essentialism (the idea that the sexes differ at a fundamental psychological level) and committed on the one hand to social constructionism (there is no objective truth "out there," only negotiable subjective representations) and on the other to extreme environmentalism (all sex differences result from factors external to the person). Neither road has taken us very far towards an accurate understanding of why men and women differ. (2002, 1; see also 2006)

Campbell presses the issue to insist that a serious engagement with evolution and biology must be a part of any acceptable consideration of male–female differences and relationships (8). Campbell explores such issues as differences between males and females with regard to sexual strategies and parental investment, aggression and violence, risk-taking and fear, competition and friendship, marriage and nurturing, and related matters.

If Hrdy, Jolly, and Campbell all deal to some extent with feminist concerns in their studies of biology and evolution, philosopher Griet Vandermassen makes it the central subject of her book *Who's Afraid of Charles Darwin?* (2005; see also 2004). Vandermassen opens by showing how she changed from a social constructionist feminist and critic of science in general to what she calls a "Darwinian feminist." She believes, for instance, that "feminism throws away a valuable tool for understanding sexism in denouncing an evolutionary approach to the human mind" (60). Explaining why she gladly draws from evolutionary science in her approach to feminism, Vandermassen comments:

> People who would never think about questioning the legitimacy of psychological or sociological research sometimes ask me what evolutionary insights are good for. I continue to be amazed by this. Is it not tremendously important to try to understand human nature? I suspect that the main reason is the false dichotomy between nature and nurture that is still preeminent in people's minds: It is good to study "nurture," since that implies we can find ways of remedying human behavior, but why should we study "nature," that fixed entity we cannot change? And aren't we cultural beings who have left the constraints of nature behind us? As I hope to have shown, things are more complex than that. (2005, 179–80)

In her conclusion, she elaborates her position that a rejection of biological and social scientific evidence by feminists will cause them to "back themselves into an embarrassingly uninformed corner" (2005, 196). Vandermassen devotes over a third of her book to a discussion of evolutionary psychology and even proposes it as "a unifying metatheory for feminism" (2005, 148).[2]

Physician Leonard Shlain's *Sex, Time and Power: How Women's Sexuality Shaped Human Evolution* (2003) is another particularly interesting approach to evolutionary theory. Shlain cites the evidence from genetics that the human species passed through a "bottleneck" somewhere between 100,000 and 200,000 years ago. At that time, there may have been no more than about 20,000 of our ancestors (probably in northeast Africa), and we are all descended from them (Olson 2002, 28). Then, he proposes that this bottleneck stemmed from the shape of the birth canal: "The death of the Unknown Mother and her unlucky baby, and the subsequent dying off of increasingly large numbers of hominid mothers and their newborns, was the stressful Factor X that precipitated the *Homo sapiens* line" (7–8). Thus, it was the human female, not the male, who passed through the evolutionary bottleneck, an event that transformed her in a way not experienced by the female of any other species. This, in turn, caused the male to adapt his behavior to this new sexual/social reality (14). Shlain's hypothesis is that "*the history of our species could be written from the perspective that males have spent the last 150,000 years trying to regain the power they so emphatically lost to females when we differentiated away from* Homo erectus" (15). Among the profound changes brought about by human female sexuality, all of which together distinguish our species from any other, are hidden ovulation, elimination of estrus, the uncoupling of sex from reproduction, female orgasm, variety of sexual positions, and prolonged sexual foreplay (15–17).

Above all, Shlain calls attention to women's massive loss of iron-rich blood, unique among the 4,000 mammals living on earth: "*By one avenue or another, a woman is always losing iron.* Over a lifetime, the average woman loses the equivalent of approximately *fifteen gallons* of iron-rich blood due to menses, pregnancy, delivery, birth trauma, placental loss, failure to consume the expelled placenta, and lactation" (30). Shlain then argues that almost every aspect of human evolution and, as a result of human social structure—including human concepts of birth, death, and paternity—can be traced back to the human female's loss of blood and iron and to other aspects of female sexuality. He places so much emphasis on women's leading role in evolutionary history that he even employs the term *Gyna sapiens* (20) rather than *Homo sapiens*. His book, like the work of Hrdy, Jolly, Campbell, and Vandermassen, should contribute to a reevaluation of the role of biology and evolution in the social sciences and humanities.

If the female of the human species was once viewed as little more than the passive receptacle and vessel of transmission of precious male genes, that day is long past. Evolution today is an equal-opportunity theory, recognizing the roles of both biological sexes, in addition to external forces of various kinds, in all processes that have resulted in our species' current state. Specifically, sex, gender, the body, the mind-brain, and all social relationships came into being as a result of the interaction and interrelationships between biology and culture. It is, again, nature *and* nurture. Concepts of gender and sex are obviously crucial to literary, cultural, and feminist studies, and sexual selection is at the same time a theory that has been neglected for so long. We hope that this section of our book will help promote knowledge of this important direction in evolutionary theory, particularly among humanists.

3 Human Evolution

Some humanist scholars hold vague and inadequate concepts of human evolution that then inform their scholarship. One of these is that at some point in evolution humans became something qualitatively different from all other species and that biology now plays no part in our cognitive processes. Another is the straight-line theory, a sort of traditional Great Chain of Being, in which humans are the direct and inevitable result of a simple, causal chain of developments that begins with the most elementary forms of life and ends with us as its crowning achievement (see Pinker 1994, 342–49). Daniel Dennett bluntly expresses a view with which we completely concur: "To put it bluntly but fairly, anyone today who doubts that the variety of life on this planet was produced by a process of evolution is simply ignorant—inexcusably ignorant, in a world where three out of four people have learned to read and write" (1995, 46). The truth is, rather, that there has been a complex and often unclear and confusing development of hominid species. *Hominids* (or *hominidae*) are human-like species. This term should not be confused with a similar one, *hominoids* (or *hominoidae*), which refers to all apes, including humans; it is thus a superordinate term. Hominids—essentially bipedal apes—are the various intermediary forms between the last common ancestor of both apes (chimpanzees and bonobos) and humans; they are "humans before humanity" (see Foley 1995, 77–79). Humanity itself begins with the emergence of *Homo sapiens sapiens* about 100,000 years ago.

Since literary scholars tend to be unfamiliar with the various species of early hominids and generally do not have a solid of understanding of when they existed or how they are related—nor did we before we started on this project!—we include here in schematic form an overview based largely on the model traced by Steven Mithen (1996, especially 24–25):

SPECIES	TIME SPAN IN YEARS AGO
Australopithecus	4.5 million–1 million
Homo habilis	2 million–1.6 million
Homo erectus	1.8 million–300,000

(continued)

(continued)

SPECIES	TIME SPAN IN YEARS AGO
Homo spaiens	400,000–100,000
Homo neandertahlensis	220,000–30,000
Homo sapiens sapiens	100,000–present

The contextualizing terminology, frequent in archeological and evolutionary literature, for these phases of hominid development can also be confusing, especially the overlapping terms Paleolithic and Pleistocene. The *Paleolithic* is essentially what is also more popularly known as the Stone Age; it is divided into three subsections:

- Early, beginning about 2,000,000 years ago, characterized by the primitive stone tools of *Homo habilis* and successors;
- Middle (also often referred to as the Mousterian period), beginning about 200,000 years ago, and characterized by the more refined stone tools used by the Neanderthals and *Homo sapiens*; and.
- Upper, beginning about 100,000 years ago, and featuring the more elaborate tools of stone and other materials used by the Cro-Magnons—modern human beings.

The *Pleistocene* corresponds essentially to the Ice Ages; it, too, is subdivided into three periods:

- Lower, beginning about 1,800,000 years ago when the glaciers first began to spread over the northern hemispheres;
- Middle, beginning about 900,000 years ago when the temperature began to fluctuate rather "rapidly" (in evolutionary terms) between cooler (glacial) and warmer (interglacial) periods; and
- Upper, beginning about 100,000 years ago as the temperature fluctuations continued but with the glacial periods much shorter (the last took place between about 24,000 and 13,000 years ago).

In both schemes, the distinction between Middle and Upper is frequently blurred, and they are often referred to together as Middle/Upper. The choice of terminology is generally determined by the writer's emphasis on climate (Pleistocene) or technology (Paleolithic). The last 10,000 or so years are often assigned new terminology: *Holocene* for climate and *Neolithic* for technology (with the unofficial and inconsistently used geologic term *Anthropocene* either overlapping with or following these periods). This period includes the entire span of recorded history, the expansion of human beings over the entire planet, and the increase in human population from a few million to nearly eight billion at this writing. The terminology designates the period in which human beings actively began to change the world's environment—and not just cope successfully with it—and it is a convenient way of distinguishing the modern world from the primitive one.

Human evolution is a history that is constantly being rewritten; all the proposed dates suggested here are subject to change. For example, a huge cache of fossils was uncovered in the old Gran Dolina railway gorge near the town of Atapuerca in northern Spain in the 1990s. At that site, and especially in the Sima de los Huesos (Cavern or Pit of Bones), evidence was found that suggests that ancestors of the Neanderthals (and perhaps also modern humans)—the name *Homo antecessor* has been proposed—lived there perhaps an astonishing three hundred thousand years earlier than any date that has previously been suggested for Europe. For a concise summary and careful consideration of the importance of these and related discoveries, see Arsuaga (2002) and Arsuaga and Martínez (2006, 184–96); see also Wills (1998, 127–50). The Museo de la Evolución Humana (Museum of Human Evolution), the only museum of its kind in the world, in nearby Burgos, Spain, opened in 2010 to feature the discoveries at Atapuerca. More prehistoric human remains have been excavated from Atapuerca than from the combined total of all other sites in the world (see Vergano 2014 for more about the "bumpy" evolution of humans in prehistoric Europe as indicated by studies of the Neanderthal skulls from this site).

An important evolutionary development is the Lower Paleolithic appearance of *Homo habilis* (also known as *Homo ergaster* and *Homo rudolfensis*, although these may be different subspecies, perhaps mid-way between *Homo habilis* and *Homo erectus*), the earliest in the line that led more or less directly to modern humans. As the variety of names suggests, these may have been various species with different degrees of relationship, but the single term *habilis* is most often used to refer to them all. *Homo habilis* was taller than *australopithecus* and had a brain volume of some 675 cc., a little more than 50% larger. There is clear evidence now of tool use, the famous handaxes, and other Oldowan artifacts.

Homo erectus was the first hominid to move out of Africa into Europe and Asia (below the ice sheets), as far as Java (famous fossils include the Peking Man and the Java Man). The discovery in 1984 of the Nariokotome boy made available the most complete skeleton of a single early hominid ever discovered (see Walker and Shipman 1996). This species has, again, increased height (about six feet) and brain size (approximately 950 cc.), over a third larger than that of *Homo habilis*. Though *Homo erectus* almost certainly did not have full spoken language, there was probably some gestural and proto-linguistic communication. *Homo erectus* was a more proficient tool-user than any ancestor, usually employing the more sophisticated Acheulian technology.

Archaic *Homo sapiens* is distinguished from *Homo erectus* again largely by a greater brain size, about 1300 cc., fully a third larger than their predecessors, and almost the size of that of modern humans. The European examples of the species are often referred to by the name of *Homo heidelbergensis*, from remains discovered in Germany. The most interesting new artifacts of the species are the much more refined Mousterian pear-shaped stone handaxes, and some remains of artifacts made of wood.

Homo neanderthalensis, the famous Middle Paleolithic cave-dwelling Neanderthals of the Ice Ages, represent a dead end in the *Homo* line, having apparently completely died out in southwestern Europe (perhaps in Gibraltar or Portugal), about 30,000 years ago. A species with a slightly larger brain size (about 1500 cc.) than humans today, the Neanderthals inhabited Europe and the Near East until modern humans replaced them. Their Mousterian stone technology resulted in relatively elegant and efficient tools. It is unlikely that the Neanderthals had any sort of language in the modern sense, and this may have been a decisive factor in their demise in competition with *Homo sapiens sapiens*, the species alongside whom they coexisted for thousands of years. Although often depicted as loutish ape-men, the Neanderthals adapted to a severe climate and lasted for nearly 200,000 years. As Ian Tattersall has commented, they should be regarded "as exponents of the most complex—and in many ways admirable—lifestyle that it has ever proved possible to achieve with intuitive processes alone" (2002, 155). For more on the perennially fascinating and often underestimated species known as Neanderthals, see Shreeve (1995), Arsuaga (2002), Mithen (2006), Finlayson (2009, 2019), Papagianni and Morse (2013), Pääbo (2014), and Sykes (2020).

Anatomically modern humans, *Homo sapiens sapiens* (or, more often, though perhaps a little confusingly, simply *Homo sapiens*), originated in

Africa, but are soon found in Europe and the Near East and then—by means of what must have been a remarkable feat of navigation—in Australia. It is now not brain size that matters, for *Homo heidelbergensis* and the Neanderthals had brains just as large or larger than the 1400 cc. average of *Homo sapiens sapiens*, which remains the approximate volume of the human brain today. Fascinatingly, humans have changed significantly little (except in features such as skin color) since first appearing in Africa some 100,000 years ago. The earliest European representatives of the species, the Cro-Magnons, date back to about 40,000 years ago. The Cro-Magnons had language and created the superb cave paintings of Spain and France, where the interplay of shape, color, and ceiling and wall surface is nothing short of brilliant. In place of the earlier stone tools, these people left behind artifacts of wood, bone, and ivory. In addition to living in caves, this species begins to construct dwellings to inhabit. Overall, however, modern *Homo sapiens* remained a hunter-gatherer species for tens of thousands of years. Only within the last 10,000 years or so have agriculture, the domestication of animals, the growth of towns and cities, and other characteristics of modern civilization begun to appear.

Human evolution in the last 5 million years or so has remarkably outpaced that of other primates,[3] perhaps because the latter has been much less nomadic. Human beings today occupy virtually the entire surface of the planet, while our primate relatives remain in the stable environment of the forest, where, sadly, we are responsible for their rapidly decreasing numbers as we destroy them (especially the orangutans, gorillas, and bonobos) and their habitats. It may be only a matter of time until the planet's only remaining species is that of *Homo sapiens sapiens*, far less wise than its proud name proclaims.

All literary theory and criticism must entail some concept of what language is, how it functions, and what we can say and write about it. All discourse about literature, as well as literature itself, is based on some concept of language. Even the absence of an explicit language theory constitutes a (naive and functionally inoperative) theory of language. No principled study of literature can afford to be based on anything less than the best possible theory of language, and no understanding of the phenomenon of language is possible without acknowledgement of the evolutionary history and biological composition of our species, with the beginning (birth, emergence, invention, development, and evolution) of language itself in the history of our species.

4 The Origins of Language

Noam Chomsky, the most important modern linguist to ground the study of language within the fields of psychology and biology, has, surprisingly, consistently maintained that, even though language is largely a biological function, there is no way to determine how it might have come about as a result of an evolutionary process. Rather, Chomsky has proposed language must have been a relatively sudden and unmotivated occurrence in early human history, possibly beginning with a fortuitous genetic modification in a single individual (Chomsky 2012, 14). On this issue, Chomsky was influenced by Eric H. Lenneberg, who was carrying out the research that led to his important *Biological Foundations of Language* (1967, 227–70), where he specifically rejects the idea that language could have developed by means of normal evolutionary processes. Primatologist David Premack agrees, stating that "[h]uman language is an embarrassment for evolutionary theory because it is vastly more powerful than one can account for in terms of selective fitness" (1986, xx). Although it might seem curious that the founder of a biology-based approach to language should deny what ought to seem logical—that is, an evolutionary origin for language—it turns out that such a position is not at all uncommon. Even our age's most popular and influential promoter of evolutionary science, Stephen J. Gould, argued for a non-adaptationist origin of language. Gould considered language an *exaptation*, or *spandrel*, that is, an accidental by-product of evolution rather than a trait selected for survival through normal evolutionary adaptation (Pinker and Bloom 1992, 452).

The precise time when linguistic abilities became an essential part of early human biology and culture and the precise anatomical and cognitive changes that bring such abilities are not crucial matters, but the continuity hypothesis is the one that seems to have the most support. The linguistic and symbolic capabilities that characterize humanity are clearly the result of evolutionary developments. So while we do not claim that the version sketched out below must be exactly what happened, our approach to the question of the origins of human language is based on the following:

- the interrelated theories of a protolanguage and a gestural basis for a mimetic phase of human culture,
- a social origin for language,
- the concept of coevolution of mind and brain,

- the development of cognitive fluidity,
- the evolution of human anatomy, and
- the linguistic capabilities (or lack thereof) of other animals.

This position is based on much of the best work in the field from a variety of disciplinary perspectives, and the findings converge in a coherent fashion.

Terrence W. Deacon makes much of the fact that no simple natural language—of the sort seen in myths and fables, cartoons, and Walt Disney movies—seems to exist. No such language is spoken either by humans who live in "primitive" cultures or by some animals such as the higher primates (1997, 12, 41–42). If no such thing as a simple language exists today, it must have existed at some time in our evolutionary past and have been replaced by full-blown natural languages of the type that exist today throughout the world (44–45). It is impossible (barring supernatural or miraculous intervention) that suddenly some species of hominids broke the linguistic barrier of all other species and started to speak as we do today. Language must have had its own evolution, with some kind of primitive and simple origins, emerging eventually into the complex linguistic reality that we know.

Derek Bickerton addresses Deacon's concern in *Language and Species* (1990) and presents a logical case for what he calls *protolanguage*. Bickerton's task is to reconstruct this protolanguage to fill in the gap while also addressing the Chomsky-Lenneberg position that we cannot know what the formal evolution of language might have consisted of. In so doing, he considers what he calls the "fossils of language," the artificial, primitive, incomplete, and ad-hoc languages—in effect Deacon's simple languages—spoken (or signed) by apes, language of children in the earliest stages of language acquisition, language of so-called wolf children, pidgin languages, and cases of language aphasia. Each of these is in different ways reminiscent of a primitive language state—a protolanguage—that characterized hominid cognition at a more fundamental level than full language. Trained apes and biologically or culturally deprived human beings can acquire this protolanguage, at least to some limited extent, because it does not require a critical period for acquisition and because it is a more basic and more robust form of human biological endowment (118). Bickerton's careful description of the characteristics of protolanguage (122–26) reveals "wide and deep" differences between it and all natural languages. Bickerton's work in focusing our attention on a model for the development of language in the natural course of human evolution is an important start, but it is incomplete.

In *Formulaic Language and the Lexicon* (2002), Alison Wray has given the theory of protolanguage a new twist. Wray posits formulaic language as a major factor in language use, and she defines this as "[w]ords and word strings which appear to be processed without recourse to their lowest level of composition" (4). Formulaic language is, she suggests, not merely an adjunct to compositional syntax, but a major, omnipresent, and defining characteristic of language. Everyday language use is filled with formulaic linguistic units, including set phrases such as "good luck," "have a nice day," "thank you very much," "you know what I mean," and so forth.

In *The Singing Neanderthals* (2006), Steven Mithen has made the most original use of this idea, arguing for holistic language communication at an early stage in human evolution. Much early human verbal communication, he suggests, was not carried out with discrete words and elementary syntax, but with sounds that represented whole ideas, concepts, or messages—utterances associated with one single meaning. Mithen describes such a communication system as "'Hmmmmm': *H*olistic, *m*anipulative, *m*ulti-*m*odal, *m*usical and *m*imetic. Its essence would have been a large number of holistic utterances, each functioning as a complete message in itself rather than as words that could be combined to generate new meanings" (172). For Mithen, elements of music such as rhythm, pitch, and motion would have been major characteristics of such vocal communication. "Hmmmmm" would thus have been especially good for communicating feelings or emotions; after all, a major difference between language and music is that music lacks referential meaning (17–23). The earliest "Hmmmmm" would have been characteristic of hominids as early as 1.8 million years ago in *Homo habilis* (or *Homo ergaster*).

For Mithen, the Neanderthals probably represent the peak of "Hmmmmm" communication and illustrate why this type of vocalization was inadequate in comparison with full-fledged language. They may have been "singing Neanderthals" (221), but they lacked the cognitive fluidity that led to language in *Homo sapiens* (232–33). *Homo sapiens* had language abilities that fostered a new way of thinking, and they subsumed "Hmmmmm" into speech, so that when Neanderthals had to coexist and compete with *Homo sapiens,* the Neanderthals lost out and became extinct. Remnants of "Hmmmmm" remain in infant-directed speech (motherese), in the use of spontaneous gestures during speech, in elements such as onomatopoeia, in sound synesthesia, and in rhythm in speech.

If once we could ignore the role of the hand in the evolution of human cognitive abilities, the publication of Frank R. Wilson's extraordinary book, *The Hand* (1998), made that impossible.[4] Wilson argues that "any theory of human intelligence which ignores the interdependence of hand and brain function, the historic origins of that relationship, or the impact of that history on developmental dynamics in modern humans, is grossly misleading and sterile" (7). In a series of compelling chapters on subjects as dissimilar as juggling, automobile repair, and piano playing, Wilson writes passionately about the role of the hand in human activities, specifically including those that tend to be considered mental rather than physical.

As the hand evolved, it enabled the brain to change in ways not previously possible; as the brain evolved, it enabled the hand to pursue further physical adaptations (146). In part, the brain grew because of the physical capabilities of the hand, while the hand became capable of more and more because of an evolving brain that was capable of conceiving of new things for the hand to do. *Coevolution* is a concept appropriate to brain and language, but also applies to what the human hand was able to accomplish during the long passage of time between the descent of the australopithecines from the trees until early modern humans painted on the walls of the caves of Altamira and Lascaux. The dialogic relationship between hand and brain, Wilson's "hand-brain complex," is yet another illustration of mind-body continuity rather than dualism.

The relationships among gesture, tool making and tool use, and language are all underscored by the prominence of the left hemisphere, especially Broca's and surrounding areas. As Michael C. Corballis suggests in *The Lopsided Ape* (1991), all of these activities involve similar processes, particularly recursion, embeddedness, and generativity. Especially the latter process, generativity—which Corballis defines as "the ability to construct an unlimited number of different forms from a finite number of elementary parts" (65)—plays a crucial role in human evolution. Because this ability is unique to our species, Corballis posits the existence of a "generative assembling device, or GAD" (219), based on and related to Chomsky's language acquisition device, or LAD. Generativity also plays an important role in mental imagery, which, perhaps surprisingly, is also very largely a left hemisphere function. The evidence of tool use in hominids as early as two million years ago with *Homo habilis* suggests to Corballis that we can logically posit the most rudimentary form of generativity at about that time. A long, slow coevolution of generativity, tool making, gesture, protolanguage, and, eventually, language fits well into the scheme proposed here. Corballis extended his study of the relationship between gesture and language in *From Hand to Mouth* (2002), where he suggests that speech fully independent of gesture may date back only to some 50,000 years ago. At that time, he proposes, autonomous speech was "invented" largely to free the hands from the burden of communication to permit them to perform other tasks while communication was being carried on.

The modern study of sign language was initiated primarily by William C. Stokoe in the 1950s and 1960s and has led to the general recognition of signed languages, such as American Sign Language (ASL), as full languages in every sense of the term and to the expansion of the concept of language well beyond a spoken phenomenon. Modern research in sign languages has added support to the thesis that gestural communication most probably preceded spoken language (or, at least, was more prominent than language at the earliest stages) in the history of human evolution. Three books by prominent scholars from this field stand out: (1) *Gesture and the Nature of Language* (1995) by David F. Armstrong, William C. Stokoe, and Sherman E. Wilcox; (2) Armstrong's *Original Signs* (1999), essentially an extension of and elaboration of the theses of the first book; and (3) Armstrong and Wilcox's *The Gestural Origin of Language* (2007).

Armstrong et al. argue, among other things, that "[a] linguistic theory must acknowledge the physically embodied grounding of language; in a very real sense, the body is in the mind (Johnson 1987)—the essence of language is bodily activity" (1995, 36). This is because, they insist, "if we are to do the biology of language at all, it will have to be done by tracing language to its roots in the anatomy, physiology and social environment of its users. Only in this way can we hope to arrive at an account of language perception and production fitted to animals rather than machines" (38). In justification of this position, Armstrong et al. cite the work of cognitive psychologist Ulric Neisser, who has used the term "articulatory gestures" for human speech (1976, 156). Grounding their work in this line of thought, which includes some of the best research in cognitive science in recent years, together with a sophisticated study of the nature of gesture, and other results from research on sign language, the authors make a strong case for a gestural basis for speech. Although Armstrong (1999, 111–32) recognizes the clear advantages of speech over signed languages, he also makes a case for a superiority of the latter because of its simultaneity, as opposed to the inherently limiting sequentiality of speech. The ability to combine gesture and speech, along with intonation and other pragmatic strategies, makes a multi-modal spoken language the most effective means of communication available to human beings. Armstrong and Wilcox continue these ideas in their 2007 book, using human fossil records to support their claims that grammar is grounded in gesture.

With his *The Origins of the Modern Mind* (1991), Merlin Donald offered one of the most original, influential, and persuasive presentations of human cognitive development. Donald traces human evolution through four stages: (1) episodic; (2) mimetic; (3) mythic; and (4) theoretic (our current evolutionary stage). Apes, and other mammals to varying degrees, inhabit an episodic cognitive culture, where they live "entirely in the present, as a series of concrete episodes, and the highest element in their system of memory representation seems to be at the level of event representation"; in effect, they are "bound to the concrete situation or episode; and their social behavior reflects this situational limitation" (149). Early humans emerged from this sort of episodic culture. And their emergence, during the period of *Homo erectus*, was, according to Donald, into a mediating phase that he calls "mimetic" culture, the missing link in the evolution of cognition. Mimesis for Donald involves much more than simple mimicry; it "rests on the ability to produce conscious self-initiated, representational acts that are intentional but not linguistic" (168) and, as such, is a kind of multi-modal representation of concepts by imitation, posture, body movement, facial expression, movements of the eyes, gesture, sign, vocalization, intonation, rhythm, and so forth. (Recall that mimesis is the final *m* of Mithen's "Hmmmmm.") Donald places the transition to what he calls a "mythic," or language-centered, culture, during the period of anatomically modern *Homo sapiens*. For Donald, following Robin Dunbar (see below), the primary function of language was social and narrative, and since the most important narratives were those purporting to explain origins, "to construct conceptual models of the human universe" (215).

For Donald, the modern mind is a hybrid structure (like a set of Chinese boxes or Russian nesting dolls) in which each of its major components—the episodic, the mimetic, the mythic, and the theoretic—is recursively embedded, one within the other. Particularly important for Donald is the increasing role of culture in humanity's evolutionary development; in effect, culture (together with its product, technology) now supplements (and complements) biology as a major factor in our continued evolution. Donald repeated and expanded on this idea in his later book *A Mind So Rare* (2001). This is not to say that culture has replaced biology or that we are culturally determined, but that our created culture (involving technology, society, language, and much more) has grown in importance with each stage of evolution. Donald's view is thus consistent with Plotkin's idea of a secondary (cultural) heuristic built upon and constrained by the primary heuristic of biology. As one of the most significant inquiries into human evolution, *The Origins of the Modern Mind* should be of interest to all literary scholars.

Donald specifically relates the origins of language to primate social practices, basing this to some extent, as indicated above, on the work of anthropological psychologist Robin Dunbar, whose book *Grooming, Gossip, and the Evolution of Language* (1996) is one of the most influential contributions to the subject. Dunbar's point of departure is the fact that research shows that the majority of human conversation is about relatively trivial social matters—in a word, gossip. Even most of what we read in books (and see in film and on television) is about the personal lives of other people, much of it imagined (i.e., fiction). Why is it, he asks, that the most powerful tool ever invented by human beings, language, is used more than anything else for what would seem to be such a trivial activity: "Here, then, is a curious fact. Our much-vaunted capacity for language seems to be mainly used for exchanging information on social matters; we seem to be obsessed with gossiping about one another" (6–7).

Explicitly and firmly rejecting the assumption that language is solely a cultural phenomenon that is "beyond the pale of biological explanation" (8), Dunbar builds on recent developments in evolutionary science to set forth his approach to the social origins of human language. Noting that all primates are, to varying degrees and in different ways, extraordinarily social animals, Dunbar calls attention to "the long sessions of grooming so peculiarly characteristic of primate societies" which he perceives as "the key to the processes that give primate societies their cohesion and sense of belonging" (35). Through an ingenious and detailed procedure that prominently involves the size of various animals' neocortex and patterns of ethology related to group size, Dunbar reaches the conclusion that perhaps "language evolved as a kind of vocal grooming to allow us to bond larger groups than was possible using the conventional primate mechanism of physical grooming" (78). (Mithen added the soothing effects of music; 2006, 136.) As he develops this hypothesis, Dunbar draws heavily on the concept of Theory of Mind (see our Chapter 10): "being able to understand what another individual is thinking, to ascribe beliefs, desires, fears and hopes to someone else, and to believe that they really do experience these feelings as mental states" (83).

Geoffrey Miller, in his book *The Mating Mind* (2000, 341–91), extends Dunbar's gossip theory to displays of fitness, status, and courting, which helped language to evolve in its complexity (353). Even public speaking allows someone the opportunity to display her "knowledge, clear thinking, social tact, good judgment, wit, experience, morality, imagination, and self-confidence" (357). Miller offers us yet another example of how sexual selection has played such an important role in the evolution of so many aspects of human art and culture.[5]

5 Coevolution of Language and Brain

Neuroscientist and evolutionary anthropologist Terrence Deacon, in his extraordinary book *The Symbolic Species* (1997), also sees language as evolving very slowly over a very long period of time. Deacon defines language as "a mode of communication based upon symbolic reference (the way words refer to things) and involving combinatorial rules that comprise a system for representing synthetic logical relationships among these symbols" (41). In this sense, Deacon holds that there is no such thing as a simple (human or non-human) language. Various forms of animal communication, such as calls, cries, grunts, songs, and so forth—which have generally been underestimated for their communicative power by previous researchers, and which humans share to a large extent—are not languages.

Key to Deacon's definition is the concept of *symbolic* reference. Using C. S. Peirce's useful terminology of icon, index, and symbol, Deacon traces the necessary hierarchy among them: Icon stands alone; index incorporates icon; symbol subsumes both index and icon. Other animals may use iconic and/or indexical means of communication, but only human beings communicate naturally at the symbolic level. Noteworthy here also is Deacon's insistence that linguistic (symbolic) communication is organized around the concept of reference to the external world: "When we strip away the complexity, only one significant difference between language and nonlanguage communication remains: the common, everyday miracle of word meaning and reference" (43). In direct opposition to the notion that language is a closed or self-contained system that can refer only to itself, Deacon (and the rest of cognitive science) is emphatic that linguistic reference is to the world.

Deacon, like Bickerton previously, postulates the existence of some sort of simple form—a "multi-modal" form, presumably something like Bickerton's protolanguage and/or Donald's mimesis (353)—that led to the development of language, and he sees language, once its development had begun, as its own prime mover. As language evolved, so did the human brain (44–45). The term Deacon uses (see his book's subtitle) for this process is the "co-evolution of language and the brain," a process that recalls Wilson's "co-evolution of hand and brain." Just as language was shaped by evolutionary changes in the brain, so the brain was shaped by evolutionary changes in language; see also Lieberman (1991, 80–82). To a large extent, Deacon writes, "an idea changed the brain" (322): as part of the evolutionary process our brains were—literally, physically—shaped by symbolic thought: "The remarkable expansion of the brain that took place in human evolution, and indirectly produced prefrontal expansion, was not the cause of symbolic language but a consequence of it" (340).

Rather than assume that the brain is adapted for language, Deacon proposes that language, in its evolution, became adapted for the brain. Whereas biological evolution is slow, taking place over thousands of years, linguistic evolution is relatively rapid, with important changes taking place within just a few years. Likening languages to living organisms, Deacon (while readily admitting the metaphoric, not literal, value of his suggestion) proposes that "[t]he most basic principle guiding their design is not communicative utility but reproduction—theirs and ours" (110); Deacon imagines languages as "an independent life form that colonizes and parasitizes human brains, using them to reproduce" (111). It is interesting that both Dunbar and Deacon compare language to a parasite inhabiting the brain, thus figuratively giving us a new and significant insight into the relationship between brain and language. Deacon holds that although the earliest, most primitive, aspects of language may have come into existence as long as two million years ago, *Homo erectus* probably had some degree of symbolic communication. But language reached its full development with archaic *Homo sapiens* (i.e., even before anatomically modern humans), which he places in the time period of 200,000 to 100,000 years ago.

Deacon's coevolution approach bridges the apparent gap between the purely biological on one side and the purely cultural on the other. This bridge can also be anchored on other solid ground, particularly that of "cognitive fluidity," as proposed by Steven Mithen. In *The Prehistory of the Mind* (1996), Mithen characterizes his work as *cognitive archaeology*, conceived as an exact counterpart to evolutionary psychology (13). Mithen starts by reviewing some of the metaphors traditionally used to describe the mind: The mind is a blank slate, written on by culture/language/ideology; it is a sponge, soaking up information; it is a computer, processing information by manipulating symbols; it consists of multiple intelligences (Gardner 1993); and, the favorite of the evolutionary psychologists, it is a Swiss army knife with a multi-modular structure. Mithen finds fault with all of these and proposes a new one: The mind is like a cathedral (1996, 65–72).

Drawing on recent work in infant cognition and developmental psychology, Mithen recognizes the following intuitive abilities of children: language, psychology, biology, and physics. But beyond this cluster of abilities for which human beings may well have specialized circuits, Mithen finds the army knife metaphor misleading, because he is convinced that the human mind also has the kind of all-purpose, general cognitive skills—another area with strong support from child and developmental psychology—that evolutionary psychologists such as Cosmides and Tooby et al. do not adequately acknowledge. His mind-as-cathedral model consists of a great edifice with four "chapels," consisting of four "intelligences"—technical, linguistic, social, and natural history—which surround a central "nave" of general intelligence and which are connected both to that nave and to each other (see his graphic model, 67). Mithen argues that the ability to move mentally within this structure gives humans a *cognitive fluidity* that is unique.

Human anatomy provides further convincing evidence of the biological nature of language. Philip Lieberman (1984, 1991, 1998, 2000) has studied both the physiology and the psychology of speech. To speak, one must have the physical mechanism for producing sounds—and only human beings have this anatomical apparatus: the supralaryngeal vocal tract. As air is expelled from the lungs, it passes through the trachea, makes the larynx (the voice box) vibrate, and then is formed into the large variety of characteristic human sounds by the velum (or soft palate), tongue, teeth, lips, and other parts of the mouth and nasal cavity. The key here is the larynx, an adaptation of human beings that has its evolutionary origins in fish, where its purpose was to keep water from entering the lungs. In *Homo sapiens*, however, the larynx is lower in the neck than in any other animal, positioned in such a way that it has room to vibrate in the fashion necessary to produce the sounds of language. But this adaptation was achieved only at the expense of another feature, the simultaneity of breathing and swallowing food. The larynx is placed in such a way in human beings that we run the risk of getting food lodged at the point of the larynx and blocking the passage to the lungs; thus, thousands of our species choke to death every year.

Obviously, some substantial differences exist among these evolutionary theorists. But some very basic similarities emerge in the totality of this convergent work from linguistics, psychology, biology, genetics, anthropology, archeology, ethology, primatology, and neuroscience that, together, make what we take to be a definitive case: Human language is a natural product of the regular adaptive and selective processes of evolution. These processes probably began in a very rudimentary way at a very early time, perhaps shortly after the branch in the evolutionary tree that definitively separated other primates from humans as australopithecines first began to walk on just two feet, thus freeing the hands for other purposes, such as gestures. Certainly by the time of *Homo erectus*, there was some sort of protolanguage and/or gestural communication—a mimetic culture—that already distanced this species from all others bound irremediably to an episodic culture. Once begun, the process of symbolic communication, originally and primarily a social activity, fed recursively back upon itself, accelerating its own development and at the same time changing the size, shape, structure, and function of the human brain and, with it, the maturational calendar for the species. This coevolutionary process also made possible a more general and generative type of thinking that can be referred to as cognitive fluidity.

This review of research in language and evolution is intended to serve at least four purposes. First, it establishes the evolutionary baseline for all theorizing about things human, specifically human creations including language and literature. Second, it introduces some concepts that will be referred to later in this book, such as coevolution, mimesis, and Theory of Mind. Third, it confirms the superiority of an approach to language that is consistent with Chomsky's theories over one founded on Saussurean principles. And fourth, it leads directly into the topic of the evolution of the human brain. The seat of our I-language, the brain is a crucial subject of interest for scholars in the humanities.

Notes

1. Plotkin discusses human knowledge in terms of two *heuristics*, a term he uses to mean "something that leads to discovery and invention" (1993, 153). Since "[t]he evolution of adaptations is the invention and discovery of how to organize living structures relative to particular features of the world" (153), and since this is both prior to and what forms the conditions for all other types of invention, Plotkin terms the evolutionary process of adaptation the *primary heuristic*. What is constructed upon this base, the *secondary heuristic*, includes "[t]he process of discovery and invention making up the adaptations that compensate for the shortcomings of those formed by the primary heuristic" (154).
2. For a summary of feminist approaches to biology by both scientists and humanists, and a comprehensive bibliography on the topic, see the entry "Feminist Philosophy of Biology" in the Stanford Encyclopedia of Philosophy (2011). https://plato.stanford.edu/entries/feminist-philosophy-biology/
3. On this topic see: (Wills 1998, 151–56, 188–89), see also Wrangham and Peterson (1996, 43–48), and Berger (2000, 23–25).
4. More recently on this topic, see Colin McGinn's *Prehension: The Hand and the Emergence of Humanity* (2015).
5. Shlain (2003, 196–206) also stresses the sex-language nexus, arguing that "language evolved primarily because men and women had to *negotiate* sex" (201).

References

Abbott, H. Porter. 2003. "Unnarratable Knowledge: The Difficulty of Understanding Evolution by Natural Selection." In *Narrative Theory and the Cognitive Sciences*. Ed. David Herman. Stanford: CSLI Publications. 143–62.

Alexander, Richard D. 1974. "The Evolution of Social Behavior." *Annual Review of Ecology and Systematics* 5: 325–83.

Andersson, Malte. 1994. *Sexual Selection*. Princeton: Princeton UP.

Armstrong, David F. 1999. *Original Signs: Gesture, Sign, and the Sources of Language*. Washington, DC: Gallaudet UP.

———, William C. Stokoe, and Sherman E. Wilcox. 1995. *Gesture and the Nature of Language*. Cambridge: Cambridge UP.

———, and Sherman E. Wilcox. 2007. *The Gestural Origin of Language*. Oxford: Oxford UP.

Arsuaga, Juan Luis. 2002 (1999). *The Neanderthal's Necklace: In Search of the First Thinkers*. Trans. Andy Klatt. New York: Four Walls Eight Windows.

———, and Ignacio Martínez. 2006 (1998). *The Chosen Species: The Long March of Human Evolution*. Trans. Rachel Gomme. Oxford: Blackwell.

Barash, David P., and Ilona A. Barash. 2000. *The Mammal in the Mirror: Understanding Our Place in the Natural World*. New York: W. H. Freeman.

Berger, Lee R., with Brett Hilton-Barber. 2000. *In the Footsteps of Eve: The Mysteries of Human Origins*. Washington, DC: Adventure P.

Bickerton, Derek. 1990. *Language and Species*. Chicago: U of Chicago P.

Bowlby, John. 1971–81 (1969–80). *Attachment*. 3 vols. New York: Basic Books.

Campbell, Anne. 2002. *A Mind of Her Own: The Evolutionary Psychology of Women*. Oxford: Oxford UP.

———. 2006. "Feminism and Evolutionary Psychology." In *Missing the Revolution: Darwinism for Social Scientists*. Ed. J. H. Barkow. New York: Oxford UP. 63–99.

Campbell, Donald T. 1974. "Evolutionary Epistemology." In *The Philosophy of Karl Popper*. Ed. Paul A. Schilpp. La Salle: Open Court. Vol. 1, 413–63.

Carroll, Sean B. 2005. *Endless Forms Most Beautiful: The New Science of Evo Devo and the Making of the Animal Kingdom*. New York: W. W. Norton.

Chomsky, Noam. 2012. *The Science of Language: Interviews with James McGilvray*. Cambridge: Cambridge UP.

Corballis, Michael C. 1991. *The Lopsided Ape: Evolution of the Generative Mind*. New York: Oxford UP.

———. 2002. *From Hand to Mouth: The Origins of Language*. Princeton: Princeton UP.

Cronin, Helena. 1991. *The Ant and the Peacock: Altruism and Sexual Selection from Darwin to Today*. Cambridge: Cambridge UP.

Darwin, Charles. 1971 (1871). *The Descent of Man, and Selection in Relation to Sex*. Adelaide: Griffin Press.

———. 1998 (1859). *The Origin of Species by Means of Natural Selection or The Preservation of Favored Races in the Struggle for Life*. New York: Modern Library.

Dawkins, Richard. 1989 (1976). *The Selfish Gene*. Oxford: Oxford UP.

Deacon, Terrence W. 1997. *The Symbolic Species: The Co-Evolution of Language and the Brain*. New York: W. W. Norton.

Dennett, Daniel. 1995. *Darwin's Dangerous Idea: Evolution and the Meanings of Life*. New York: Touchstone Books.

Dissanayake, Ellen. 1992. *Homo Aestheticus: Where Art Comes from and Why*. New York: Free P.

Dobzhansky, Theodosius. 1973. "Nothing in Biology Makes Sense Except in the Light of Evolution." *American Biology Teacher* 35: 125–29.

Donald, Merlin. 1991. *The Origins of the Modern Mind: Three Stages in the Evolution of Culture and Cognition*. Cambridge: Harvard UP.

———. 2001. *A Mind so Rare: The Evolution of Human Consciousness*. New York: W. W. Norton.

Dunbar, Robin. 1996. *Grooming, Gossip, and the Evolution of Language*. Cambridge: Harvard UP.

Egan, Kieran. 1997. *The Educated Mind: How Cognitive Tools Shape Our Understanding*. Chicago: U of Chicago P.

Ehrlich, Paul R. 2000. *Human Natures: Genes, Cultures, and the Human Prospect*. Washington, DC: Island P.

"Feminist Philosophy of Biology." 2011. *Stanford Encyclopedia of Philosophy*. Online.

Finlayson, Clive. 2009. *The Humans Who Went Extinct: Why Neanderthals Died Out and We Survived*. Oxford: Oxford UP.

———. 2019. *The Smart Neanderthal: Bird Catching, Cave Art, and the Cognitive Revolution*. Oxford: Oxford UP.

Foley, Robert. 1995. *Humans before Humanity: An Evolutionary Perspective*. Oxford: Blackwell.

Gardner, Howard. 1993 (1983). *Frames of Mind: The Theory of Multiple Intelligences*. New York: Basic Books.

Geary, David C. 1998. *Male, Female: The Evolution of Human Sex Differences*. Washington, DC: American Psychological Association.

Gould, Stephen J. 1977. *Ever Since Darwin: Reflections in Natural History*. New York: W. W. Norton.

Hamilton, William D. 1964. "The Genetical Evolution of Social Behavior." *Journal of Theoretical Biology* 7: 1–52.

Hrdy, Sarah Blaffer. 1977. *The Langurs of the Abu: Female and Male Strategies of Reproduction*. Cambridge: Harvard UP.

———. 1999a (1981). *The Woman that Never Evolved*. Cambridge: Harvard UP.

———. 1999b. *Mother Nature: A History of Mothers, Infants, and Natural Selection*. New York: Pantheon Books.

———. 2009. *Mothers and Others: The Evolutionary Origins of Mutual Understanding*. Cambridge: Belknap P of Harvard UP.

Humphrey, Nicholas K. 1976. "The Social Function of Intellect." In *Growing Points in Ethology: Based on a Conference Sponsored by St. John's College and*

King's College, Cambridge. Ed. P. P. G. Bateson and Robert A. Hinde. Cambridge: Cambridge UP. 303–17.

Johnson, Mark. 1987. *The Body in the Mind: The Bodily Basis of Meaning, Imagination, and Reason*. Chicago: U of Chicago P.

Jolly, Alison. 1999. *Lucy's Legacy: Sex and Intelligence in Human Evolution*. Cambridge: Harvard UP.

Lenneberg, Eric H. 1967. *Biological Foundations of Language*. New York: Wiley.

Levine, George. 2006. *Darwin Loves You: Natural Selection and the Re-Enchantment of the World*. Princeton: Princeton UP.

Lieberman, Philip. 1984. *The Biology and Evolution of Language*. Cambridge: Harvard UP.

———. 1991. *Uniquely Human: The Evolution of Speech, Thought and Selfless Behavior*. Cambridge: Harvard UP.

———. 1998. *Eve Spoke: Human Language and Human Evolution*. New York: W. W. Norton.

———. 2000. *Human Language and Our Reptilian Brain: The Subcortical Bases of Speech, Syntax, and Thought*. Cambridge: Harvard UP.

Lopreato, Joseph, and Timothy Crippen. 1999. *Crisis in Sociology: The Need for Darwin*. New Brunswick, NJ: Transaction Publishers.

Mayr, Ernst. 1997. *This Is Biology: The Science of the Living World*. Cambridge: Belknap P of Harvard UP.

———. 2001. *What Evolution Is*. New York: Basic Books.

McGinn, Colin. 2015. *Prehension: The Hand and the Emergence of Humanity*. Cambridge: MIT P.

Miller, Geoffrey. 2000. *The Mating Mind: How Sexual Choice Shaped the Evolution of Human Nature*. New York: Doubleday.

Mithen, Steven. 1996. *The Prehistory of the Mind: The Cognitive Origins of Art, Religion and Science*. London: Thames and Hudson.

———. 2006. *The Singing Neanderthals: The Origins of Music, Language, Mind and Body*. Cambridge: Harvard UP.

Neisser, Ulric. 1976. *Cognition and Reality: Principles and Implications of Cognitive Psychology*. San Francisco: W. H. Freeman.

Olson, Steve. 2002. *Mapping Human History: Genes, Race, and Our Common Origins*. Boston: Houghton Mifflin.

Pääbo, Svante. 2014. *Neanderthal Man: In Search of Lost Genomes*. New York: Basic Books.

Papagianni, Dimitra, and Michael A. Morse. 2013. *The Neanderthals Rediscovered. How Modern Science Is Rewriting Their Story*. London: Thames & Hudson.

Patai, Daphne, and Noretta Koertge. 1994. *Professing Feminism: Cautionary Tales from the Strange World of Women's Studies*. New York: Basic Books.

Pinker, Steven. 1994. *The Language Instinct: How the Mind Creates Language*. New York: William Morrow.

———. 1997. *How the Mind Works*. New York: W. W. Norton.

———, and Paul Bloom. 1992 (1990). "Natural Language and Natural Selection." In *The Adapted Mind: Evolutionary Psychology and the Generation of Culture*. Ed. J. H. Barkow, L. Cosmides, and J. Tooby. Oxford: Oxford UP. 451–93.

Plotkin, Henry. 1993. *Darwin Machines and the Nature of Knowledge*. Cambridge: Harvard UP.

Premack, David. 1986. *Gavagai! or the Future History of the Animal Language Controversy*. Cambridge: Bradford Books/MIT P.

———, and Ann Premack. 2003. *Original Intelligence: Unlocking the Mystery of Who We Are*. New York: McGraw-Hill.

Russell, Robert Jay. 1993. *The Lemur's Legacy: The Evolution of Power, Sex, and Love*. New York: Jeremy P. Tarcher/Putnam.

Searle, John R. 1992. *The Rediscovery of the Mind*. Cambridge: MIT P.

Shlain, Leonard. 2003. *Sex, Time, and Power: How Women's Sexuality Shaped Human Evolution*. New York: Viking.

Shreeve, James. 1995. *The Neandertal Enigma: Solving the Mystery of Modern Human Origins*. New York: Avon.

Small, Meredith F. 1993. *Female Choices: Sexual Behavior of Female Primates*. Ithaca: Cornell UP.

Smith, John Maynard. 1974. "The Theory of Games and the Evolution of Animal Conflict." *Journal of Theoretical Biology* 47: 209–21.

Sykes, Rebecca Wragg. 2020. *Neanderthal Life, Love, Death and Art*. London: Bloomsbury.

Tattersall, Ian. 2002. *The Monkey in the Mirror: Essays on the Science of What Makes Us Human*. San Diego: Harvest Book.

Thelen, Esther. 2000. "Grounded in the World: Developmental Origins of the Embodied Mind." *Infancy* 1.1: 3–28.

Thompson, Evan. 2007. *Mind in Life: Biology, Phenomenology, and the Sciences of Mind*. Cambridge: Belknap P of Harvard UP.

Trivers, Robert L. 1971. "The Evolution of Reciprocal Altruism." *Quarterly Review of Biology* 46: 35–57.

———. 1972. "Parental Investment and Sexual Selection." In *Sexual Selection and the Descent of Man, 1871, 1971*. Ed. Bernard Grant Campbell. Chicago: Aldine de Gruyter. 1972. 136–79.

———. 1974. "Parent-Offspring Conflict." *American Zoologist* 14: 247–62.

Vandermassen, Griet. 2004. "Sexual Selection: A Tale of Male Bias and Feminist Denial." *European Journal of Women's Studies* 11.1: 9–26.

———. 2005. *Who's Afraid of Charles Darwin? Debating Feminism and Evolutionary Theory*. Lanham: Rowman and Littlefield.

Varki, Ajit, and Danny Brower. 2013. *Denial: Self-Deception, False Beliefs, and the Origins of the Human Mind*. New York: Twelve.

Vergano, Dan. 2014. "Bonanza of Skulls in 'Pit of Bones' Changes View of Neanderthals." *National Geographic*, 19 June (Online).

Walker, Alan, and Pat Shipman. 1996. *The Wisdom of Bones: In Search of Human Origins*. London: Weidenfield and Nicolson.

Wills, Christopher. 1998. *Children of Prometheus: The Accelerating Pace of Human Evolution*. Reading, MA: Perseus Books.

Wilson, Edward O. 1975. *Sociobiology: The New Synthesis*. Cambridge: Harvard UP.

Wilson, Frank R. 1998. *The Hand: How Its Use Shapes the Brain, Language, and Human Culture*. New York: Vintage.

Wrangham, Richard, and Dale Peterson. 1996. *Demonic Males: Apes and the Origins of Human Violence*. Boston: Mariner Books.

Wray, Alison. 2002. *Formulaic Language and the Lexicon*. Cambridge: Cambridge UP.

Zahavi, Amotz. 1975. "Mate Selection: A Selection for a Handicap." *Journal of Theoretical Biology* 55: 205–14.

CHAPTER 6

The Brain

More important, the human brain is adaptive and context-dependent, and it has enormous generalizing powers.

—Robert L. Nadeau

The major difference between plants and animals is movement: plants do not have brains because they never move; animals move, so they have brains. Human beings are animals. We move about and reproduce, so we need brains to do these things. In addition, we have added some chores to the list of things we expect the brain (and mind and body) to do: build bridges and rockets, engage in politics, dance and sing, and write and read literature, to name just a few. The mind-brain is intimately involved in the reading and understanding of literature, so anyone working with literature on a critical or theoretical level should have a basic familiarity with the evolution and development of the physical brain, along with some idea of the basic structures and functions of the brain. To ignore the physical reality of the brain and to claim authority in a discipline intimately concerned with that organ's function is rather like claiming authority in the subject of automobiles with no understanding of how an internal combustion engine works. A contextualized approach to literary theory entails a basic comprehension of the embodied reality of the human animal and, more specifically, a basic comprehension of the

H. Mancing and J. Marston William, *Restoring the Human Context to Literary and Performance Studies*, Cognitive Studies in Literature and Performance,
https://doi.org/10.1007/978-3-030-89078-0_6

brain. This does not mean that we have to be brain surgeons to do literary theory, but we should know the frontal cortex from the brainstem and the corpus callosum from the amygdala. In this chapter, we will: take a brief look at the evolution of the brain; present an overview of the brain's structure; review some of the brain's characteristics; and consider research on the differences between the male brain and the female brain.

1 Evolution of the Brain

Movement necessarily involves the decision as to when, why, and how to move; brains are designed to decide what to do next; see Sheets-Johnstone (1999) and Berthoz (2000). A classic illustration of this is the life cycle of the small animal known as a sea squirt. The sea squirt swims about in its youth, using its rudimentary brain to coordinate its movements and to respond to light and vibrations. When mature, the sea squirt makes a major change in lifestyle and attaches itself to a rock or piece of coral, whereupon it stops moving about for the rest of its life and consumes its now useless brain. Neuroscientist Susan A. Greenfield, after citing the above example, concludes that a brain is unnecessary for stationary life forms, and thus "the brain, in whatever shape, size, and degree of sophistication, is somehow connected in a very basic way to ensuring survival as both a consequence and a cause of movement" (1997, 34).

Evolution works by a process of *bricolage*, of tinkering with material already on hand, of modifying and updating, of adding on and getting by. Every brain of every organism living today can be traced back to the first organisms that ever lived on earth, because every brain is the result of the long series of modifications that have taken place since then. Given the reality of evolution, the concept of a brain as a blank slate or a universal information-processing device is absurd (Plotkin 1993, 163). We cannot make direct comparisons between the brain of modern *Homo sapiens* and ancient hominid brains, because no examples of ancient brains remain, but we can speculate about and consider the probable make-up of those more primitive hominid brains, as Steven Mithen (1996) did with his cognitive archeology. We can also compare the human brain to the brains of other animals alive today.

For a long time, it was assumed that the brain was a single, uniform structure, and an all-purpose learning device. But as the study of anatomy developed through recent centuries, we came to recognize a maze of complex textures and shapes within the brain, and as we began to understand how different sections (such as Broca's and Wernicke's areas) play different roles in different brain functions, we had to modify that straightforward, simple notion of the brain. Furthermore, it also became evident that while the human brain shares some structures and configurations with animals as different from us as the snake, the bird, the rat, the horse, and the chimpanzee, other parts of our brains are unique to our species. Eventually we came to realize that some parts of the brain reflected earlier stages of animal evolution, while others must be the product of more recent developments.

Paul MacLean set forth a theory about what he called the *triune brain* over a period of time in the 1970s (but see MacLean 1990 for the fullest expression). MacLean proposes that the innermost, and most primitive, part of the brain—what he calls the *reptilian brain*—is encapsulated by an evolutionarily newer *old mammalian brain*, which is also known as the limbic system. All of this, in turn, is wrapped in the *neo-mammalian brain*, called the neocortex. The concept of the triune brain is useful for non-specialists, although it is a simplification of what the brain is and how it functions.

The innermost part of the brain, and the part with the oldest evolutionary ancestry, is what MacLean calls the reptilian brain. The reptilian brain is much like the brain of reptiles that lived in eons long past and of reptiles today. The most basic and elemental part of the brain, it encompasses the sensory-motor system and the nonconscious processes essential to our drives to eat, reproduce, and move. The *cerebellum*, for instance, located at the back of the head, coordinates muscle movement, especially movement not requiring conscious thought; it also integrates signals from the skin, muscles, and sense organs and sends signals of various kinds to other parts of the brain. At the base of this core brain, located at the juncture of the neck and the head, is the *brainstem*, which connects the brain in the head with the rest of the central nervous system, especially the spinal cord. The central nervous system and the spinal cord form the *autonomic nervous system*, regulating a number of largely unconscious bodily functions, as well as coordinating communication between body and brain. The spinal cord consists of two pathways, one for basic sensations such as pain and temperature, and the other for more precise signals related to the sense of touch. This most primitive part of the brain oversees many of the basic, largely nonconscious, functions involved in the monitoring of heart rate, digestion, and breathing, as well as movement and other basic functions.

The next layer of the brain is what MacLean calls the old-mammalian brain or *limbic system*, and this is the part that deals directly with emotions. Essentially enveloping the reptilian brain, but located underneath the more modern cortex, the subcortical old-mammalian brain includes structures that reflect the evolution of early mammals beyond the simple existence of reptiles. The old-mammalian brain lends the brain as a whole more direct responsiveness and higher levels of awareness than reptiles muster. The limbic system forms the heart of most structures and processes involved in olfactory processes, sexual ecstasy, and the emotions in general. The more primitive, instinctual feelings of pleasure and pain, the straightforward approach-avoidance impulses, and the basic drives to eat and reproduce that characterize the reptilian brain are all modified, nuanced, extended, and elaborated in the limbic system. This is the seat of the emotional bonds between mother and infant, male and female, even individual and society, bonds that are present in many mammals and particularly prominent in human beings. The old-mammalian layer is also the primary locus of our bodies' immune and endocrine systems, which regulate health and homeostasis. It is further involved in dreaming, fantasy, intuition, and other thought processes that are more nonconscious than conscious. The limbic system is intimately connected both with the reptilian brain and the neocortex.

Three subsections of the limbic system are particularly important: the amygdala, the hippocampus, and the thalamus and hypothalamus. The *amygdala* is a small almond-shaped organ that receives information from all the perceptual systems and is intimately involved in emotions. The amygdala is one of the areas of the brain most fully developed (hardwired) during gestation. Adjacent to the amygdala is the *hippocampus*, a seahorse-shaped organ also involved in emotions and in the workings of memory. The largest of the subcortical areas of this part of the brain is the *thalamus*, which relays various kinds of information from the outside world to the cortex. Immediately below the thalamus is the *hypothalamus*, which though quite small (it occupies less than one percent of the brain's volume), is a powerful area of the brain, as it is involved in homeostasis, the organization of metabolism, circulation of hormones, regulation of body temperature, heart rate, blood pressure, pulse rate, glandular activity, ovulation in females, internal states, sexuality, and much more. The hypothalamus is the main link between the brainstem and the limbic system. Overall, the limbic system is in many ways the heart of the brain, the glue that holds it all together, and the grand central station where connections are made.

Blood pressure offers a unique illustration of the interrelatedness of mind and body, body and environment, nature and nurture. Physician James J. Lynch (1985) has studied the medical practice of taking blood pressure, which has traditionally been conceived as a simple mechanical, hydraulic measurement. But Lynch was surprised when he realized that when a patient talks, the blood-pressure reading is higher than when she is silent. The implications of this are substantial: "No other hydraulic or hydrostatic system in physics was known to be influenced by simple conversations" (49). Lynch's further investigation of this process led him to conclude that "[t]he belief that mind and body live together in some sort of self-contained dual existence, apart from everything and everyone else, has obscured the fact that human beings also live in and are biologically linked to other natural environment as well as to other human beings" (13). Ultimately, Lynch came to recognize that "the entire body is influenced by human dialogue. Thus, it is true that when we speak we do so with every fiber of our being" (3). Further, "[t]he entire body, even down to the microscopic levels of circulation and the exchange of blood gases in individual tissues, is involved in human dialogue" (173). Bakhtin's dialogism thus becomes a prototype of embodied cognition in the most literal sense.

The evolutionarily newest part of the brain, which is far more developed in primates than in other mammals and even further developed in humans than in other primates, is the neo-mammalian brain, usually referred to as the *cortex*, and the *cerebral cortex* or *neocortex*. The cortex is about five times more voluminous than the other two layers combined. It is about 2–5 mm thick and consists of six layers, within which different types of cells are arranged in vertical columns. The cortex alone contains some ten to fifteen billion *neurons* (often called brain cells) and many billions of *glial cells* (there are between ten and fifty times more glial cells than neurons), which have the important functions of supporting, insulating, and protecting the neurons. The brain as a whole, including both the cortex and the subcortical areas, contains something on the order of a hundred billion neurons and a trillion glial cells. The neocortex is the seat of language, intelligence, thinking, and reasoning, as well as sympathy, empathy, compassion, and love. It is this part of the brain that enables us to tell stories, invent myths and scientific theories, reason and do mathematics, gossip, lie and deceive, imagine possible worlds, make music, and create art. Virtually nothing done in the neocortex, however, is done without the participation (both direct and indirect) of the two more primitive (and more basic) layers. Nothing in the brain is simple and direct; rather, everything is connected to and dependent on everything else in ways that we have only recently begun to appreciate fully.

2 Structure of the Brain

The brain is not a blank slate, nor is it a carbon-based computer with a central processing unit and a number of programs that run on its general architecture, nor yet a large memory storage device that processes information. Perhaps it would be more accurate to say that the brain is not an organ but a collection of organs that function as a *system*. As we will see, the brain is not even confined to the head, but is in some very literal ways distributed throughout the body. Further, brain and body are part of a system, such that their functions are integrated in everything done by both.

If the brain is not a *tabula rasa*, how is it organized? The reigning thesis is that it is modular in nature. But *modularity* has degrees, and the term has a variety of connotations. In the early formulation of Jerry Fodor (1983) and others, the modules were seen as separate compartments of the brain that functioned in independence of other modules and did not communicate directly with them. Chomsky, at least in his earlier formulations, seems to have had such a concept of modularity in mind when he stressed the independence of the language organ in the acquisition and generation of language.

Today, however, while the modularity thesis remains a frame of reference and a topic of discussion, and the term modularity has become standard, almost no one sees the brain as divided into discrete sections to carry out separate processes. As Joseph LeDoux says, the functions of the brain are "properties of integrated systems rather than of isolated brain areas" (1996, 78). Robert Blank stresses that "[u]nder the modular brain theory, the brain is a complex system of linked structures, with an emphasis on 'linked'" (1999, 47). Susan Greenfield, too, reminds us, "there is no simple one-to-one matching between a function and a particular part of the brain" (2000, 6). Neuroscientist Joaquín Fuster describes the situation as a "welcome change," namely "the transition from the modular model to the network model of cognition" (2003, x). The stress is on the interrelatedness and interconnectedness of multiple parts of the brain; brain processes are seen as parallel and distributed, rather than sequential and localized; any given part or section of the brain is likely to be involved in several functions. As Fuster stresses, "[n]o cognitive function has a fully dedicated cortical or network" (56). For Antonio Damasio, it is "a fundamental notion of cognitive neuroscience" that "[a]ny complex mental function results from concerted contributions by many brain regions at varied levels of the central nervous system rather than from the work of a single brain region conceived in a phrenological manner" (2003, 73). Finally, in the words of renowned neuroscientist V. S. Ramachandran, "Sure enough, brains are made up of modules, but the modules are not fixed entities; they are constantly being updated through powerful interactions with each other, with the body, the environment, and indeed with other brains" (2011, 126).

It has been estimated that the brain carries out some 200 trillion (200,000,000,000,000) simultaneous operations per second (Campbell 1989, 12). It is thus more useful to talk of neural networks (circuits and systems), neuronal groups (or assemblies), maps, webs, association areas, and convergence zones than of modules, since these are places where neural functions from different locations come together and are integrated (Damasio 1994, 102). The mind-brain is more a society than a monolith, more like a shopping mall than a prison, more an orchestra than a bunch of soloists. If the occipital lobes deal primarily with visual processing, the frontal lobes, and neural subsets within them, are involved in multiple and varied functions from one millisecond to the next. Rather than a discrete number of relatively independent modules, the brain is better conceived as consisting of a "many billions of functionally relevant subdivisions, each responsive to a broad but proprietary range of highly complex stimuli" (Churchland 1989, 178).

The neocortex is (artificially, arbitrarily, and for convenience) divided into four general regions on each side of the brain referred to as *lobes*. The *occipital lobes* occupy the majority of the back of the skull and are the primary areas in which vision is processed. They are somewhat more self-contained and discretely modular than the other lobes and make up the largest area of the devoted primarily to a single (complex) function, which is at least part of the reason why vision is the most powerful source of knowledge in the human being brain. But vision is not processed in the occipital lobes alone; it has been estimated that up to 60% of the cerebral cortex is involved in visual processing in one way or another (Bownds 1999, 185–86). The *parietal lobes* cover most of the back top of the brain and are the primary location of functions that involve the senses such as weight, size, shape, and feel (the senses except olfaction, that is, which is the evolutionarily oldest of all senses and has a subcortical base). The *temporal lobes* occupy much of the sides of the left and right hemispheres and are involved in language, hearing, and memory. And the *frontal lobes* occupy the front part of the brain, behind the forehead. They are quite large, occupying some 25–33% of the total mass of the cortex. The frontal lobes are most implicated in voluntary muscle movement, social behavior, working memory, narration, reasoning, curiosity, anticipation, imagination, sympathy, and empathy. All the lobes have *association areas* that analyze, interpret, and coordinate various sensory experiences. These are gross generalizations; virtually everything that takes place in the brain is mediated by processes occurring both in rapid sequence and in parallel in other regions.

Two more brain areas could be considered lobes: the *insula* and the *cingulate*. These are usually excluded from discussions of the neocortex because they mediate, connect, and coordinate between limbic and cortical activities. But Louis Cozolino, for example, argues that "the most important cortical areas are no longer those on the surface of the brain but those hidden from view" and that "the cingulate and insula cortices could, and perhaps should, be considered as the fifth and sixth cortical lobes" (2006, 50). The cingulate is particularly important in the coordinating of "maternal behavior, nursing, and play" (104; see 104–7). Cozolino's discussion of the role of the insula (206–9) calls attention to this structure as "an interface between our sense of self and our self-control capabilities" (207), its role in mediating extremes of emotion, and so forth (see his summary chart on 209).

Much has been made in the popular press in recent decades about brain lateralization, the division of the brain into left and right hemispheres, and hemisphericity, different cognitive modes in the two sides of the brain. It is true that *left hemisphere* functions tend to be more involved in cognitive processes that are analytic, sequential, and abstract, while *right hemisphere* functions tend to be more holistic, simultaneous, synthesizing (Hellige 1993). But most of what we read in popular accounts of left-brained, logical, linear, analytic, linguistic men, and right-brained, emotional, intuitive, holistic, artistic women, is misleading at best and downright wrong at worst. We *do not ever* "think" with just one side of the brain; individuals are not right-brained or left-brained.[1] Some brain functions are essentially localized, but overall "*There is almost nothing that is regulated solely by the left or right hemisphere*" (Ornstein 1997, 68). For example, most of the mechanical functions involved in the production and comprehension of speech are located primarily in the left hemisphere (in about 95% of all people), and damage to areas of the left hemisphere tends to result in different types of language aphasia. But the right hemisphere also participates importantly in acts of language, for damage to this side of the brain can result in an individual's inability to comprehend the tone, inflection, or prosody of speech (which are, interestingly, elements crucial to a Bakhtinian concept of language—and absent in the Saussurean account). To a large extent, the right hemisphere is the contextualizing part of the brain (99–115)—the Bakhtinian complement to the Saussurean and Chomskyan structural, mechanical, and computational left hemisphere.

When it was learned that certain kinds of serious epilepsy could be cured by completely severing these inter-hemispheric connections—an operation called a *commissurectomy*—it also became possible to conduct a series of psychological experiments on individuals whose left and right hemispheres did not communicate. Studies involving these "split-brain" patients by Roger Sperry and Michael Gazzaniga, especially as described by the latter, have become classic illustrations of the complicated and subtle relationships that exist among the various functions of the two hemispheres. The presentation of information to the visual field of one hemisphere and of different information to the other hemisphere reveals some ways that the two hemispheres deal with things on their own without direct and constant communication. Most significant, according to Gazzaniga (1985, 27–99; 1992, 95–137), is the inevitable reaction by the left hemisphere to make up stories about what was perceived. This non-veridical self-reporting is generally called *confabulation*, the reporting of imaginary events as though they were truthful; see Hirstein (2005, 2012). Based on these experiments, Gazzaniga posits as a primary function of the brain a mechanism that he calls the "interpreter," the brain's capacity to draw interpretive inferences, to confabulate, to tell tales about ourselves and our motivations (1992, 112–37; 1998, 133–75).

Not all of the brain is contained within the head. Michael D. Gershon (1998) documents what he calls the "second brain," the brain in the gut. It is only in recent years that we discovered a large number of neurons in the bowel, which gives a new and literal meaning to traditional talk about "gut feelings." Gershon reminds us that we may have a brain in our head, but the "gut has a mind of its own" (xiii). The existence of a primitive but extensive neural system, called the *enteric nervous system* (see also Nuland 1997, 285–86), in the lower intestinal tract allows the basic bodily functions of digestion and absorption to take place in an almost continuous process without the involvement of any higher cognitive functions, allowing us to go about the more important activities of finding food, reproducing, defending ourselves and our families, and reading novels. This earlier research is by now a staple in popular cure-all literature of the twenty-first century; see, e.g., Sonnenburg and Sonnenburg (2015), *The Good Gut: Taking Control of Your Weight, Your Mood and Your Long-Term Health.*

Further, the enteric nervous system is quite extensive; according to Gershon, "We have more nerve cells in our gut than in the entire remainder of our peripheral nervous system" (1998, xiii). In addition to the second brain's neuronal structure, an extensive production and storage of neurotransmitters exist in the enteric nervous system. One's perspective on brain and body begins to change when one thinks of our intestines as having greater (nonconscious) cognitive capability than many other animals.

3 The Emotional Brain

One of the foundations of Western thought has been the conviction that what most distinguishes human beings from the animals is that, while other animals purportedly react largely on the instinctual basis of emotion, we humans are characterized by reasoned, logical, intellectual cognitive processes. Traditionally, the binary conceptual set, and variants of it, has been one of the means by which patriarchy has perpetuated itself:

- man vs. woman
- mind vs. body
- objectivity vs. subjectivity
- logic vs. emotion.

Once again, however, neuroscience has proven this binary oppositional stance invalid: no thought takes place without emotion. Indeed, since

1987, there has existed an entire scientific journal devoted to this intimate interconnection, titled *Cognition and Emotion.*

The first thing to note about emotions is that without them we would not be here (LeDoux 1996, 40; Damasio 1999, 53–56). In the early stages of evolution, at least through the time of the reptiles, emotions played little (if any) role. But mammalian brains, which are larger and more complex than reptile or amphibian brains, are clearly characterized by emotions. In the triune brain system, the major parts of the brain related to emotions are located in the limbic system, the part of the brain that humans share (to a large extent) with other mammals. An animal's ability to perceive, recognize, and react to possible dangers with maximum speed obviously enhances the possibility of survival.

Neuroscientist Joseph LeDoux, one of the most influential students of human emotions and author of *The Emotional Brain* (1996), shows that emotional and defensive behaviors in response to danger evolved before conscious feeling (128). Most of the brain systems for dealing with emotions are therefore older, more primitive, and thus more powerful than the cognitive systems, which developed slowly and much later. LeDoux proposes that "emotion and cognition are best thought of as separate but interacting mental functions mediated by separate but interacting brain systems" (69). According to LeDoux, our brain/body starts reacting emotionally to a perception before our conscious cognitive powers are even aware of it or can begin to evaluate it.

Since reason and emotion have traditionally been cast as diametrically opposed and mutually exclusive, it is important to see what contemporary neuroscience has to say about the relationship between the two, and this turns out to be very different from what we have traditionally believed. Neurologist Antonio R. Damasio has studied this relationship extensively, especially in his influential book *Descartes' Error* (1994; see also 1999). The title refers to Cartesian dualism, described by Damasio as "the separation of the most refined operations of mind from the structure and operation of a biological organism" (1994, 249–50). A mind–body continuity of the sort that Damasio, LeDoux, and many others cited in this book advocate blurs all the traditional dualistic distinctions between mind and body. Neat separation between reason and emotion must be called into question; see also the very good book by Devlin (1997) on Descartes's legacy.

Damasio illustrates his point that what we call logical thinking never takes place in the absence of emotion and feeling by describing the experiences of people whose brains were damaged in ways that make this non-separation explicit. Phineas Gage, for example, was a man whose entire social being was destroyed when he lost his emotional capabilities after an accident in 1848 in which a steel pipe was literally blasted through his head, carrying away sections of his brain. Damasio's accumulation of evidence from clinical cases, neuroanatomy, results from various kinds of brain surgery, and laboratory experiments make an overwhelmingly convincing case for his position that human reason evolved with help of, and to this day still depends on, the biological mechanisms of emotion and feeling (1994, xii). His conclusion is that emotions and feelings are every bit as cognitive, in the most literal sense of that word, as are visual and other types of perception and the supposed higher cortical functions; he defines feelings as "*the cognition of our visceral and musculoskeletal state* as it becomes affected by preorganized mechanisms and by the cognitive structures we have developed under their influence" (159). Damasio accords feelings "a truly privileged status" because of their representation and influence at multiple levels (159–60).

Damasio makes an important distinction between *emotion* and *feeling*: an emotion is "a collection of changes occurring in both brain and body, usually prompted by a particular mental content," while a feeling is "the perception of those changes" (270 n1). Damasio also distinguishes between primary "early" emotions and secondary "adult" emotions. The primary (or innate) emotions are those fully (or almost fully) functional, or hard-wired, in the subcortical regions of the brain at birth. The amygdala is an important component in the construction and processing of these emotions, as is evident in its role in the fear circuit. Primary emotions, of which fear is the prototype, are strong, automatic, and rapid. Secondary emotions, such as love or hate, are more conscious and depend more on cortical functions than on the limbic system. Damage to the frontal cortex can have a negative influence on the ability to experience secondary, but not primary, emotions. Many feelings are generated by primary emotions, but others do not originate in and are not related to emotions. Feelings such as happiness, sadness, and disgust are clearly based on primary emotions and are among those most easily seen in bodily responses. Subtler varieties of these, such as euphoria, melancholy, and shyness, are more tuned by temperament and experience. The secondary emotions and their variants are probably the most interesting for students of literature, as these are the kinds of emotions that are most obviously and easily generated by the reading of a novel or poem.

Finally, there is a category that Damasio calls "background feeling" because "it originates in 'background' body states rather than in emotional states" (150). We experience it during the (quantitatively greater) periods between the experience of emotion-related feelings, so background feeling corresponds to "our image of the body landscape when it is not shaken by emotion" (150–51); it also corresponds roughly to the idea of "mood." Background feeling is not associated with any particular part of the body nor is it exhibited in any specific bodily symptom, but it is more of an overall state of the body as a whole. The importance of background feeling is, Damasio suggests, that "[o]ur individual identity is anchored on this island of illusory living sameness against which we can be aware of myriad other things that manifestly change around the organism" (155).

Many neuroscientists limit the study of emotions to human beings, on the (basically behaviorist) belief that we cannot directly detect and measure emotions in other animals as we can in ourselves. The person who has argued most strongly against this idea is Jaak Panksepp in his massive, detailed, and demanding text *Affective Neuroscience* (1998). Panksepp believes that while ethology, behaviorism, the cognitive sciences in general, various strains of contemporary psychology (cognitive, evolutionary, clinical), and even psychiatry have all made some attempt to understand the relationship between brain and emotion, they are also all lacking something. Panksepp proposes that "a missing piece that can bring all these disciplines together is a neurological understanding of the basic emotional operating systems of the mammalian brain and the various conscious and unconscious internal states they generate" (5). Panksepp calls this new perspective "affective neuroscience." Further, he believes that "it is only through a detailed study of animal emotions and the brain substrates that a satisfactory foundation for understanding human emotions can emerge" (10). Specifically, he proposes that "it is correct to assume that primary-process affective feelings in humans (i.e., 'raw feels') arise from distinct patterns of neural activity that we share with other animals, and that these feelings have an important role in controlling behavior, especially conditionally" (13). This approach leads Panksepp to presume that "the major evolutionary differences within the subcortical operating systems are matters more of emphasis than of kind" (15). Panksepp makes use of MacLean's triune brain concept, grounding basic innate actions and matters of basic survival to the oldest part of the brain, the reptilian brain (the brainstem and basal ganglia). The old-mammalian brain (the limbic system) is at the core of the processes Panksepp studies, the subjective feelings and emotions that we share with other mammals. The new mammalian brain (the neocortex) is the locus of symbolic capabilities and the area that coordinates perception. All areas interact constantly, but at the heart of emotions is the old-mammalian brain.

Panksepp describes seven emotional "systems" that have their origin in the limbic system: SEEKING, RAGE, FEAR, PANIC, LUST, CARE, and PLAY (first described and summarized 51–54; Pankseep uses capitals for these terms). His pages on sleep, anticipation, rage, anger, fear, anxiety, sexuality, nurturance, sorrow, grief, and play, plus much more, make it evident that our evolved brains, and especially the neurochemicals that course through them, play an active role in virtually everything that characterizes our species. For Panksepp, it is unfortunate that

> social constructivist frameworks have too commonly disregarded the vast amount of behavioral and physiological evidence for specific emotional response patterns, as well as the wealth of neuroscientific evidence suggesting that there are genetically provided affective infrastructures for different emotions within the brain. The great advantage of constructive approaches is the full recognition that language is our most important social instrument … The disadvantage is that this view finds it so easy to overlook the universe of neurobiology that exists independently of our vast, and often deceptive, linguistic abilities. (45)

It is one thing to study neurons, which tend to stay in place and remain available for detailed examination by microscope after death. But how can we study the chemicals that flow in ever-changing currents throughout our bodies and brains? The thousands of chemicals that work within the brain leave no trace for postmortem examination; the very existence of many of them, as well as their complicated roles, have only recently come to attention. Brain chemistry is a much younger field of inquiry (with its origins in the 1970s and 1980s) than is brain anatomy. As Candace Pert notes, it is only in relatively recent times that the "chemical brain" has been able to be studied scientifically (1997, 26). But we are very far from knowing enough about the formation, distribution, excitation, transmission, and effects of the amino acid transmitters (glutamate, GABA), the biogenic amines (serotonin, norepinephrine, dopamine, acetylcholine, histamine), the many peptides (opioids, endorphins, oxytocin, vasopressin), or the various steroids (cortisol, testosterone, estrogen, progesterone). And we still know even less about other systems. The influence of these constantly flowing chemicals is pervasive and constant. As Steven Johnson puts it, "Right now, as you read these words, you are under the influence of chemicals that are, molecularly speaking, almost indistinguishable from drugs that could get you arrested if you consumed them openly in a public place" (2004, 137). Future advances in this area will dramatically modify our understanding of the brain.

Pert's *Molecules of Emotion* (1997) provides a good, accessible explanation of how neurotransmitters function in the brain (and most of the rest of the body). Protein molecules called *receptors* are attached to neurons; there may be millions (!) of receptors on the surface of a single neuron (23). The three types of ligands (which Pert calls "information molecules") are neurotransmitters (dopamine, serotonin, and others mentioned above) that carry information across the synaptic gap, steroids (including the sex hormones), and peptides (molecules consisting of strings of amino acids) that do not act between synapses but attach to molecules as described above (25). Pert estimates that as much as 98% of neuronal communication takes place away from the synapse (139). The ligand-receptor system is virtually a second nervous system spread throughout the body (26). The processes occurring within this evolutionary ancient distributed system are as important to cognition—particularly with respect to sensations, feelings, and emotions—as anything happening within our bodies. Pert's suggestion that most of our neuronal communication does not take place between neurons, in the synaptic clefts so extensively studied and discussed in neuroscience, can be staggering. If the majority of our research to date has dealt with perhaps as little as 2% of what is going on in our brains, we see how little we know and how far we are from understanding how the brain really functions.

4 The Sexual Brain

- Men are from Mars and women are from Venus.
- Men are aggressive and violent while women are emotional and nurturing.

- Men's brains are more lateralized and therefore they are better at math.
- Women's brains are more equally balanced between left and right hemispheres and therefore they are better at verbal tasks.

Are these and other recent popular expressions of essential male–female biological (and brain) differences true, partly true, or just the latest expression of the traditional (male) tendency to see women as different and therefore inferior? New information about the role of sexual selection in evolution, along with recent studies on sexuality, mate selection strategies, jealousy, violence, and other related subjects in the field of evolutionary psychology, together with discoveries in neurochemistry, have all substantially reframed the discourse on gender differences. THEORY—particularly much influential feminist discourse—has consistently made assumptions and drawn conclusions that must be substantially overhauled from an evolutionary perspective.

Men and women are, of course, different biologically. Women can give birth to then breast-feed children, while men cannot (but see Diamond 1997, 41–62); men and women generally differ in size, strength, shape, and in the timbre to their voices; men and women have a different balance of neurotransmitters and hormones in their bodies. Throughout evolution, men and women developed different roles in hunting and gathering, giving birth to and raising children, and in social relationships in general. Surely these biological and evolutionary differences between men and women extend to the brain. After the genitals, the brain is the organ that demonstrates the greatest sexual differentiation; the heart, liver, or lungs are characterized by no difference whatsoever (except perhaps size) according to sex. (It is worth noting that the default plan for all mammal fetuses is female. It is only when testosterone is released in large quantities during the second trimester of human pregnancy that the gonads take the form of testes in males and ovaries in females.) Recent research in neuroscience and brain imagery should be able to pinpoint differences between male and female brains with a degree of precision never before possible. And they have—to a limited extent. Scottish psychologist Stuart J. Ritchie and his team have conducted extensive brain-imaging research in this area, as they state, "the largest single-sample study of neuroanatomical sex differences to date" (Ritchie et al. 2018, 11).

Ritchie et al. report on biological differences in, e.g., brain volume and cortical thickness, but overall the similarities between the brains of men and women outweigh the differences: "Overall, for every brain region that showed even large sex differences, there was always overlap between males and females, confirming that the human brain cannot—at least for the measures observed here—be described as 'sexually dimorphic'" (8). Past research has indicated some physical differences in the hypothalamus, where much sexual and emotional regulation takes place (see Bownds 1999, 147, for a brief summary). Similarly, some researchers have found differences between men and women in the corpus callosum (particularly in the part called the *splenium* and the related *anterior commissure*). These structures tend to be slightly larger in women than in men, which would imply a somewhat greater degree of hemispheric connectivity in women—a feature that might account for the often-reported greater verbal fluency of women (147–48). There is also some strong indication that there is a detectable difference in visuospatial abilities between men and women (Hines 2004, 12–14, 165–82). But, to date, there is no convincing evidence that these small physical and cognitive distinctions indicate any meaningful difference between the nature or function of the brains of men and women (in Ritchie et al.'s study, the male participants did slightly better on two cognitive tests, but the authors note these results could be due to the sample or types of tests performed, assessing specific skills rather than general ability; see Ritchie et al. 11). Furthermore, calling attention to potentially minor structural differences could obscure the overwhelming structural similarity between the two. As Lise Eliot notes at the conclusion of *Pink Brain, Blue Brain* (2009), studies confirm "the argument that boys and girls need different educational experiences because 'their brains are different' is patently absurd" (305).

The day may come when neuroscience can irrefutably identify truly significant differences in structure, neuronal development, or (more likely) hormonal and other chemical functions in the brains of men and women. The desperate need to identify defining biological differences between men and women is part of the hunger to simplify our lives by dividing up the world into mutually exclusive binary oppositions: self-other, nature-nurture, male-female. Tavris's plea to move beyond simple binarism in this context is compelling:

> As long as the question is framed this way—"What can we do about *them*, the other, the opposite?"—it can never be answered; no matter which sex is being regarded as "them." The question, rather, should be this: What shall we do about us, so that our relationships, our work, our children, and our planet will flourish? (1992, 333)

As far as the brain goes, what men and women have in common is much more important than what divides them: "it is sometimes forgotten how very much alike, rather than different, all the members of our species are" (Mandler 1997, 119). Deborah Blum describes the contrasts between the male and female brain as "too tiny, and still far too mysterious, to suggest that these are profoundly different organs. They may do some things in different ways, but the basic repertoire is the same" (1997, 63).

Pert's work illustrates the principle of brain and body being united. The nervous system (the brain and other nerve cells in the body), the endocrine system (the body's glands and hormones), and the immune system (the defense system of the body) all work together constantly; they are "joined together in a wonderful system coordinated by the actions of discrete and specific messenger molecules" (1997, 172). The functions of chemicals in the brain also illustrate the inseparable nature-nurture relationship and spark considerations about the effect of experience and emotion on those chemicals. For example, testosterone surges as a man strains to beat an opponent at tennis—but winning can cause testosterone levels to become elevated. Any talk of just one factor, whether biological or cultural, determining behavior or traits is mistaken. Both human male and human female brains are profoundly influenced by the different hormones present in the brain at any given time, by the nature and timing of these hormone flows, and by hormone activity related to the biology of sexuality and reproduction.

Women, especially, have a much greater ebb and flow of more kinds of hormones and other neurochemicals between the age of puberty and that of menopause than do men at any time in their lives except at puberty. But this doesn't *determine* essential and non-changeable differences between women and men, and contemporary biologists are not attributing male-female differences exclusively to hormones, genes, or gender. Ultimately, neurochemicals can be considered one important means by which sexual selection has operated throughout evolutionary history to modify both the body and the psychological traits that make men and women different in many ways (Miller 2000, 226–30).

What the brain is, how it came to be through the eons of evolution, and how it is structured is only half of the story. How the brain of each individual person grows and develops in a specific cultural context, and the periods of its growth, are equally important. We turn to these questions in Chapter 7.

Note

1. At the same time, recent research points to studies of brain asymmetry—also in non-human animals—as being the key to understanding brain dysfunctions. Cognitive neuroscientist Onur Güntürkün notes, "There are almost no disorders of the human brain that are not linked to brain asymmetries" (Cell Press 2017; and see Güntürkün's co-authored research article with Sebastian Ocklenburg in *Neuron*, April 2017).

References

Berthoz, Alain. 2000 (1997). *The Brain's Sense of Movement*. Trans. Giselle Weiss. Cambridge: Harvard UP.

Blank, Robert. 1999. *Brain Policy: How the New Neuroscience Will Change Our Lives and Our Politics*. Washington, DC: Georgetown UP.

Blum, Deborah. 1997. *Sex on the Brain: The Biological Differences between Men and Women*. New York: Viking.

Bownds, M. Deric. 1999. *The Biology of Mind: Origins and Structures of Mind, Brain, and Consciousness*. Bethesda: Fitzgerald Science P.

Campbell, Jeremy. 1989. *The Improbable Machine: What the Upheavals in Artificial Intelligence Research Reveal about How the Mind Really Works*. New York: Simon and Schuster.

Cell Press. 2017. "Why Animals Have Evolved to Favor One Side of the Brain." *ScienceDaily*, 19 April (Online).

Churchland, Paul M. 1989. *A Neurocomputational Perspective: The Nature of Mind and the Structure of Science*. Cambridge: MIT P.

Cozolino, Louis. 2006. *The Neuroscience of Human Relationships: Attachment and the Developing Social Brain*. New York: W. W. Norton.

Damasio, Antonio R. 1994. *Descartes' Error: Emotion, Reason, and the Human Brain*. New York: Grosset/Putnam.

———. 1999. *The Feeling of What Happens: Body and Emotion in the Making of Consciousness*. New York: Harcourt Brace.

———. 2003. *Looking for Spinoza: Joy, Sorrow, and the Feeling Brain*. Orlando: Harcourt.

Devlin, Keith. 1997. *Goodbye, Descartes: The End of Logic and the Search for a New Cosmology of the Mind*. New York: Wiley.

Diamond, Jared. 1997. *Why Is Sex Fun? The Evolution of Human Sexuality*. New York: Basic Books.

Eliot, Lise. 2009. *Pink Brain, Blue Brain: How Small Differences Grow into Troublesome Gaps—and What We Can Do About It*. Boston: Houghton Mifflin Harcourt.

Fodor, Jerry A. 1983. *The Modularity of Mind*. Cambridge: MIT P.

Fuster, Joaquín M. 2003. *Cortex and Mind: Unifying Cognition*. Oxford: Oxford UP.

Gazzaniga, Michael S. 1985. *The Social Brain: Discovering the Networks of the Mind*. New York: Basic Books.

———. 1992. *Nature's Mind: The Biological Roots of Thinking, Emotions, Sexuality, Language, and Intelligence*. New York: Basic Books.

———. 1998. *The Mind's Past*. Berkeley: U of California P.

Gershon, Michael D. 1998. *The Second Brain: The Scientific Basis of Gut Instinct and a Groundbreaking New Understanding of Nervous Disorders of the Stomach and Intestine*. New York: HarperCollins.

Greenfield, Susan A. 1997. *The Human Brain: A Guided Tour*. New York: Basic Books.

———. 2000. *The Private Life of the Brain: Emotions, Consciousness, and the Secret of the Self*. New York: Wiley.

Güntürkün, Onur, and Sebastian Ocklenburg. 2017. "Ontogenesis of Lateralization." *Neuron* 9.2: 249–63.

Hellige, Joseph B. 1993. *Hemispheric Asymmetry: What's Right and What's Left*. Cambridge: Harvard UP.

Hines, Melissa. 2004. *Brain Gender*. Oxford: Oxford UP.

Hirstein, William. 2005. *Brain Fiction: Self-Deception and the Riddle of Confabulation*. Bradford Books/MIT P.

Johnson, Steven. 2004. *Mind Wide Open: Your Brain and the Neuroscience of Everyday Life*. New York: Scribner.

LeDoux, Joseph. 1996. *The Emotional Brain: The Mysterious Underpinnings of Emotional Life*. New York: Simon and Schuster.

Lynch, James J. 1985. *The Language of the Heart: The Body's Response to Human Dialogue*. New York: Basic Books.

MacLean, Paul D. 1990. *The Triune Brain in Evolution: Role in Paleocerebral Functions*. New York: Plenum.

Mandler, George. 1997. *Human Nature Explored*. New York: Oxford UP.

Miller, Geoffrey. 2000. *The Mating Mind: How Sexual Choice Shaped the Evolution of Human Nature*. New York: Doubleday.

Mithen, Steven. 1996. *The Prehistory of the Mind: The Cognitive Origins of Art, Religion and Science*. London: Thames and Hudson.

Nuland, Sherwin B. 1997. *The Wisdom of the Body*. New York: Alfred A. Knopf.
Ornstein, Robert. 1997. *The Right Mind: Making Sense of the Hemispheres*. New York: Harcourt Brace.
Panksepp, Jaak. 1998. *Affective Neuroscience: The Foundations of Human and Animal Emotions*. Oxford: Oxford UP.
Pert, Candace B. 1997. *Molecules of Emotion: Why You Feel the Way You Feel*. New York: Scribner.
Plotkin, Henry. 1993. *Darwin Machines and the Nature of Knowledge*. Cambridge: Harvard UP.
Ramachandran, V. S. 2011. *The Tell-Tale Brain: A Neuroscientist's Quest for What Makes Us Human*. New York: W. W. Norton.
Ritchie, Stuart J., et al. 2018. "Sex Differences in the Adult Human Brain: Evidence from 5,216 UK Biobank Participants." *Cerebral Cortex*. 28.8: 2959–975.
Sheets-Johnstone, Maxine. 1999. *The Primacy of Movement*. Amsterdam: John Benjamins.
Sonnenburg, Justin, and Erica Sonnenburg. 2015. *The Good Gut: Taking Control of Your Weight, Your Mood and Your Long-Term Health*. New York: Penguin.
Tavris, Carol. 1992. *The Mismeasure of Woman*. New York: Simon and Schuster.

CHAPTER 7

Development of the Brain

Our brains use context to shape our perception of relationships.

—M. Deric Bownds

The human infant is born more prematurely—at an earlier stage of development—than the offspring of other primates largely because of the size of the human head. As a result, about three-quarters of human brain development takes place after birth, and thus in circumstances where individual, family, and social factors all vary enormously, influencing brain development in radically different and often profound ways. In this chapter, we will: review the basics of brain composition as they are seen in infancy; look at what happens in childhood and at puberty, with an exploration of how modern technology affects the developing brain; consider how learning to read changes the brain; comment on what happens as the brain ages; and close the chapter by introducing the field of neurolinguistics.

H. Mancing and J. Marston William, *Restoring the Human Context to Literary and Performance Studies*, Cognitive Studies in Literature and Performance,
https://doi.org/10.1007/978-3-030-89078-0_7

1 The Brain in Infancy

Every individual human brain, although conforming to the overall species pattern, is unique in literally millions of ways. Psychologist Henry Plotkin describes the process by which an organism develops from birth to maturity as "an immensely complex series of interactions between the different parts of the genetic constitution of that individual; and also between its genes, its developing parts and its environment" (1993, 122). When a human infant is born her brain is about 350 cubic centimeters (cc.) large, and within two years, it has nearly tripled in size, to about 1000 cc. This overall growth in size is accompanied by an extensive development in complexity, as dendrite and axon growth, along with myelination and the creation of new synapses, progress with astonishing speed during this period. By the time a child is about fourteen years old, her brain has quadrupled in size to nearly 1400 cc, the approximate size at which it will remain for the rest of her life (in contrast, a chimpanzee achieves virtually full adult maturity in brain size in less than two years). This pace of growth is by far the fastest of any part of the human body, which is also why the final closure of the human skull is delayed longer than in any other mammal.

But this does not mean that the brain as a single entity increases in size during this period, for the various subsections of the brain do not develop at a uniform rate. Nor does this fast early pace of brain growth mean that development stops at puberty, for the brain continues developing actively through puberty and may continue, albeit at an ever-decreasing pace, throughout life. The term used to describe the brain's ability to grow, develop, and modify itself is *plasticity* (or *neuroplasticity*). The more evolutionarily ancient, innermost parts of the brain are the parts already most developed at birth, and these are also the parts most involved in basic bodily functions that need to operate from the beginning for life to be viable such as heart rate, breathing, and so forth. The next most developed parts are the other sub-cortical areas, while the cortex itself is the least developed part of the brain at birth. All sections of the brain continue to develop at different rates from birth on. The prefrontal cortex is among the last to be completed, at about the age of puberty.

William Greenough and his colleagues (1987) make an important distinction between *experience-expectant* and *experience-dependent* aspects of brain development; see also Bruer's (1999, 107–11, 156–58) helpful summary of this distinction. Experience-expectant aspects are those functions such as vision, hearing, movement, and language, all of which develop normally after birth with no special effort on the part of either the individual child or the child's caretakers. It is as if evolution had prepared the body for a certain essential fine-tuning of basic functions that are almost always carried out automatically by the fact of living in any normal human environment and culture, no practice required.

The experience-dependent aspects are, however, quite different in that they depend entirely on experience: reading, playing baseball, composing a symphony, or deconstructing a text. Neither evolution nor biology specifically prepares for any of these activities, although biology makes them possible. The great majority of human beings who have lived quite successful lives have never done any of these things. Experience-dependent activities are thus very much like what Plotkin (1993) calls the secondary heuristic: culture. No one learns to see, but everyone who reads has learned, through conscious effort, to do so. No illiterate person has ever had reading grow in her brain, just as no one who is blind has ever been able to see through conscious effort alone. Critical periods are important in experience-expected development, but they play almost no role in experience-dependent learning.

Humans are born prematurely and helpless, and we require more time than other animals to develop our brains to full potential. This is not a handicap but an essential key to our humanity. Because the brain of the human neonate is so undeveloped, it has the opportunity (and the need) to build itself and to adapt to its environment. The brains of babies are "learning machines" (Eliot 1999, 8), and from day one are actively involved in this, their major task. One reason why human babies are born with so much more subcutaneous fat than other primates is that this energy store is diverted to continued rapid brain development after birth; see Cunnane (2005).

A newborn baby has about 100 billion neurons, generated at the staggering rate of about 250,000 per minute during pregnancy. More neurons continue to develop during the first year and a half of life, after which very few new neurons are created. In general, neurons that die are not replaced, so at no time is the number of neurons in an individual brain greater than at (or very soon after) birth, and that number decreases as the brain matures. Unlike cells in all other organs of the body, neurons are not basically all alike: each one is unique in its shape and function.

The brain grows as it loses neurons, however, because the neurons that remain undergo years of development in size and complexity.

Eliot describes the wiring of the brain as "an intricate dance between nature and nurture" as experience contributes to a customized brain circuitry (1999, 29). However, the exact nature and extent of this circuitry are not entirely fathomable.[1] What is easier to grasp, backed up by modern science, is that biology and culture are inseparable parts of the same process of brain growth and development: "The world wires up the mind" (Ornstein 1991, 122). Louis Cozolino (2006) further blurs the distinction between biology and culture, nature and nurture, when he stresses that "when we interact, [we are] impacting each other's internal biological state and influencing the long-term construction of each other's brains" (5). We need each other's brains: "Without mutually stimulating interactions, people and neurons wither and die" (11). The person we become is in part a result of our environment to be sure, but we can't forget that this environment includes biological entities always interacting corporeally at some level.

2 The Brain in Childhood and at Puberty

The development of the brain is largely complete in its major configurations by about the time of puberty, usually between eleven and seventeen years of age. A spurt of synaptogenesis and a general cleaning-out of the brain occurs around puberty, when undeveloped neural networks are eliminated, while those that have been myelinated are impervious to this cleansing. The process of myelination continues, though more slowly, in some parts of the brain after puberty. New synaptic connections can continue to be made throughout the life span, and other minor adjustments can take place, but, overall, the majority of the brain is basically wired up for life by this time. Speaking in gross terms, one could say that the human animal's life divides into three periods. The first is the uphill period of brain growth and development, which lasts from conception to puberty (in the context of brain development, the act of birth itself might be considered a relatively minor incident in the development of the brain); the second is the steady-state period from puberty through maturity; and the third is the slow and ill-defined downhill stage of old age when neuronal death and other types of brain degeneration increase.

Puberty is the major dividing line in the course of a human life, with significant changes in the body and brain at that time. It is almost as though the early years are designed to prepare the brain to cope with the physical maturity that comes on at that time. When puberty arrives, brain growth and development slow down, allowing the body to make use of the previous years' neural work. Before puberty, when there is an excess of synapses, plasticity is at its greatest, and the brain can grow, develop, and adjust in a number of ways. Plasticity for different areas of the brain, and thus for different cognitive abilities, passes through a critical period at various times, as we sequentially lose the ability, for example, to wire up the brain for vision within the first half year, or the ability to become a native speaker of a language by about the age of two (the disputed "critical period hypothesis"). Most of this plasticity, however, is lost at puberty and the brain is on its own: it must do with what it has; as Eliot says, "there's no trading up for a faster computer" (1999, 38). Or is there? New research with mice suggests the possibility that the chemical messaging process in the brain may be manipulated in humans to store juvenile plasticity, enabling auditory learning as needed for acquiring languages or musical skills (Blundon et al. 2017). Still, while the existence of some kind of a "window" for acquiring native languages seems to be taken as a universal, the exact extent to which individuals' environments have an influence on how slowly this window shuts, or whether it can be reopened, is still unknown.

It is commonly thought that when conception takes place, an individual's genetic inheritance is fixed, as a chain of hereditary developments begins to unfold. But in reality, the way genetics play out depends on a multitude of environmental factors. As Eliot has written, "[g]enes and environment are both important, but the fact is that we can do very little about our genes, and a great deal about the kind of environment we provide for our children" (1999, 9). Asbury and Plomin (2014) argue that educational institutions still falsely assume a "blank slate" and do not take genetics into account, to the detriment of the children (the authors are based in the UK, but their arguments certainly transfer to the American system). The process of "genotype-environment correlation" that they outline highlights how an understanding of genetics can help us tailor instruction effectively rather than assuming, as many school systems do, a one-size-fits-all approach.

3 Reading and the Frontal Lobes

There is a great difference between perceptual cognition, which involves perceiving something directly (as when one watches television or other electronic media), and symbolic cognition such as reading. French neuroscientist and reading expert Stanislas Dehaene argues for a "theory of reading" that

> postulates that the brain circuitry inherited from our primate evolution can be co-opted to the task of recognizing printed words. According to this approach, our neuronal networks are literally "recycled" for reading. The insight into how literacy changes the brain is profoundly transforming our vision of education and learning disabilities. (2009, 2)

In *Proust and the Squid* (2007), Maryanne Wolf brings the study of literacy into the age of neuroscience as never before. Wolf invites her readers "to ponder the profoundly creative quality at the heart of reading words. Nothing in our intellectual development should be less taken for granted at this moment in history, as the transition to a digital culture accelerates its pace" (ix). Wolf traces the history of reading in terms of the intellectual evolution of the human brain: she is concerned with how learning to read changes the human brain from what it had been throughout all of our long evolutionary history into something very different. Wolf first examines the history of reading and of writing systems from the Sumerians through the time of the ancient Greeks. She then details how the brain is physically reconfigured through five phases of reading development: (1) the emerging pre-reader; (2) the novice reader; (3) the decoding reader; (4) the fluent comprehending reader; and (5) the expert reader (114–62). This section of Wolf's book is compelling and wholly relevant to literary scholars. Wolf concludes with a consideration of the problem of dyslexia and other reading disabilities in the context of both cognitive and evolutionary science.

Seeing is as prewired as any of our cortical activities. We do not learn to see any more than we learn to digest food; it is an inheritance of our evolutionary history and a universal human capability. The same cannot be said for reading: we have no DNA and no neural network predisposing us biologically for reading. The vast majority of human beings who have lived have not known what it is to read. Yet we can learn, with determination and practice, to read in a way that can seem almost as effortless as seeing the world (or watching television or a YouTube video). Perhaps as much as language acquisition itself, reading illustrates the flexibility of our brains and once again demonstrates the fallacy of the nature and nurture dichotomy. We cannot learn to read without the ability to use our infinitely complex neural structures to comprehend creatively the possible significance of the symbols printed on the page (nature). Nor can we learn to read without the social contexts of home and school, where we are taught what reading is and how to go about doing it, and ideally, why it's important (nurture). The brain makes possible the social and psychological activities involved in reading, while the social and psychological activities involved in learning to read change the physical structure of our brain. Again: 100% nature and 100% nurture.

Merlin Donald has commented on the way learning to read profoundly changes brain structure: "The literacy brain is a cultural add-on to the normal preliterate state of the brain. It determines a great deal about how the operations of the conscious mind are carried out. It affects the relative growth and synaptic richness of certain regions and structures of the brain" (2001, 302–3). What we do in life has direct influence on how our brains are structured and how they function. The literate brain is different from the illiterate brain, and it is also different from the brain profoundly influenced by electronic media from an early age.

N. Katherine Hayles has described a "Generation M," a media-oriented generation of the twenty-first century, whose preferred cognitive style is what she terms "hyper attention": "switching focus rapidly among different tasks, preferring multiple information streams, seeking a high level of stimulation, and having a low tolerance for boredom" (2007, 187). It is clear to Hayles (as it should be to anyone who looks objectively at the realities of brain development), that "[c]hildren growing up in media-rich environments literally have brains wired differently from those of people who did not come to maturity under that condition" (192). The Generation M cognitive style stands in contrast to what Hayles calls "deep attention": "the cognitive style traditionally associated with the humanities, [which] is characterized by concentrating on a single object for long periods …, ignoring outside stimuli while so engaged, preferring a single information stream, and having a high tolerance for long focus times" (187).

In his book *The Shallows*, journalist Nicholas Carr makes the case against the Internet within a broad historical context, warning that the flexible and imaginative literary mind "may soon be yesterday's mind"

(2010, 10). Cognitive scientists are, of course, pursuing this topic and finding indications that the increased use of screen time is impacting not only social interactions but also potentially mental health and brain health; see Firth et al. (2019) on "the online brain" and indications of how cognition is changing due to the Internet. As Mancing has argued before (2006), symbolic reading and perceptual watching are two very different cognitive activities. We read books—novels, stories, poems, essays—but we perceive (watch) other media—theater, television, film, computer screens. Reading requires imaginative and creative participation in ways that perceiving does not, and reading is harder than perceiving. In learning to read, the brain is changed physically and conceptually. Literature demands a literate brain actively involved in the understanding of symbolic language, and literary theories that do not recognize this fact are doomed to failure. The authors of this book are by no means Luddites—we are writing this manuscript on our laptops with our smart phones nearby—but the fact is that teachers and scholars of literature are tasked not only with exposing the younger generations to the joys of reading great literature, but also with explaining its benefits to their cognitive development and to the betterment of society. While many students (whose frontal lobes are still developing in most cases) are resistant to reading, their instructors need to find ways to adapt to the digital age and understand that while we may prefer they read a print copy of original literature, we may need some technological bells and whistles to maintain students' attention and to incite the desire and motivation to read.[2]

The brain is a complex system, and brain development is a complex function, one bordering on chaos. In chaos theory, small initial incidents can have major, even catastrophic, effects later on. This is the so-called *butterfly effect*. The term was sparked by MIT meteorologist Edward Lorenz in the early 1960s, "after Lorenz suggested that the flap of a butterfly's wings might ultimately cause a tornado. And the butterfly effect, also known as 'sensitive dependence on initial conditions,' has a profound corollary: forecasting the future can be nearly impossible" (Dizikes 2011). Choices made, or not, and actions taken, or not, throughout childhood and youth, choices that are simultaneously biological and cultural contribute enormously to the development of the person who enters university classrooms and is asked to read literature. The ability to read and understand a story or a poem, to empathize with both fictional characters and real-life peers, depends to a very large degree on these developmental processes. As the evidence in this section suggests, college students at the beginning decades of the twenty-first century may not be in possession of a mind-brain that is as capable of creatively and imaginatively understanding written texts as their equivalents in past years had.

Are we trying to teach literature to a generation that has a diminished imaginative capacity to understand a written text? Is technology contributing to a reversal of the evolutionary trends toward intelligence, symbolic abilities, and imagination that have been characteristic of humans? Again, we aren't doomsayers, but some warning signs are hard to deny. On the very bright side, humanities scholars are uniquely qualified to help reverse some of these negative trends and we can try to shape the learning environment accordingly, if we stay apprised of research on technology's impact on developing brains.

4 The Mature Brain

Neurons die every day of our lives. In infancy and childhood, this is essential to the formation of strong and healthy neuronal assemblies, dendritic and axonal growth, creation of synapses, and myelination. But by the time we are adults, these activities become less of a factor in our lives. At a certain point during the normal lifespan, our brains begin to get smaller; we lose about 5% of total brain volume and weight by the age of 60, and about 20% by 90 (Greenfield 1997, 118). The *ventricles* (the cavities in the interior of the brain that contain cerebrospinal fluid) and the *sulci* (the spaces between the folds of the cortex) become larger; the space between neurons increases, and our synapses become less dense; blood flow and thus the supply of oxygen to the brain decreases.

These physical changes are not distributed evenly throughout all parts of the brain. The structures that are last to develop (especially the frontal lobes) are the ones most affected by aging: "we have come to recognize the maturation of the frontal lobes as the central theme of cognitive development and their decay as the central theme of cognitive aging" (Goldberg 2005, 161). In addition, there can be a significant loss of myelin in certain areas as the brain ages, which slows down brain operations. The hippocampus experiences a loss of synaptic connections and a change in the neurochemicals it receives (Gazzaniga 2005, 24). Patricia Churchland describes what she calls the "grim reaper death" of the normal aging process, "which fells about a thousand neurons per day in the adult brain after forty—a rather appalling statistic given the lack of replacements" (1986, 40). But this is not to say that change and development ever stop. Brain plasticity and even *neurogenesis* (the creation of new neurons) continue throughout life, and even the oldest of dogs can sometimes be taught new tricks. The brain is constantly remaking itself throughout life: "it turns out that the brain is an ongoing construction site" (Schwartz and Begley 2002, 128). Furthermore, as Louis Cozolino (2006) suggests, the older we get the better we may be at integrating cognitive processes, utilizing at the same time the functions "of both hemispheres in processing the information about the self and the world" (46). Neuroplasticity continues to be a major research focus, with indications of how exercise, friendship and love, stress levels, and various mental stimulation can affect levels of cognitive decline and of neurogenesis (see, e.g., Shaffer 2016). Again here, we stress both the genetic and environmental factors that make human brain development hardly a clear path.

5 Neurolinguistics

Neurolinguistics, which may be defined as "the study of how the brain ('neuro') permits us to have language ('linguistics')," is at the heart of the modern study of language as a biological and psychological phenomenon (Obler and Gjerlow 1999, 1). Like most work done today in the study of cognition, neurolinguistics draws on a large variety of more specifically delimited disciplines, such as behavioral neurology, aphasiology, studies of the deaf and the blind, aging research, brain imaging techniques, speech-language pathology, artificial intelligence, animal (especially ape) cognition, neuropsychology, second-language learning, sign-language research, psycholinguistics, developmental psychology, and cognitive science. Since ethical considerations rule out the possibility of certain types of research, such as the surgical removal of specific sections of the normal human brain or keeping children from acquiring language during their critical period, many of the basic findings have depended on the study of abnormal brain situations resulting from injury, disease, and abnormal development.

Although neurolinguistics is characteristically a new discipline that has gained prominence in the wake of the Chomsky-led cognitive revolution and enormous recent advances in neuroscience, its roots extend into the nineteenth century, when Paul Broca and Carl Wernicke made their foundational discoveries of the relationships between specific parts of the brain and specific language functions. Two areas of the brain are named after these pioneers: Broca's area is in the left frontal lobe and Wernicke's in the left temporal lobe. It was long thought that language functions were located in these places. Further research, however, has made it clear that language is indeed crucially dependent on a somewhat larger language area of the left hemisphere that includes both of these areas (essentially the left perisylvian region, which is the region surrounding and including the Sylvian fissure, the single most prominent sulcus, which separates the temporal lobe from the parietal lobe), but even this is not *the* site where language is both produced and comprehended. Rather, language is a function of the brain as a whole, with important participation of various parts of the neocortex and various sub-cortical areas, the inner, older, more widely shared biological structures (see Lieberman 2000). Language is not just a left-brain function, since the right hemisphere also contributes in important ways to both language production and language comprehension. Anderson and Lightfoot provide a definition of the "language organ"[3]: "it is a biologically determined aspect of certain tissue (primarily in the brain), rendering it uniquely sensitive to linguistic events in the environment, and allowing the development of a highly specialized capacity as a consequence of that environmental interaction" (2002, 221). And note that although these authors use the term "biologically determined," they also refer twice to the fact that nothing happens without "environmental interaction." We are "biologically determined" only in that we are dependent upon biology, which always interacts with the environment (society, culture, family structures, etc.). Mehler et al. (1984) warned already in the very first issue of *Cognitive Neuropsychology* against studies that focus exclusively on the biological processes behind language, promoting a "flexible approach to the relationship between psychological processes and the underlying physiology" (85).[4] Twenty years later, Schumann et al. (2004) warned of basically the opposite trend in language acquisition studies, namely that of dismissing neuroscience altogether (ix). Clearly, a consideration of both sides is the only approach that makes sense.

Neurolinguistics provides conclusive proof that language is a function of both biology and psychology. Languages exist only in and as a function of human mind-brains, and not in society at large. Saussure's *langue* does not exist; it is as valid a scientific explanation as is phlogiston (a chemical substance once thought to be released by combustion), the flat-earth theory, or divine creationism; it is totally inconsistent with the biological reality of language. Concepts of language in which it plays a major part are mistaken, and approaches to literary theory in which Saussurean concepts of language are prominent are fatally flawed. In contrast, Chomsky's innate language capacity, UG, LAD, and I-language are all reasonable approximations of the facts of human language, basically consistent with what is known about the biology of language. But, as Chomsky repeatedly notes, language is a matter of both biology and psychology. It is to that latter field that we now turn.

Notes

1. Attempts to illustrate the staggering complexity of a single brain's neuronal structure, involving new analogies or metaphors for comparison and/or enormous numbers made up of countless zeros, have been frequent among those who write on the brain (see, e.g., Churchland 1993, 66; 1997, 286; Flanagan 1992, 37; Greenfield 1997, 79, 83; and Hobson 1994, 28). See Edelman's book *Bright Air, Brilliant Fire* (1992) for a detailed and widely recognized explanation of neural growth and development.
2. Comparative literature professor Ellen McCracken has recently made the case for enhanced e-books to this end, for example, in the case of *Digital Dubliners*: "Available in iBooks for reading on tablets and Mac computers, this edition of Joyce's book includes video clips with students introducing each story, video 'office hours' with Joyce scholars, maps, 3-D visuals, memorabilia, photos, glossed words and phrases, and many other augmentations. Collectively written and available as a free iBook, this superb digital text shares academic knowledge with a wider public and uses the 21st-century cultural vernacular to draw readers to a century-old literary masterpiece" (McCracken 2017). If you can't beat them, join them.
3. Some scholars have taken issue with this language organ concept even if agreeing with other of their principles; see, e.g., Everett (2006).
4. See also Moreno and Giró (2014) who distinguish between natural languages and cultivated languages, the former of which develops more spontaneously while the latter (an extension of natural language) needs to be learned.

References

Anderson, Stephen R., and David W. Lightfoot. 2002. *The Language Organ: Linguistics as Cognitive Physiology*. Cambridge: Cambridge UP.

Asbury, Kathryn, and Robert Plomin. 2014. *G Is for Genes: The Impact of Genetics on Education and Achievement*. West Sussex: Wiley.

Blundon, Jay A., et al. 2017. "Restoring Auditory Cortex Plasticity in Adult Mice by Restricting Thalamic Adenosine Signaling." *Science* (30 June): 1352–56.

Bruer, John T. 1999. *The Myth of the First Three Years: A New Understanding of Early Brain Development and Lifelong Learning*. New York: Free P.

Carr, Nicholas. 2010. *The Shallows: What the Internet Is Doing to Our Brains*. New York: W.W. Norton.

Churchland, Patricia Smith. 1986. *Neurophilosophy: Toward a Unified Science of the Mind-Brain*. Cambridge: MIT P.

Churchland, Paul M. 1993. "On the Problem of Truth and the Immensity of Conceptual Space." *Realism and Representation: Essays on the Problem of Realism in Relation to Science, Literature, and Culture*. Ed. George Levine. Madison: U of Wisconsin P. 44–69.

———. 1997 (1989). "On the Nature of Theories: A Neurocomputational Perspective." In *Mind Design II: Philosophy, Psychology, Artificial Intelligence*. Ed. John Haugeland. Cambridge: Bradford Book/MIT P. 251–92.

Cozolino, Louis. 2006. *The Neuroscience of Human Relationships: Attachment and the Developing Social Brain*. New York: W. W. Norton.

Cunnane, Stephen C. 2005. *Survival of the Fattest: The Key to Human Brain Evolution*. Hackensack: World Scientific.

Dehaene, Stanislas. 2009. *Reading in the Brain: The Science and Evolution of a Human Invention*. New York: Viking.

Dizikes, Peter. 2011. "When the Butterfly Effect Took Flight." *MIT Technology Review*, 22 February (Online).

Donald, Merlin. 2001. *A Mind so Rare: The Evolution of Human Consciousness*. New York: W. W. Norton.

Edelman, Gerald M. 1992. *Bright Air, Brilliant Fire: On the Matter of the Mind*. New York: Basic Books.

Eliot, Lise. 1999. *What's Going on in There? How the Brain and Mind Develop in the First Five Years of Life*. New York: Bantam Books.

Everett, Daniel L. 2006. "Biology and Language: Response to Anderson & Lightfoot." *Journal of Linguistics* 42.2: 385–93.

Firth, Joseph, et al. 2019. "The 'Online Brain': How the Internet May Be Changing Our Cognition." *World Psychiatry* 18.2: 119.

Flanagan, Owen. 1992. *Consciousness Reconsidered*. Cambridge: Bradford Books/MIT P.

Gazzaniga, Michael S. 2005. *The Ethical Brain: The Science of Our Moral Dilemmas*. New York: Ecco.

Goldberg, Elkhonon. 2005. *The Wisdom Paradox: How Your Mind Can Grow Stronger as Your Brain Grows Older*. New York: Gotham Books.

Greenfield, Susan A. 1997. *The Human Brain: A Guided Tour*. New York: Basic Books.

Greenough, William T., James E. Black, and Christopher S. Wallace. 1987. "Experience and Brain Development." *Child Development* 58.3: 539–59.

Hayles, N. Katherine. 2007. "Hyper and Deep Attention: The Generational Divide in Cognitive Modes." In *Profession 2007*. Ed. Rosemary G. Feal. New York: Modern Language Association. 187–99.

Hobson, J. Allan. 1994. *The Chemistry of Conscious States: Toward a Unified Model of the Brain and the Mind*. Boston: Back Bay Books.

Lieberman, Philip. 2000. *Human Language and Our Reptilian Brain: The Subcortical Bases of Speech, Syntax, and Thought*. Cambridge: Harvard UP.

McConachie, Bruce, and F. Elizabeth Hart, eds. 2006. *Performance and Cognition: Theatre Studies and the Cognitive Turn*. London: Routledge.

McCracken, Ellen. 2017. "Enticing Students to Read Again." *The Chronicle of Higher Education*, 31 May (Online).

Mancing, Howard. 2006. "See the Play, Read the Book." In McConachie and Hart, 189–206.

Mehler, Jacques, et al. 1984. "On Reducing Language to Biology." *Cognitive Neuropsychology* 1:1: 83–116.

Moreno, Juan-Carlos, and José Luis Mendívil Giró. 2014. *On Biology, History and Culture in Human Language: A Critical Overview*. Sheffield: Equinox Publishing Limited.

Obler, Loraine K., and Kris Gjerlow. 1999. *Language and the Brain*. Cambridge: Cambridge UP.

Ornstein, Robert. 1991. *The Evolution of Consciousness: Of Darwin, Freud, and Cranial Fire: The Origins of the Way We Think*. New York: Prentice-Hall.

Plotkin, Henry. 1993. *Darwin Machines and the Nature of Knowledge*. Cambridge: Harvard UP.

Schumann, John H., et al. 2004. *The Neurobiology of Learning: Perspectives from Second Language Acquisition*. Mahwah: Erlbaum.

Schwartz, Jeffrey M., and Sharon Begley. 2002. *The Mind and the Brain: Neuroplasticity and the Power of Mental Force*. New York: ReganBooks.

Shaffer, Joyce. 2016. "Neuroplasticity and Clinical Practice: Building Brain Power for Health." *Frontiers in Psychology* 7: 1118.

Wolf, Maryanne. 2007. *Proust and the Squid: The Story and Science of the Reading Brain*. New York: HarperCollins.

PART III

Psychology and the Development of the "Literary Mind"

Study of the mind in context has gained increasing interest in cognitive and developmental psychology in recent years.

—Katherine Nelson

The only difference is the context.

—Marco Iacoboni

CHAPTER 8

The Mind at Work

The contextual element is crucial: mind cannot arise in isolation.
—Ian Stewart and Jack Cohen

The mind is what the brain does.

This sentence captures the mind-brain relationship and focuses our attention onto the mind as activity rather than as entity. One reason the mind-body binary is fundamentally wrong-headed is that it compares apples of physical structure and the oranges of an ongoing activity. The functions of the brain (the object of study in neuroscience) are the activities of the mind (the object of study in psychology). Many of the assumptions about human psychology that are most prominent in THEORY in particular are inconsistent with the scholarly received concept of the mind-brain. This chapter contains sections on the nature of consciousness and the sense of self; perception (especially visual perception); sleep and dreams; and the cognitive unconscious.

1 Consciousness and Self

The earliest articulation of the notion that the mind is what the brain does, is, as best we can tell, that of Marvin Minsky, who blurs mind–body dualism by stating: "*Minds are simply what brains do*" (1986, 287).

H. Mancing and J. Marston William, *Restoring the Human Context to Literary and Performance Studies*, Cognitive Studies in Literature and Performance,
https://doi.org/10.1007/978-3-030-89078-0_8

W. Deric Bownds makes the following comparison: "the mind is what brain/body does—in the same sense that digesting our food is what the gut does" (1999, 4). Antonio Damasio uses the term "mind" to encompass all brain operations, both conscious and nonconscious: "It refers to a *process*, not a thing. What we know as mind ... is a continuous flow of mental patterns" (1999, 337 n7).

It seems appropriate to use a term like *mind-brain* to indicate that there are two separate concepts simultaneously, one biological (brain) and one psychological (mind). These two are so inextricably intertwined that, although they can be discussed separately on some occasions, it is often necessary to make clear that we are referring to both of them at once and as one. Some might be tempted to see the mind-brain relationship as comparable to the familiar hardware-software relationship from computer science, but as Rudolfo Llinás states clearly and correctly: "this type of language usage is totally misleading. In the working brain, the 'hardware' and the 'software' are intertwined in the functional units, the neurons themselves" (2001, 3). Brain may be compared to hardware (or, as some prefer, wetware), but mind is not at all like software. A mind is an activity, while software is a program, a tool used by human beings. We do not *use* our minds; rather, we *are* our minds.

The mind-brain is shaped by the coordinated factors of genes and personal experiences. Daniel J. Siegel begins his excellent book *The Developing Mind* by stating succinctly, "The mind emerges from the activity of the brain, whose structure and function are directly shaped by interpersonal experience" (1999, 1). Culture, society, and temperament, together with genes, nutrition, and neurochemicals, all in context, all in conjunction, all at the same time, determine the nature, structure, and function of the mind-brain.

Gerald Edelman's theory of Neural Darwinism provides a point of departure for a plausible concept of consciousness, differentiating between *primary consciousness*, which is inextricably bound together with the present, with no concept of time, self, or context, and *higher-order consciousness*, which allows us to be aware of our own consciousness and in terms of past, present, and future (1992, 12). Primary consciousness is much like Donald's episodic culture, the worldview or degree of awareness that many animals have and from which human mimetic culture developed. Edelman proposes that only human beings have higher-order consciousness, a product of both "socially constructed selfhood" and "biological individuality" (133).

Damasio proposed a very similar approach to the continuum of consciousness in *The Feeling of What Happens* (1999), distinguishing between core and extended consciousness. *Core consciousness* "provides the organism with a sense of self about one moment—now—and about one place—here. The scope of core consciousness is the here and now" (16). *Extended consciousness*, on the other hand, "provides the organism with an elaborate sense of self—an identity and a person, you or me, no less—and places that person at a point in individual historical time, richly aware of the lived past and of the anticipated future, and keenly cognizant of the world beside it" (16).

Babies are born with a minimal primary consciousness, and higher-order consciousness emerges in them as they grow and develop. That is, even higher-order consciousness is not an all-or-nothing proposition, but something that grows and slowly emerges within each individual and which is in constant flux within the brain. Consciousness flashes into existence in brief states, occurring in rapid succession, always moving and changing. Different parts of the brain are in continual dialogue as consciousness moves flickeringly and transiently about within it. Consciousness thus depends on multiple factors: the electric signals coursing among the brain's neurons; the neurochemicals that flow through the brain; perception of things and events in the world; and memories of past experiences.

The baseline definition of consciousness is the experiencing of sensations. On top of this comes the higher conscious functions, including language. Being conscious, feeling sensations, is something we *do*, an action rather than a quality or a thing: the mind is what the brain does. Nicholas Humphrey, who defines consciousness in terms of "the having of sensations," says, "[f]eelings enter consciousness, not as events that happen *to us* but as *activities* that we ourselves engender and participate in—activities that loop back on themselves to create the thick moment of the subject present" (1992, 217). For Humphrey, to have sensations, you must have a body; consciousness by definition is embodied consciousness (203–4). Philosopher Michelle Maiese also underscores the importance of recognizing that consciousness is embodied: "consciousness is not simply something that happens within our brains, but rather something that we do through our living bodies and our lived, bodily engagement with the world" (2011, 1). Susan Greenfield concurs: "One thing for sure is that consciousness always entails some sort of feelings" (2000, 2).

Edelman, Damasio, Humphrey, and Greenfield (and many others) use terms such as reentry, bootstrapping (the process of making something out of nothing, of pulling yourself up by your own bootstraps), self-referential feedback, and emergence when discussing the concept of consciousness. *Emergence* is the key term here, one that has gained popularity and status in recent research in the sciences of complexity and chaos. This view of consciousness necessarily implies a lack of central place in the brain where it occurs; in the felicitous phrase of Daniel Dennett, there is no "Cartesian Theater," the supposed "place where 'it all comes together' and consciousness happens" (1991, 39). In place of the Cartesian Theater, Dennett proposes what he calls the "Multiple Drafts model of consciousness," according to which "all varieties of perception—indeed, all varieties of thought or mental activity—are accomplished in the brain by parallel, multitrack processes of interpretation and elaboration of sensory inputs. Information entering the nervous system is under continuous 'editorial revision'" (111). What Dennett's model makes apparent is the multi-track, parallel, distributed, simultaneous, and incredibly rapid nature of the activity of mind. Thoughts may only last for milliseconds and occur simultaneously in many parts of the brain. So consciousness is everywhere, always in flux, never complete.

Alva Noë has explored the embodied, contextualized approach to consciousness, writing, "The locus of consciousness is the dynamic life of the whole, environmentally plugged-in person or animal. Indeed, it is only when we take up this holistic perspective on the active life of the person or animal that we can begin to make sense of the brain's contribution to conscious experience" (2009, xiii). The take on the concept of consciousness presented here is by no means the only one found in the cognitive sciences today. Others abound, but no attempt can be made here to review them all; for a summary, see the *Blackwell Companion to Consciousness*, edited by Max Velmans and Susan Schneider (2007; second edition 2017), and Jonathan D. Cohen and Jonathan W. Schooler's *Scientific Approaches to Consciousness* (2014).

If consciousness is not a thing but a process or an activity, it stands to reason there is no such thing as a *self*. And yet, like the Cartesian Theater, it seems almost impossible for us to think or talk without the assumption that we have a self or without referring to our self.[1] If consciousness is an essential aspect of the activity that is mind, something the brain does, then what do we speak of when we talk of our self? Such talk seems to slip the homunculus of dualism in by the back door, as the self who does the mind's work—the "real me" of folk psychology, the "mind's I," the "ghost in the machine," the Christian "soul," the Kantian "transcendent, noumenal self," the Freudian "ego." But surely this cannot be—and it is not, and yet *something* is there.

The idea that we are (or have) multiple selves that we exhibit (or inhabit, or that inhabit us) at different times and in different places for different reasons has long been explored in psychology, literature, and the popular imagination. Psychologists often speak of distinguishing among our professional, family, public, private, spiritual, and even future selves. What is often meant is that it is impossible to identify one unified, stable, recognizable, individual self for everyone. In the more simplistic terminology of THEORY, the self is decentered; that is, there is no single, unitary, controlling (self-centered) self, but only a socially constructed subject. But, characteristically, this binary moves the self from the stable center to the unstable margin; to call a self decentered clarifies nothing.

Although there is no such thing as a self that we have, we do have the *sense* of having a self; that is, we have a *sense of self*. Consciousness is self-referential. There must be some way to talk of that-which-we-perceive-to-be-us-within-ourselves. Evolution, if nothing else, demands that we be aware of, and take actions to ensure the continuation of, our self. Feminist psychologist Ellyn Kaschak describes our sense of self as "a metaphor, an organizing concept," and notes the "emerging sense of self is a set of abstract symbols and, at the same time, an embodiment of the abstract" (1992, 154). Kaschak's views are wholly consistent with the ecological mind-brain that we promote in this book, with her stress on context as well as the emergent and narrative nature of our sense of self.

Because we experience our lives in narrative terms, we understand how we both change and remain the same person through time. Marc D. Hauser summarizes how this happens:

> What makes me different from you is something about the continuity of my own personal experiences. I can share experiences with others in terms of seeing, tasting, and smelling the same thing, but they are my experiences. What makes me the same person over time is that although my views may change, it is my experience of the world that is changing them. I own my experiences, even if they are foisted on me by someone else, and even if I am unaware of what I own. This characterization of the self is universally preserved, even though other aspects vary across cultures. (2006, 183)

In the same way that the physical development of the brain involves a complex interplay between biology and culture, genes and environment, so do consciousness and our sense of self emerge from constant, simultaneous, and self-identical nature/nurture interactions. Human beings are animals, and the biology of our species is an integral, defining, and irreducible aspect of consciousness. We are also social beings, and the social context—the dialogical, culturally and historically situated, interpersonal context—is every bit as integral, defining, and irreducible.

Language is the single most important cognitive tool available to higher level consciousness. This does not mean, however, that consciousness depends on or is defined by language. Damasio's review of evidence for consciousness without language summarizes and illustrates the issue very well (1999, 107–25, 184–89) and aligns with the consensus of thought in contemporary cognitive science. The notion that consciousness is dependent upon and defined solely by language and that language is an autonomous social agent that determines subjectivity does not resonate in the intellectual world of the twenty-first century. Merlin Donald is merciless in his commentary on this position:

> The Postmodernists adhere to this idea in its most exclusive and censorious form. They define awareness entirely in terms of an imaginary ongoing text that characterizes a particular culture. This idea is a return to the old canonical definition of consciousness, in which it is once again conscripted into the service of rabid a priori tribalism, in this case anthropocentrism. Taken to its logical conclusion, this text-bound definition denies consciousness not only to animals but to many humans as well. Pity the poor preliterate Mesolithic humans; they were unconscious no less! They struggled with survival in conditions that the typical modern person couldn't negotiate for an hour, and they did this unconsciously! Remarkable. These are the opinions of ideological high priests in secular garb. So, who was the first conscious human? Foucault? Derrida? The mind boggles. (2001, 119)

Later, he returns to the idea and reiterates that "thought is the arbiter, and language is the child of thought, invented in the service of thought, employed forever as the amplifier and mediator of thought. The human mind is infinitely wider and more supple than all its languages and symbols" (277). But if our self-concepts are not formed exclusively by and through language, our (Bakhtinian) dialogical relationships within our unique social contexts are undoubtedly a major factor in the process (Hermans and Kempen 1993). Or, as Chris Frith says, "Meaning arises from the interactions between minds" (2007, 187).

2 Perception

According to much THEORY, we perceive the world solely by way of signs, which we read as we do the paradigmatic sign system—language—meaning that all cognition is, in effect, linguistic. Vision itself, the primary human perceptual system, provides us with just another means of reading (and misreading) the textualized world. This concept contradicts what we know about evolved, situated, multi-modal, embodied human cognition.

The modern understanding of perception owes more to James J. Gibson than to any other individual. During World War II, Gibson served in the Air Force and studied actual problems in the visual perception of pilots. Working with perceptual realities in real-world situations led him to re-conceptualize what vision consists of, to realize the value of the work done by the Gestalt psychologists, and to conclude that the study of vision should be based on real-world, or ecological, considerations rather than on static drawings and pictures. The result was his book *The Perception of the Visual World* (1950).

Gibson's concern is the world as visually perceived by upright, forward-looking, terrestrial beings whose perceptual field is bounded by earth and sky and is seen from the position of the contextualized individual. He describes the visual world as being "extended in distance and modeled in depth; it is upright, stable, and without boundaries; it is colored, shadowed, illuminated, and textured; it is composed of surfaces, edges, shapes, and interspaces; finally, and most important of all, it is filled with things which have meaning" (3). Before Gibson, virtually no one was talking of vision in these humanly meaningful terms. Previous philosophers and psychologists tended to discuss vision in terms of spaces and lines, two-dimensional sensations, visual illusions, mental constructs based on fallible perceptions, and, above all, cultural norms that determine visual understanding. It was Gibson's innovation to remind everyone that "all human beings, everywhere, probably see the ground and the sky the same way" (212).

Visual illusions are still commonly used to demonstrate the unreliability of human visual perception. Gibson's definition of a visual illusion is "a perceptual judgment or estimate which is consistently not in agreement with measurements of the object giving rise to the perception" (181). We are very rarely deceived in our perception of simple sizes and shapes in the real—ecological—world, but quite easily misled by simple drawings and visual on-screen tricks. Modern cognitive researchers and philosophers agree. Alva Noë, for example, writes,

> Our perceptual skills have evolved for life on earth, not life in an environment in which objects materialize and vanish at the whim of supernatural deceivers (or engineers). So the fact that we are vulnerable to deception—in the psychology lab or at the movies—just reveals the context-bound performance limitations of our cognitive powers. It does not show that our cognitive powers are radically deluded! (2009, 142)

Rather than building a whole perceptual theory on two-dimensional drawings, Gibson has taught us, we should build our theories on what happens in the world in which we live as embodied creatures. Using his work on visual perception as the springboard, Gibson then went on to develop a complete theory of perception.

In *The Senses Considered as Perceptual Systems* (1966),Gibson showed that human perception is not a matter of the brain's passively receiving stimuli from an external environment and then processing them. Rather, Gibson suggests, we actively seek out information from the world, and our understanding of the world is a result of a constructive, ecological relationship with it: "The classical concept of a sense organ is of a passive receiver, and it is called a receptor. But the eyes, ears, nose, mouth, and skin are in fact mobile, exploratory, orienting" (33). All forms of perceiving, like all forms of understanding language, are always active, and this has been a consistent theme in cognitive studies; see Damasio (1994, 225), Berthoz (2000, 1), Gallagher (2005, 172), and Thompson (2007, 366).

Rather than the traditional five senses, Gibson (1966, 49–50; see also Gibson 1979, 244–46) proposes five *perceptual systems*:

(1) the orienting system, basic to all the others, and whose primary activity is general orientation;
(2) the auditory system—which involves listening;
(3) the haptic system—touching;
(4) the taste–smell system—tasting and smelling; and
(5) the visual system—looking.

The occipital lobe is the largest sector of the brain dedicated almost exclusively to a single function, but in fact, multiple, parallel functions are involved in vision (Zeki 1993). In the course of evolution, the percentage of the brain devoted to other senses, especially to the more primitive olfaction, decreased considerably more in humans than in other animals (Watson 2000). As a result, the human sense of smell is far less important to us than to many other animals. Meanwhile, vision occupies a greater percentage of the brain than all other senses combined. The result is that, as the traditional proverb has it, our eyes are the windows on the world. Citing this proverb, Ann Marie Seward Barry adds that vision is "[t]he last of our senses to evolve and the most sophisticated, they [our eyes] are our main source of formation about the world, sending more data more quickly to the nervous system than any other sense" (1997, 15). It needs to be noted duly, though, that the vision-impaired can know the world just as intimately as those possessing sight, and as is well known, the other perceptual systems tend to become enhanced to compensate for the lack of vision (see Cohut 2019).

Gibson often uses the term "veridical" to describe the "invariances" picked up from the environment by our active perceptual systems. By this, he means not that we see the world "exactly" as it is, for there is no sense in which such a concept has any meaning. We perceive the world, for the most part, in reliable and relevant ways that allow us to take advantages of the *affordances* the world makes available. The idea of affordances is a neologism central to Gibson's thought and has potential for our understanding of the act of reading: "The *affordances* of the environment are what it *offers* the animal, what it *provides* or *furnishes*, either for good or ill" and his very deliberately selected term "implies the complementarity of the animal and the environment" (1979, 127). An affordance is a pragmatic concept and exemplifies the inextricable, contextual relationship between any organism and its environment.

The importance of Gibson's approach to perception cannot be overstated. As his biographer Edward Reed has noted, "Gibson was replacing an assumption with an hypothesis. The idea that perceptions are mental constructs based on inadequate sensations is centuries old and was no longer considered an active topic of debate—until Gibson attacked it" (1988, 144). The appearance of Gibson's book during an era of COGNITIVIST hegemony made it a complete anomaly; it was greeted with incomprehension and was either criticized—for example, by Fodor and Pylyshyn (1981) and Pylyshyn (1984)—or ignored. The same is true of the subtle book by Rudolf Arnheim (1969), which made some of the same points with respect to visual perception. Within a decade, however, it was the subject of intense debate (Reed 1988, 216–17). The work of Ulric Neisser in the 1970s (see Neisser 1976) was crucial in bringing Gibson to the attention of a far wider field. In Chapter 14, we consider further the implications of Gibson's theories of perception on embodied cognition and its applicability to literary studies.

Gibson begins his last major book, *The Ecological Approach to Visual Perception* (1979), by noting that although we normally say that vision depends on the eye, which in turn is connected to the brain, it is more accurate to say that "natural vision depends on the eyes in the head on a body supported by the ground, the brain being only the central organ of a complete visual system" (1). As with his book on visual perception a quarter of century earlier, Gibson is interested in the way we perceive the real world in real life, and this is the reason he uses the term "ecological" to describe his approach.

Language, for Gibson, is not a literal, virtual, or analog experience as is all visual experience, whether direct or mediated. "Perceiving precedes predicating" (260), states Gibson, that is, awareness comes before verbalizing. Gibson contrasts perceptual and verbal meaning as follows:

> *Perceptual meaning* has
>
> - an Environmental source, which via
> - a Physical law is related to
> - a Stimulus invariant, which has
> - a Psychological resonance, which results in
> - a Percept; while
>
> *Verbal meaning* has
>
> - a Referent, which via
> - a Social convention is related to
> - a Symbol or word, which has
> - a Psychological association, which results in
> - a Thought. (1966, 244)

It is not our *perception* of something that is significant; what matters is our *experience* or our *understanding* of that thing perceived. Damasio makes this fact fundamental to his consideration of the neurobiology of consciousness: "You and I can have an experience of the same landscape, but each of us will generate that experience according to our own individual perspective. Each of us will have a separate sense of individual ownership and individual agency" (1999, 306). This distinction is of fundamental importance, for example, when we read a work of literature. We all read (see, perceive) the same thing, but what we understand by what we read may be something else entirely. The failure to take this distinction into consideration, the subsequent conflation of perception and understanding, and the illogical conclusion that the varied nature of understanding means nothing can be known, are often implicit (and sometimes explicit) in much contemporary THEORY.

3 Sleep, Dreams, and the Unconscious

Fish and amphibians do not sleep; they rest at times, but never sleep. But sleep is a defining characteristic of birds and all mammals (including dolphins, who sleep with only one cerebral hemisphere at a time)—for them it is a necessary life function (reptiles fall at various places along the continuum between fish and mammals). Sleep is a time for restoration and repair of vital bodily organs and, in human beings, an opportunity for the consolidation of memories. But what is the role of dreams? Since ancient times, dreams have been the object of wonder, speculation, and symbolic interpretation. But only in the twentieth century did we begin to understand the physical processes involved in dreaming. The modern era of dream research began in 1953 with the discovery—largely accidental—of rapid eye movements (REM) during sleep by graduate student Eugene Aserinsky who was working under the direction of Nathaniel Kleitman (Hobson 1988, 139–46). Since then, sleep and dreams have been analyzed in great detail and it is now possible to describe with some accuracy how it is we dream. *Why* we dream, however, is less certain, although there is no shortage of plausible, as well as absurd and implausible, suggestions (see, e.g., Greenfield 2000, 148–49). It is often assumed that we dream only during REM states. However, we dream virtually all the time we are asleep, but we almost never recall our non-REM dreams. Most nightmares, terrifying dreams, and sleepwalking take place during delta sleep, the deepest of the four phases of sleep (Panksepp 1998, 129).

J. Allan Hobson's book *The Dreaming Brain* (1988) is the standard presentation of the *activation-synthesis hypothesis*, perhaps the most popular and influential modern theory of sleep and dreams. Domestic cats are often used in sleep research because about two-thirds of their life is spent catnapping (13 to 16 hours daily), with over 200 minutes a day in REM sleep (Rock 2004, 15). Building on the work with cats by French neurosurgeon Michel Jouvet that showed a relationship between brainstem activity and dreaming during REM sleep (146–54), Hobson (who had worked in Jouvet's laboratory) and his colleagues developed a research plan that led to the conclusion that during sleep, some crucial neurons in the brainstem turn on the mind while others turn off both the perceptual systems and the motor-command system of the brain stem and spinal cord. This permits the brain-mind (as Hobson calls it) to generate and process information both without the interference of sounds and other distractions from the outside world, and without our attempting to react physically to or participate in our dreams. During this process, the neurochemical mix in the brain undergoes a dramatic shift from a normal waking brain, with a notable reduction in the levels of serotonin and norepinephrine and a higher level of acetylcholine. The simultaneous internal activation and sensorial disconnect allow the images, activities, and narrative of dreams to occur but keep us from acting upon our dream experiences. Thus, in the activation-synthesis hypothesis, the "now auto-activated, disconnected, and auto-stimulated brain-mind" (207) can generate and interpret its own signals in terms of the individual's memories, experiences, concerns, and so forth.

Hobson traces the history of dream research through the ages, explicitly compares his own approach based on the modern understanding of what the brain is and how it functions with Freud's theory of dreaming, and criticizes the latter thoroughly. Basically, Hobson's point—more recently, as we will see, called into doubt—is that if it is the primitive brainstem, together with other areas of the midbrain, that activates dreams, they obviously are not generated by areas of the cortex, particularly the frontal lobes, and would not then have any of the symbolic connotations attributed to them by Freud. His critique of Freud is so intense that Hobson has been called "the anti-Freud" (Rock 2004, 17). In 1976, Hobson made a presentation of his activation-synthesis hypothesis at the annual meeting of the American Psychiatric Association, where a vote was taken on opinions about the validity of Freudian theory in light of Hobson's approach. The result went strongly against Freud in this group of professionals basically sympathetic to psychoanalysis, which suggested that for many Freud was passé (Solms and Turnbull 2002, 190).

Hobson continues his polemic against Freud and psychoanalysis in *The Chemistry of Conscious States* (1994). Here he grounds his concept of brain-mind activity within a version of chaos theory, in which the brain is the result of self-organization and an emergent quality. Within this paradigm, Hobson identifies three conscious states (as opposed to nonconscious states such as coma): waking, sleeping, and dreaming. Much of Hobson's work in this area is of great interest to humanistic, and especially literary, scholars, for example, his comparison of dreaming with insanity; his discussion of how sleep and dreams positively influence both physical and psychic health; and the role of emotion and narrative in dreams.

In his lavishly illustrated *Consciousness* (1999), Hobson presses even more his view of the brain-mind paradigm, significantly stressing that unlike many neuroscientists, he does not take the approach that reduces everything to brain function. Hobson describes a contextual view of consciousness as an emergent phenomenon that involves degrees of consciousness, based on feelings or awareness, and he describes this explicitly with respect to his dream theory. He further elaborates (157–215) his unified model of mind and brain from his earlier books, what he calls the "AIM (Activation energy-Information source-Modulation) model" (1994, 64–77). To make this model more accessible, the book is illustrated with three-dimensional cubes. But Hobson's work is not without its critics, and Freud still lingers in the mind of several other researchers.

As Hobson was mounting his frontal attack on Freud's dream theories, Jonathan Winson attempted to reconcile aspects of Freudian theory and neuroscience in *Brain and Psyche* (1985). Winson's detailed description of the role of REM sleep, the importance of critical periods in brain development, and the function of the hippocampus and other brain regions in memory consolidation during dreaming remain largely accurate by today's standards. He summarizes his conclusions as follows: "*I believe that the phylogenetically ancient mechanisms involving REM sleep, in which memories, associations, and strategies are formed and handled by the brain as a distinct category of information in the prefrontal cortex and associated structures, are in fact the Freudian unconscious*" (209). Winson's version of the unconscious, however, is not purely Freudian, as he sees the unconscious as something more cohesive and active than Freud's "caldron of untamed passions and destructive instincts held in check by repression" (245).

The most substantial effort to reconcile Freud and neuroscience has been made by South African psychoanalyst Mark Solms (Solms and Turnbull 2002). Solms criticizes Hobson's activation-synthesis hypothesis (186–98), highlighting research showing that the brainstem is not the sole point of origin for REM and other dreaming activity. Patients with certain kinds of parietal aphasias and others who had some parts of their frontal lobes surgically cut out (to remove a tumor, for example) reported that they no longer had dreams. This meant that at least a large part of the brain functions involved in dreaming had nothing to do with brainstem activity. Solms does not deny any role to the brainstem, for it is clearly involved in REM activities, but he calls attention to the fact that it alone cannot account for all Hobson credited it with. Furthermore, the major chemical difference between dreaming and waking states is dopamine, rather than the ones Hobson described. Solms concluded REM sleep is one thing and dreaming is another: parallel but separate activities, with different brain regions involved in each. Finally, whereas Hobson sees similarities between delirium and dreams, Solms believes that dreams are more like schizophrenia.

One has the impression that Solm was trying too desperately to reconcile Freud and neuroscience. For example, he enlists Nobel-Prize-winning memory researcher Eric Kandel as an ally. Kandel was born in Freud's Vienna and has been fascinated by psychiatry all his life, studying the discipline in depth and almost becoming a psychiatrist. Solms cites the following 1998 statement by Kandel: "psychoanalysis still represents the most coherent and intellectually satisfying view of the mind that we have" (Solms and Turnbull 2002, 304). But in his more recent autobiography, Kandel describes his flirtation with psychoanalysis and his firm belief that its radical departure from late-twentieth-century brain science made it less viable (2006, 35). Kandel's repeated remarks of this sort—see the chapter dedicated to psychoanalytic thought (363–75), as well as other comments throughout—make it clear he no longer believes psychoanalysis to be a viable project.

Literary theorist Norman Holland (1988, 1992, 2009) has campaigned longest and hardest for the reconciliation between neuroscience and psychoanalytic theory Solms proposes. Holland consistently cites Solms's work in support of his own approach. His meticulous work, so influential in the formation of relationships between cognitive science and literary theory, must be taken seriously into consideration by anyone wanting to maintain a link between Freud, or psychoanalytic theory in general, and literary theory and criticism.

In some ways, the Solms-Freud vs. Hobson-anti-Freud neural turf wars cast more heat than light on the subject of dreams. G. William Domhoff (2005) has summarized their work and shown that they have much in common. Both have stressed the role of emotion in dreams, have called attention to the bizarre nature of dreams, and conceive of dreams as being related to some form of psychosis (delirium for Hobson, schizophrenia for Solms). Their dispute over exactly which areas of the brain have more influence on dreaming and their disagreements about the relative involvement of neurochemicals are issues that still await the results of further research. Especially Hobson's early work is becoming ever more dated as other neuroscientists investigate sleep and dreams. What Domhoff accentuates, however, is how both Hobson and Solms have either ignored or rejected out-of-hand research calling into question some of their similar assumptions and conclusions.

Most research on dream content, including Domhoff's own book, *The Scientific Study of Dreams* (2003), does not support their shared conviction that dreams are normally bizarre, exotic, and emotion-laden. His argument that the scientific study of dreams leads to the conclusion that the elements of bizarreness and disunity considered so important in earlier theories, Freudian as well as those of Hobson and Solms, are mistaken. Domhoff's "neurocognitive model" of dreaming makes the whole process more mundane and unexciting. Most of our dreams are "reasonable simulations of the waking world inhabited by the dreamer" (168) and are without symbolic significance or characteristically bizarre or suggestive of psychotic elements. Rather, "much dream content is still not understood and may turn out to be the product of freewheeling improvisation of little import" (169). In sum, "it now seems likely that dreaming is a comprehensible cognitive process with many similarities to waking thought" (170). See cognitive psychologist Richard Schweickert's research on cognitive social networks (2007; and Han et al. 2016) as illustrations of how the unconscious dream world and the waking world have been compared scientifically.

Owen Flanagan has proposed in *Dreaming Souls* (2000; see also 1996, 32–52),that dreams are *spandrels*, making use of a term first suggested by Gould and Lewontin (1979), to suggest that dreams "have never been subjected to biological selection pressures themselves. They are free riders that come with sleep. Dreams are the spandrels of sleep" (2000, 22). Sleep is essential to biological function, and, as such, was selected for in evolution. But dreams need not have any explanation in natural selection. Importantly, both Domhoff (2003, 267) and Foulkes (1999, 139) concur that dreams are probably no more than an evolutionary by-product of sleep and conscious cognitive processes. Even Hobson concedes that dreams could be no more than "an epiphenomenon, a secondary and unnecessary outcome of an underlying process" (1994, 282).

Flanagan, like Hobson, Foulkes, and Domhoff, rejects the deep significance accorded to dreams by Freud since no neurological evidence supports such a theory:

> Forget the views that all dreams are meaningful or the work of a bad poet inside you. Here is a better view: Dreams are produced by activity originating in the brainstem that awakens stored or semi-stored thoughts and memories that are then put into some sort of narrative structure by higher brain sectors that are designed to make sense of experience by light of day, but continue to work, less efficiently, when the lights go out. (2000, 127)

In spite of their non-functional role in evolutionary terms, however, Flanagan, like Domhoff, still maintains that many dreams are self-expressive. Dreams contribute to our sense of self, our identity. The (partly, perhaps largely) fictional sense of self that we construct, and constantly revise, for ourselves in our self-narratives includes aspects taken from our dreams. So, even if our dreams are not a royal road to anywhere, they are not meaningless; they "can be used to shed light on mental life, on well-being, and on identity" (140), and we incorporate them into the ongoing meaning-making project that is living. "Life is a dream," wrote Calderón de la Barca in the seventeenth century. Well, no; but dreams are an important part of our lives, and, as such, they are meaningful to us. What we are, our plans for the future, and our memories always necessarily include our dreams. Literary writers frequently include dreams in their narratives—and sometimes with intentionally symbolic meaning! This reality check about dreams isn't meant to take the fun out of analyzing literary dreams, but should give us pause about automatically interpreting them through a Freudian, psychoanalytic lens.

Finally, Andrea Rock's *The Mind at Night* (2004) is an accessible account of many of the important issues involved in sleep and dreaming (e.g., see 41–59 for her good summary of the Solms-Hobson feud), including much that is not addressed here; her summary of major results from the study of dreaming over the last few decades (187–89) is particularly useful. She concludes that reality is not that different from dreaming, "a con job beautifully carried out by neural circuitry of astonishing complexity" (199–200). While there have not been major developments in the past decade or so in terms of defining the nature of dreams—what they actually "are" and mean—scientists continue to use the latest technology to study them. Neuroscientists Benjamin Baird and Giulio Tononi have monitored brain waves of sleeping subjects, identifying which brain areas are associated with which types of dream topis (per self-reporting of the subject), as a means of better understanding human consciousness (Lewis 2017).

We perceive consciousness as serial and narrative, but that is only the smallest part of cognition. The activities of the mind-brain are going on all the time, distributed throughout the brain, on multiple levels, in parallel, mostly outside of our conscious awareness. The term "cognitive unconscious" was coined by psychologist John Kihlstrom (1987) to refer to all the parallel, distributed mental processes that never cease and about which we are unaware. It is a useful term, particularly in that it avoids any possible confusion between the unconscious processes that cognitive scientists study and the Freudian dynamic unconscious. LeDoux distinguishes clearly between the two concepts: "The term cognitive unconscious merely implies that a lot of what the mind does goes on outside of consciousness, whereas the dynamic unconscious is a darker, more malevolent place where emotionally charged memories are shipped to do mental dirty work" (1996, 29–30). Lakoff and Johnson describe the cognitive unconscious as including "not only all our automatic cognitive operations, but also all our implicit knowledge" and serving as a "'hidden hand' that shapes how we conceptualize all aspects of our experience" (1999, 13).

All those "gut feelings" (intuitions, hunches) that are a part of our embodied cognition are also included within the cognitive unconscious. Gerd Gigerenzer uses all these terms to refer to any kind of judgment

(1) that appears quickly in consciousness,
(2) whose underlying reasons we are not fully aware of, and
(3) is strong enough to act upon. (2007, 16)

We often make decisions or form opinions and beliefs by means of these nonconscious processes and then later rationalize the how and why of our decisions and beliefs. We use a number of heuristics to reach these decisions: "Much of intuitive behavior, from perceiving to believing to deceiving, can be described in the form of these simple mechanisms that are adapted to the world we inhabit" (40). The process by which we reach these rapid conclusions, using the "evolved capacities of the brain," may be a "Darwin Machine" (Calvin 1990), which spins out in fractions of milliseconds multiple speculative scenarios about the future consequences of our actions.

Damasio, who, like LeDoux, prefers the term "nonconscious" to describe all the mind-brain activities that take place outside of consciousness, lists the following important aspects of our "not-known" cognition:

1. all the fully formed images to which we do not attend;
2. all the neural patterns that never become images;
3. all the dispositions that were acquired through experience, lie dormant, and may never become an explicit neural pattern;
4. all the quiet remodeling of such dispositions and all their quiet renetworking—that may never become explicitly known; and
5. all the hidden wisdom and know-how that nature embodied in innate, homeostatic dispositions. (1999, 228)

He then adds, simply: "Amazing, indeed, how little we ever know."

The cognitive unconscious stands, obviously, in clear opposition to the dynamic unconscious that is a cornerstone of Freudian psychology. It is time to ask explicitly, along with Todd Dufresne, "So why do we keep referring to Freud as though he was essentially correct about human psychology when, arguably, he was trivially or incidentally correct?" (2003, viii).

If nearly all of Freud's concepts are undermined by cognitive science, Lacan's version of Freudian thought—including his concepts of the mirror stage, *jouissance*, desire, the Phallic signifier/phallus, the Other, the Name-of-the-Father, and the Imaginary, Symbolic, and Real Orders—meets an even worse fate. His famous dictum that "the unconscious is structured like a language" is beyond comprehension. The sort of Freudian dynamic unconscious he refers to has no place in any contemporary approach to the mind-brain; the idea of a structure to the nonconscious activities of the mind-brain makes no sense; and language in the Saussurean sense in which he understands it has little or nothing to do with language as it functions in embodied, contextualized human beings. Furthermore, that cornerstone of Lacanian thought, the "mirror stage" (which was developed in the late 1930s and the 1940s) bears no relationship to what actually happens between babies and mirrors (Billig 2006; Jay 1993, 341–46; Holland 1992, 200–3; Evans 2005). Given Lacan's outdatedness, it is surprising that he is still employed in writing published in respected scholarly journals, such as in an otherwise interesting recent article performing a Lacanian analysis on a work that came out a century before Lacan started his career as a psychoanalyst (Cunningham 2017).

Is Freud irrelevant to literary studies today? Aside from the fact that his historical importance should be noted, we do not see why or how Freud should play a major role in literary theory and analysis well into the twenty-first century. If a contemporary of Freud was clearly influenced in his or her writing by Freud, as for instance Franz Kafka and Arthur Schnitzler were, then it can be of interest to point out such intellectual connections and perhaps analyze the work with that in mind (see, e.g., William 2002 and de Bont 2017). Otherwise, a "Freudian reading" of a literary work in today's world makes little sense.[2] Like Saussurean linguistics, Freudian psychology is a once-important psychological theory that has little, if any, currency in the twenty-first century; it is a historical relic, something of a curiosity. The burden of proof is on the followers of Freud to demonstrate how and why we should give credence to any of the details of his nineteenth-century fantasy in the age of twenty-first century cognitive science.

4 Memory

Memory has been a source of human concern since the oral cultures of ancient times when knowledge about history, art, and science was not stored in books, libraries, and digital files. The ancient and medieval tellers of tales, singers of songs, rhetoricians, and politicians often relied on mnemonic devices such as meter, rhyme, and techniques such as the "house of memory" to help them recall with (relative) precision (see Yates 1966; Carruthers 1990; Leverage 2010). In our own day, memory has evoked particular interest in psychologists and neuroscientists in a different way: How do we remember? Where in the brain is memory located? How accurate are our memories?

In the 1950s, neurosurgeon Wilder Penfield conducted a series of studies on some of his patients during surgery by stimulating different parts of their exposed brains (where there are no nerves sensitive to pain) with a mild electrical current. When he did so, the fully conscious, non-anesthetized patients reported sudden, vivid recall of episodes from the distant past. This was taken as proof that the brain stored (presumably immutable and veridical) memories in specific parts of the brain. All we had to do was locate those places, patterns, or neural imprints—find the *engram*, the brain's neural record of an event—and we could recover all the memories of our entire lifetimes. Alas, the search proved fruitless. The psychologist who spent much of his career attempting to locate the engram was Karl Lashley. Penfield's studies were not easily replicated; it turned out that many of his and subsequent researchers' patients did not experience comparable vivid memories, and many of the supposedly remembered scenes turned out to be unverifiable and probably more invented than recalled. The popular notion of memory—held by many psychologists even today, by Freudian psychoanalysts, by many poets and novelists, by many who believe that the mind-brain is like a computer, and by the public in general—is that it's all in there, tucked away someplace in the brain, and with luck or the right set of circumstances or cues, we can recall each and every event in our lives from earliest childhood to the present. But this popular notion is wrong.

In 1932, Frederic C. Bartlett published an extraordinarily original and important book on human memory called *Remembering* (1995) in which he penetrated to the heart of human memory in a way that behaviorists, Freudians, engram seekers, and AI researchers consistently ignored. Bartlett's method of studying memory was to tell stories and then have them retold to see how well they were recalled. The stories he used were taken from Native American folklore and involved tales of witchcraft and conflict. The retellings sometimes included material not in the original version, and they were often modified in particular ways that were significant to the individual re-teller. Details were omitted, invented, rearranged, and otherwise modified, but the gist of the tale, if not the specifics, was usually fairly accurate. The subjects who remembered the stories they heard did not reproduce them, but recreated them—each in an individual way. Memory, Bartlett concluded, is not reproductive but constructive.

The term Bartlett adopted was *schema*, which he defined as "an active organisation of past reactions, or of past experiences, which must always be supposed to be operating in any well-adapted organic response" (201). The concept's relevance to memory is clear: "Remembering obviously involves determination by the past. The influence of 'schemata' is influence by the past" (202). And the notion of a "trace" of memory in the brain is not accurate: "though we may still talk of traces, there is no reason in the world for regarding these as made complete at one moment, stored up somewhere, and then re-excited at some much later moment. The traces that our evidence allows us to speak of are interest-determined, interest-carried traces" (211–12).

Bartlett argued that the personal schemata that make our unique (re)construction of individual memories possible are a function of our consciousness. He realized, however, that this position was radically at odds with the then-hegemonic approach to psychology that denied the very existence of consciousness:

> There is an active school [i.e., behaviorism] in current psychological controversy which would banish all reference to consciousness. It is common to try to refute this school by asserting vigorously that of course we know that we are conscious. But this is futile, for what they are really saying is that consciousness cannot effect anything that could not equally well be done without it. That is a position less easy to demolish. If I am right, however, they are wrong. (214)

He was right; they were wrong. Behaviorists sometimes criticized his work on memory, but much more often they dismissed and ignored it. It had little influence in psychology until a revival in recent years; it reads remarkably well today, and behaviorist approaches to memory do not. Modern cognitive psychology has repeatedly and elegantly validated Bartlett's basic conclusions. While the notion of *engram*, understood at this point in time as "enduring physical or chemical changes to populations of neurons that are triggered by new information and experiences" (University of Chicago Medical Center 2017), as well as the metaphor of "memory storage," are still referred to, the newest studies using optogenetic tools are pointing to how the brain's synaptic and intrinsic plasticity are working together in memory processes.

Searching for Memory (1996), by psychologist Daniel Schacter, is a remarkably comprehensive, inclusive, and persuasive trade book on human memory; it is highly readable to laypersons and indispensable to anyone interested in any aspect of human memory. Schacter acknowledges the importance and originality of Bartlett's work and alludes in the title to the memory work of Penfield and Lashley. Schacter not only reviews and evaluates past concepts of memory (classical, Freudian, behaviorist, pop psychological, and others), but also elucidates nearly all aspects of the most current understanding of the multiple processes we group under the general category of memory. Schacter investigates other important areas of memory's functions including amnesia, memory issues related to aging, the nature of autobiographical memory, the anatomy of memory, confabulation and narrative invention, the life-review, external memory aids (e.g., photographs), collective memory, imagination and visual imagery, and literary and artistic expressions of memory (including, of course, Proust). Schacter describes the different, and conceptually overlapping, ways that psychologists have described memory systems:

- *long-term* (memories that can be recalled even though much time has passed since the event) and *short-term*, or *working* (small amounts, usually about seven items we can remember for brief periods of time);
- *explicit* (specific information and experiences that we can recall) and *implicit* (past experiences not specifically recalled but that influence actions, thoughts, and perceptions);
- *episodic* or *autobiographical* (recollections of specific past events that are unique to each of us); and
- *semantic* (dealing with knowledge that is conceptual and factual), and *procedural* (having to do with the acquisition of skills and habits).

Different types of memory are handled differently in the mind-brain. The hippocampus is the single most important area of the brain involved in memory, particularly for long-term memories of various kinds (explicit and implicit; episodic, semantic, and procedural). Not surprisingly, the amygdala, located adjacent to the hippocampus, is involved in the emotional tone of memories. Memory, like cognition in general, functions on a continuum that extends between the anchor points of image and proposition (as in Paivio's dual-code theory, see *Mental Representations*, 1986). The former is more analog, sensory, affective, and often nonconscious, while the latter is more symbolic, linguistic, purposeful, and conscious, and there is strong evolutionary (as well as biological) support for multiple memory systems (Pillemer 1998, 104–8).

One thing is abundantly clear about the results of recent memory research: as Bartlett anticipated, no memory is really "stored" in the brain. Rather, each memory is constructed (or re-constructed) from the ground up each time. A wide variety of cues, including visual, aural, and olfactory, can instigate the reconstruction of a memory. As a result, all memory is subject to decay, modification, fictionalization, and manipulation. The remembered past is grounded in our remembering present. Not everything we remember is accurate as a photograph, videotape, or audio recording might be. Memories exist on a spectrum ranging from very accurate to totally fanciful.

Eyewitness testimony is notoriously unreliable, as we have learned from many flawed judicial processes. When compelling and dramatic testimony is made that contradicts other comparably compelling recollection, the truth may be impossible to find (Gazzaniga 2005). When we seek to remember specific actions and motives from our own past, we can be particularly creative. One famous case of recall that illustrates the nature of memory is John Dean's sensational testimony before the Senate Watergate Committee in 1974, when he described an important meeting with President Richard Nixon in which it was clear that the President knew about the Watergate burglary and was actively involved in the cover-up—one of the key pieces of evidence supporting the case for impeachment and leading to the President's resignation. When the audio tape of this meeting was later discovered and made public, it became clear that Dean, like many of the subjects in Bartlett's study, was absolutely wrong in many of the specifics of his detailed recollection of the scene and that his version of his own statements was self-serving (Neisser 1982b). But the gist, or the basic content, of the memory was clearly very accurate: Nixon clearly was aware of and actively involved in the cover-up of the illegal acts. We could say that Dean's memory was basically truthful, if factually flawed (as well as self-serving). Many of our memories can probably be described in this way.

Similarly, the subject of *flashbulb memories* has been studied with surprising results. A flashbulb memory is the recollection of how, when, and where we personally experienced a profoundly moving (usually public) event. Such memories are supposedly seared into our memory, as though they had been photographed and stored in an album. Standard examples of flashbulb memories are when people learned of the assassination of John F. Kennedy, the Challenger space shuttle explosion, or the horrific scenes of the destruction of the World Trade Center's twin towers on September 11, 2001. We are so moved by such an experience that it feels like we form an immediate, accurate, and permanent record of it, like taking a snapshot of the moment. We recite these memories as if they were indisputable. Brown and Kulik (1977), who coined the term flashbulb memories, considered them highly accurate when recalled after even a lengthy passage of time, but later researchers have found the opposite to be the case. An experiment conducted following the Challenger disaster indicated that the majority of people who recalled vividly how they had learned of the event just three years earlier held memories that were very substantially at odds with the version they wrote down the day after the event took place (see Neisser 1982a; Winograd and Neisser 1992; Luminet and Curci 2009; Hirst et al. 2009).

Many of our memories are of events we have not personally experienced. Children's earliest memories are often constructed in conversations with others, especially family members. Many memories throughout life are constructed in just the same way—in conversations and in social contexts (Engel 1999, 31). Information garnered from the imagination, our fantasies, books, films and other media, gossip, and other sources constantly supplement and shape our memories. John Kotre recounts a famous case of *cryptomnesia* ("you remember *what* someone told you but you forget *that* you were told"):

> Former President Ronald Reagan used to tell the story of the commander of a bomber who was posthumously awarded the Congressional Medal of Honor during World War II. His plane had been hit and a young gunner who couldn't get out was panic-stricken. "Never mind, son," said the commander, "we'll ride it down together." The commander stayed with the plane, accompanying the young man to his death. Reagan told the story during his campaigns of 1976 and 1980, and again in 1983. But then a journalist did some checking and found that no such medal of Honor had ever been issued. He also discovered a World War II movie called *A Wing and a Prayer* in which a pilot goes down with a wounded radioman, saying, "We'll take this ride together." A *Reader's Digest* story contained the same words: "We'll take this ride together." (1995, 36)

It is not surprising that a former actor might remember something from a movie as though it were an episode from life. (It is surprising, however, that he apparently repeated the story for years without anyone's wondering how it was possible to know the pilot's words since he was killed in the crash.) Similarly, memories implanted by others—false memories (also known as *paramnesias*)—either incidentally or purposefully, can seem as real to us as authentic recollections, although they never happened (Loftus and Ketcham 1996, 73–101; Kotre 1995, 49–53). What is clear to cognitive scientists, in any case, is that a memory is not simply deposited in a particular spot in the brain for storage until it is retrieved from that spot in the same form. Memories are enhanced or modified or combined with other similar memories to form a composite. The nature of memory as described by cognitive researchers is something about which humanists should keep apprised: memory studies are now a significant part of literary studies and cultural studies due to its relevance to, e.g., Holocaust literature and other trauma narratives. A basic understanding of how memory functions in the brain is an essential underpinning to such studies.

Forgetting is an important component of memory, and a common theme in literature. A perfect memory of all we have seen, felt, done, read, and heard would be absolutely debilitating, as is beautifully illustrated by the short story "Funes the Memorious" by Jorge Luis Borges.[3] Forgetting is adaptive (Schacter 1996, 81) and, for the most part, we remember what we must as we carry on our lives. Often we retain partial memories, fragments (usually those most important to us), or traces of memories. Memory fades in certain ways as we age, particularly certain details of "what happened, when it happened, and who said what" (289). Implicit and semantic memory hold up better than specifics of recent experiences; we are more inclined to review and relive—and retell—memories from the past. This is also why the progression of forgetting in Alzheimer's disease tends to be from most recent to most remote.

We construct many of our earliest and most fundamental memories (especially self-memories) in a social collaborative context. Parents and other family and friends participate intimately in children's construction of stories about their past by engaging them in expansive talk about past experiences, helping evoke and refine personal memories. Robin Fivush (1991, 1994), grounding her work in the psychological theories of Lev Vygotsky, has studied the role of parents who are more elaborative (rather than merely repetitive) in the way they engage children in memory construction to show the importance of such adult guidance. Her research also reveals that, at least in the United States, boys and girls are often dealt with differently in this regard, with girls receiving more interactive and affective attention, while boys are treated more as individuals and with facts more than feelings (see also Pillemer 1998, 177–212).

This factor is undoubtedly important in research that finds women talking in a "different voice" (Gilligan 1993) and in "women's ways of knowing" (Belenky et al. 1986). These approaches do not maintain an essential difference between men and women, either as a biological or socially constructed determinant. Rather, they illustrate some of the basic tenets of a realistic theory of contextualized cognition: the early experiences of the maturing mind-brain as a part of a complex, dynamic, autopoietic system can lead to radically different outcomes by using the same basic biological architecture and uniquely individual contextual experiences. Both Gilligan and Belenky and colleagues have been criticized as endorsing a retrograde essentialist approach to gender differences (e.g., by Pinker 2002, 342), but this is a misreading of their work, as none of the writers involved suggest that women are determined either by biology or irresistible social factors to be nurturing, affective, collaborative, etc. In a passage often overlooked (or conveniently ignored), Gilligan clarifies her work:

> the different voice I describe is characterized not by gender but theme. Its association with women is an empirical observation and it is primarily through women's voices that I trace its development. But this association is not absolute, and the contrasts between male and female voices are presented here to highlight a distinction between two modes of thought and to focus a problem of interpretation rather than to represent a generalization about either sex. (1993, 2)

Human memory is not static, but dynamic, responding constantly to new experiences and new contexts. Roger Schank, an AI researcher who has turned from computers to human beings, discusses the dynamic nature of human memory: "A dynamic memory is one that can change its own organization when new experiences demand it. A dynamic memory is by nature a learning system" (1999, 2). Schank stresses the importance—and complexity—of reminding for all learning and remembering (21). We need (embodied) experiences over time to remind us of past failures and successes. The stories others tell, and that we tell in return, often enable the reminding that makes dynamic memory possible (89–106).

To no small extent, our sense of self is dependent on memory; we define ourselves largely by means of our autobiographical memories: "we are our memories, and without them, we are nothing" (LeDoux 2002, 97). Each of us has a unique set of autobiographical memories. Life consists of acquiring and maintaining this memory set, and the verifiable accuracy of these is less important than the fact that they are ours. Our own personal memories largely make us who we are (Schacter 1996, 15). As a character in Ernesto Sábato's 1948 novel *El túnel* [*The Tunnel*] notes: "It is curious, but living is a process of constructing future memories" (1988, 54).

Certain events in our lives have particular importance in the creation and maintenance of our sense of self. In *The Remembered Self* (1993), Jefferson A. Singer and Peter Salovey explore the concept of "self-defining" memories, which "give shape to and are shaped by our lives, memories of our proudest successes and humiliating defeats, memories of loves won and lost—memories that repetitively influence our manner of intimacy or our pursuit of power—the memories that answer the question of who we are" (10). We reconstruct the past "through the filters of the present … In other words, our current theories about ourselves and the world can structure and recast the content of autobiographical memories" (7). Singer and Salovey's central thesis is: "*What we continue to feel about events long after they have occurred may be a function of how relevant these remembered events are to the attainment of long-term goals*" (51). Our self-defining memories are woven together in our self-narratives: "the self engaged in remembering says to itself and the world around: I was this, I am this, and soon, perhaps, I will be this" (218).

The final words of Schacter's book provide good closure for our brief consideration of human memory:

> Memory is a central part of the brain's attempt to make sense of experience and to tell coherent stories about it. These tales are all we have of our pasts, and so they are potent determinants of how we view ourselves and what we do. Yet our stories are built from many different ingredients: snippets of what actually happened, thoughts about what might have happened, and beliefs that guide us as we attempt to remember. Our memories are the fragile but powerful products of what we recall from the past, believe about the present, and imagine about the future. (1996, 308)

Memory, in all its forms and manifestations, is one of the most complicated subjects in cognitive science. There are fascinating case studies of persons who cannot remember, especially the classic case of HM, who after brain surgery to cure epilepsy, became unable to form any permanent memory; see Hilts (1995). On the other hand are cases of extraordinary abilities to remember almost everything; see Price (2008).

Memory is central to the stories we tell, and the telling of stories is a defining characteristic of the human animal.

5 Schema Theory and Categorization

The types of memory described above are among the major ways we sort and classify the elements of our cognitive life, but this section deals briefly with two others: schema theory and categorization. Schema theory precedes cognitive psychology, but it has been revitalized in the contemporary cognitive paradigm. Categorization is a concept that goes as far back as the ancient Greeks, but it has undergone a revolutionary reconceptualization in modern cognitive science.

As we have seen, F. C. Bartlett introduced the notion of the schema as a cognitive organizing structure into modern psychology in his 1932 book on memory (1995). The concept was ignored for years during the era of behaviorism, but psychologist Ulric Neisser brought it to attention again in the 1970s. Neisser's *Cognitive Psychology* (1967), one of the first books on the subject, was largely influenced by the early COGNITIVIST orientation. This book was widely used as a text and had a great deal of influence. But then Neisser became increasingly influenced by James J. Gibson's ecological approach to perception and substantially re-oriented his approach to cognition. This new Gibsonian orientation is clear in his *Cognition and Reality* (1976), as Neisser explains in his preface and introduction (xi–xiii).

Neisser was convinced that the laboratory and computation orientations of psychology were sterile efforts that would inevitably turn the discipline into "a narrow and uninteresting specialized field" (7). In place of this, he proposed what he called a "realistic turn" for psychology, dealing with actual human beings in real social contexts. Neisser thus turned to Bartlett's idea of schema, which, as he saw it, "accepts information as it becomes available at sensory surfaces and is changed by that information; it directs movements and exploratory activities that make more information available, by which it is further modified" (54).

Utilizing a Gibsonian view of perception, Neisser describes perception as a constructive process in which perceivers anticipate what to do based on what is seen in the available optic array. As new information becomes available, the schema is revised and the process is repeated (20–21). Schemata thus help us organize and understand things and events in the world. They are constantly in a state of flux, potentially changed by any new experience, always sensitive to context. Neisser (20) shows a figure illustrating the perpetual cycle of schemata: a schema directs exploration, which samples an object (available information), which modifies the schema, which directs exploration … and so forth, in a never-ending circle. Later (112), he embeds this figure into another, more complete, accurate, and contextualized one. In this new figure, a schema of present environment (which is part of an individual's cognitive map of the world and its possibilities) directs perceptual exploration, consisting of locomotion and action, which in turn samples the actual present environment (available information), which itself is part of the actual world (potentially available information), which modifies the schema, which directs … and so forth. The cycle is flexible, continuous, and self-modifying. It is, in effect, an autopoietic or dynamical system (see Sect. 2 in Chapter 9). It is simultaneously fully a biological mind-brain activity and fully a culturally embedded activity. Schemata are among the foundational concepts in cognitive science and include, importantly, our self-schema and our body-schema; see Gallagher (2005) and Noë (2004, 2009).

A second major contribution to our understanding of schemata is *The Construction of Reality* (1986) by Michael Arbib and Mary Hesse. For these authors, a schema "is a 'unit or representation' of a person's mind" (13). They concur with Neisser that schema theory is necessarily an aspect of embodied cognition; a human being is an "essentially embodied subject" (38). Interestingly, they find basic similarities between the way schemata function and traditional hermeneutics, and they relate schema theory to Gadamer's hermeneutic circle (181). Seen in these terms, schema theory becomes an essential tool for literary scholars; see also Armstrong (2013, 54–90).

Schemata are rather like memory itself, never present in a specific location in the brain, not immutable, and in need of reconstruction at a moment's notice. For example, we all have a classroom schema, but we evoke a different version of it when we are in a small graduate seminar in literature than when we are in a large undergraduate lecture in sociology. Certain aspects of it—teacher, learning, attendance, note-taking, grades—may be the same, but others—interpersonal relationships, degree of formality, seating arrangements—will be quite different. As we age, we often tend to lock in certain schemata (Mandler 1997, 92), even though the human brain is capable of generating new neurons, synapses, and schemata at any age (see Chapter 7).

One of the most important of all cognitive activities is categorization: separating one thing from another, associating one thing with another, seeing similarities or differences between two things, drawing lines of division and circles of inclusion. Often we recognize the relationships among things because of the context in which they occur, and sometimes we fail to see certain similarities because of a difference in context. Cognitive science has been the locus of a radical reevaluation of the concept of categorization in a way that has had a major impact on linguistics, philosophy, and psychology.

The classical—Aristotelian—concept of a category is of something that objectively exists in the world. If a particular item meets certain necessary and sufficient conditions, it is considered a member of that category; if not, it is excluded. A thing cannot be A and not-A at the same time. It is an all-or-nothing proposition, one that illustrates the dualism many people have perceived throughout time. In the 1970s, cognitive psychologist Eleanor Rosch challenged this essentialist tradition and, in a series of elegant experiments (see, e.g., 1973, 1977, 1978, 1981; Rosch and Mervis 1975), worked out an alternative, much more powerful, model. Her influential work has been described as a paradigmatic example of cognitive science research (Gillespie 1992, 165–66). Categories, Rosch showed, are much more like Wittgenstein's *family resemblances* than the result of clearly drawn lines of division and like what mathematicians call *fuzzy sets* (Zadeh 1965; Kosko 1993). That is to say, we can look at a group of persons or things and perceive a general resemblance or relationship among them, without being able to specify any set of necessary and sufficient conditions they all meet. In a family, a group of cousins may all be clearly related, but their physical characteristics differ considerably within the group and often share aspects of persons not related. Yet, a loose set of traits tends to be more or less characteristic of the family. It is as though we had an idea of what a prototypical member of the family might look like and found that all of the cousins approximated that model and that most others from outside the family did not. The concept of the *prototype* is at the heart of Rosch's approach to categories.

An illustration of how Rosch approached the problem is: a number of individuals are asked to close their eyes and think of a bird. The mental image most people conjure up is of a robin, or perhaps a sparrow, a bluebird, or a wren. But it is less likely that they picture a turkey, a pelican, or an owl, and it is almost certain they do not conjure up a mental image of a penguin or an ostrich. Certain birds capture better than others the essence of what being a bird is all about. The small, flying, chirping, familiar birds of our backyards are closer to the prototype. These birds possess more "birdiness" than other kinds of birds. Most human categories are creations of our practical thinking and experience.

Our categories do not have clearly defined boundaries, but often have what Lakoff (1987, 91–114) calls a *radial structure*. The category of birds, for example, can be conceptualized as consisting of a series of concentric circles. Very near the center, close to the prototype of the category, is our familiar robin. Nearby are other similar birds: sparrows, cardinals, orioles, finches, and canaries. Farther out are turkeys and hummingbirds, chickens, toucans, and eagles. And still farther away are ostriches and penguins. Perhaps nearby, but off the map, are bats. Placement into this series of concentric circles varies by individual, circumstance, culture, and much more. Categories are not stable, but relative, contextual concepts. For someone from Antarctica, a penguin might be closer to the bird prototype than a robin. Categories per se do not exist in nature, but are perceived by individuals situated in context. Recalling Chomsky's E-language/I-language distinction, which differentiates between a concept of language as something that exists in society, an aspect of culture, objective and external to any individual on the one hand (E-language) and an internalist, intentional, and individual type of language on the other (I-language), we could think of prototypes as involving "I-categories."

Lakoff has developed Rosch's model in most detail and with most originality. His book *Women, Fire, and Dangerous Things* (1987) is still considered the foundational text for the modern study of categorization, both conceptual and linguistic. The extravagant title of Lakoff's book comes from the Australian aboriginal language of Dyirbal, in which there is a category of things called *balan*, which includes women, fire, and things that are dangerous, along with water and fighting. This example, studied meticulously by Lakoff (92–104), illustrates the importance of categorization in the study of both cognition and language: "there is nothing more basic than categorization to our thought, perception, action, and speech" (5). Lakoff stresses the following concepts as themes under a "cognitive model":

- *family resemblances*
- *centrality*
- *polysemy (the quality of having many meanings) as categorization*
- *generativity as a prototype phenomenon*
- *membership gradience*
- *centrality gradience*
- *conceptual embodiment*
- *functional embodiment*
- *basic-level categorization*
- *basic-level primacy*
- *reference-point or "metonymic" reasoning*. (12–13)

Central to his argument is the idea "that we organize our knowledge by means of structures called *idealized cognitive models*, or ICMs, and that category structures and prototype effect are by-products of that organization" (68).

Categories are even more complex than this though: they can also be ad hoc or contingent. We can easily make up categories when needed, and we often do in organizing our everyday lives; see Barsalou (1983). Anthropology and developmental psychology provide strong evidence that we tend toward essentializing from earliest childhood. Paul Bloom has defined essentialism as "the notion that things have an underlying reality or true nature that one cannot observe directly and it is this hidden nature that really matters" (2010, 9). We are, Bloom claims, "natural-born essentialists" (xii). Whether this is a matter of folk psychology (Atran 1990) or inherent mind-brain tendencies (Gelman 2003) is open to debate. But surely Susan A. Gelman is correct in making the argument that "the prototype is not the full story" (110). That many people maintain such essentialist beliefs throughout their lives is one reason that stereotypes of race, gender, and class are so common: "essentialism is a myth invented by the powerful to convince people that these social categories are natural and immutable" (Bloom 2010, 15). Humanists tackling the evils of prejudice and racism in their work need to understand the psychological basis for essentialism, obviously not to defend it, but to understand how it can best be combatted.

6 Narrative and the Literary Mind

Jerome Bruner (1986, 11–44) has proposed two ways of knowing—a good story and a good argument; in other words, a narrative mode and an analytic or theoretical (paradigmatic, logico-scientific) mode. Traditionally we have privileged the latter over the former, seeing logical analysis as the quintessential means, if not the very definition, of cognition. But, suggests Bruner, to take such a stance is to undervalue narrative, for much of what we know of the world comes to us, or is conceived by us, in narrative form: "we organize our experience and our memory of human happenings mainly in the form of narrative" (1991, 4; see also Frye 1986).

Recall Merlin Donald's tracing of the phases of human culture from the episodic to the mimetic to the mythic to the theoretic in Sect. 5 of Chapter 5. Phase three, the mythic, is that period of human evolution when language, and with it narrative, became the predominant means of knowing the world. Moving beyond protolanguage and incorporating (and building upon) mimesis, the human mind capable of true symbolic thought becomes categorically and radically different from (not just an advanced version of) the episodic mind. Donald goes so far as to propose that "[t]he narrative mode is basic, perhaps *the* basic product of language" (1991, 257). The ability to communicate linguistically and to tell stories (to invent myths, for example) radically reshapes the mind as it makes possible "a *wholly new system for representing reality*" (259). Donald posits a "mimetic controller" (189–96) as the central function of the mimetic mind, and then proposes that this is subsumed within the "linguistic controller," which he describes as "a representational process whose products are narrative models" (259).

In an extraordinarily important series of essays, evolutionary psychologist Michelle Scalise Sugiyama (1996a, b, 2001a, b, c, 2003, 2004, 2005, 2008, 2017) has proposed that, uniquely among the arts, narrative probably qualifies as a human evolutionary adaptation. Part of her argument is that narrative appears in all cultures and that it draws upon a complex of cognitive skills, such as language itself, reasoning abilities, and Theory of Mind (2001a, 222). More importantly, she finds that narrative allows individuals to "take advantage of others' experience and acquire information at second hand" (226). Thus, she contends, it is "plausible that storytelling was selected as a means of disseminating information—with the opportunities for education, deception, and manipulation that such simulations afford—after which storytelling skill became a reliable cue of verbal and protean intelligence" (235). In sum, Scalise Sugiyama concludes that it is probable that "storytelling is a sufficiently ancient phenomenon to have evolved through the process of natural selection and that storytelling might serve an adaptive function" (2005, 177). The reason for this, she suggests, is that, unlike more nonverbal media (such as painting, dance, or music), "narrative is uniquely well-suited to the task of simulating human experience" (191). Recall Robin Dunbar's (1996) thesis that the first impetus for language was grooming, the need for social cohesion, and the pragmatic use of language for gossip—the telling of stories about other people. The mind is inherently narrative as an aspect of its evolutionary heritage. Although there have been attempts to show the adaptive value of other arts (see Dissanayake 1992, 2000; Dutton 2009), none is as convincing as the case for narrative.

But narration is not just a matter of our evolutionary history; it is also a matter of some of the most fundamental functions of our mind-brains. William Calvin discusses a function that he calls the brain's "narrator." He personifies and describes this defining feature of human consciousness as "the conductor of our cerebral symphony, who contemplates the past and forecasts the future, makes decisions about relative worth, plans what to do tomorrow, feels dismay when seeing a tragedy unfold, and narrates our life story" (1990, 3). Further, Calvin has proposed that, above all, the human brain is a "Darwin Machine," a device for spinning speculative scenarios about the future consequences of our actions—of telling ourselves stories about what might happen if we take a certain course of action (261–63).

One proposed reason for why we have brains is to decide what to do next, and, for Calvin, it is by means of narratives that we do this. Recall also that Michael Gazzaniga (1985) has come to conclude, as a result of his studies with split-brain patients, that the left hemisphere is the source for an "interpreter," a constant function of the brain in explaining (even if often rationalizing and confabulating) what we do and what happens around us. Louis Cozolino further suggests that narrative has an important integrative function for both emotions and neural networks (2006, 304). By all reasonable evolutionary and biological accounts, narrative is the primary mode of symbolic cognition. The inability of artificial intelligence to explain symbolic cognition further confirms this; see Chapter 11.

The view of consciousness sketched out in the last chapter features narrative as central to human conscious and nonconscious cognitive processes. Dennett's (1991) multiple-drafts model uses the metaphor of revising a narrative text as we go along, positing the idea of the self as the "Center of Narrative Gravity," the theme of the webs of narrative that we constantly spin: "the story we tell others—and ourselves—about who we are" (418). Telling stories about and to ourselves is a necessary function of being conscious (in the sense of higher consciousness) animals. Flanagan draws on both Calvin's Darwin Machine and Dennett's multiple drafts to characterize the conscious mind-brain as "the most powerful anticipation machine ever built" (1992, 42). We are, for Flanagan, "inveterate storytellers" (198).

In *The Literary Mind* (1996), Mark Turner's central idea is that the normal way in which the mind works is by *parable*. That is, we tell *stories* about certain people and events to make a point about other people and situations, and then we *project* the actions of the input story onto a target space. This, he insists, is the primary cognitive mode of human beings:

> *Story* is a basic principle of mind. Most of our experience, our knowledge, and our thinking is organized as stories. The mental scope of story is magnified by *projection*—one story helps us make sense of another. The projection of one story onto another is *parable*, a basic cognitive principle that shows everywhere, from simple actions like telling time to complex literary creations like Proust's *À la recherche du temps perdu*. (v)

Narrative is inherent to the biology of human beings, as science continues to confirm (making cross-disciplinary research involving literature scholars and scientists a "no-brainer"!). Members of the human species can't be understood without consideration of narrative cognitive processes and narrative epistemology. The final quarter of the twentieth century witnessed an almost universal recognition of narrative's role in our lives. Jonathan Gottschall has argued in the meantime for the importance of narrative in evolution, the universality of narrative thinking, and the centrality of stories in all aspects of our lives in his book *The Storytelling Animal* (2012): "Story—sacred and profane—is perhaps *the* main cohering force in human life" (138).

Theodore S. Sarbin (1986c) was a pioneer in making narrative the governing epistemology for psychology. One of the earliest and strongest supporters of a contextualist approach to psychology, Sarbin (1986a, ix;

1986b) flirts first with the metaphor of the theater for human psychology (the world is a stage and we are all actors; we assume various roles at different times and in different contexts), but, much influenced by Hayden White's concept of *emplotment* in history, eventually opts for life as a story we tell, and proposes what he calls "the narratory principle: that human beings think, perceive, imagine and make moral choices according to narrative structures" (1986c, 8). Dan McAdams (1988, 2006) has also been a leading figure in narrative psychology, and more recently, Julia Vassielieva (2016) has furthered the field with emphasis on its ethical aspects.

Today the novel continues its reign as the most popular literary genre. In the early decades of the twenty-first century, the term "literature" is usually shorthand for narrative fiction, as even theater and poetry are often understood as variants of (prose) narrative. The major mixed media of our time—film, TV, video games, virtual reality—depend largely on narratives, though naturally they are also more than that. In addition, we now have "tweets" and other social media narratives that "go viral"—they are truncated in form and probably not to be considered great literature in most cases, but they are narrative in nature, nonetheless.

Turner (1996) has made the case for the role of narrative (parable) in everyday thinking. In other words, the human mind is a literary mind, and the telling of stories and projecting relevant meaning from these stories is the most fundamental feature of human cognition. Turner also drew on both the cognitive theory of metaphor and Gilles Fauconnier's earlier work on mental spaces (1994) to develop the idea of conceptual integration, or blending. What happens in *conceptual blending* is that we take elements from one input space—say, human beings facing a dilemma—and project them onto another space—say, barnyard animals—blending the ability to think and talk with domesticated creatures that have four legs. The result: talking animals (whose actions we then project back onto a relevant human situation). The story takes place not in the mental space of speaking and dilemma-solving human beings, nor in the mental space of non-linguistic donkeys and pigs, but in the blended space in which animals talk. This process is easy, natural, and parallel to what we do in using metaphors in our everyday thinking. And multiple mental spaces can be combined into complex blended spaces (e.g., talking animals on a spaceship headed for Mars and the dilemmas they need to solve on the way).

Turner then teamed up with Gilles Fauconnier to write *The Way We Think* (2002), where the authors develop the idea of conceptual integration much further (and Turner himself went on to write a more accessible introduction to blending, *The Origin of Ideas*, 2014). The key to our cognitive ability to blend mental spaces, they contend, is ***imagination***: "The products of conceptual blending are always imaginative and creative" (6). Fauconnier and Turner relate the imaginative operations of conceptual integration to our species' evolved cognitive fluidity, a term coined by Steven Mithen (1996); see Sect. 5 in Chapter 5. Conceptual blending is a process that had not previously been addressed in cognitive psychology and one that holds enormous possibilities for the reading, understanding, teaching, discussing, and writing about literature. Literature may be the primary laboratory and showcase for conceptual blending. Fauconnier and Turner's extended presentation of what conceptual blending is and how it works makes their book an important contribution to the study of the mind and of literature.

Turner has extended the claims for blending, particularly ***double-scope stories***, by proposing that our species' ability "to activate two conflicting mental structures ... and to blend them creatively into a new mental structure" (2003, 117) is distinctively human. This ability, that no other animal has, is "the defining mental ability of cognitively modern human beings and the source of our creativity and knowledge. These mental operations ... are instances of the basic mental operations that make us cognitively modern" (132–33). While clearly remarkable, conceptual blending would, as Fauconnier and Turner originally suggested, most logically seem to be a skill that is subordinate to the more general capacity of imagination (though Turner may prefer to state it as blending makes imagination possible in the first place, i.e. it is the process underlying imagination; see his outline of the concept in *The Origin of Ideas*, 2014). Rather than *the* thing that defines the species though, we would place conceptual blending together with language, Theory of Mind, and similar capacities, as part of the total package that makes humans unique. In addition, a focus on conceptual blending theory exposes what cognitive theater scholar Amy Cook (see Chapter 15) has called the "dead end of poststructuralist conceptions of language" which maintains "problematic narratives and paradigms" (2018a, 25; see also Shaughnessy 2013, 88).

Mental imagery plays virtually no role in behaviorism, for the processes that take place in the black box of the mind are neither examinable, explainable, nor of interest. Neither does imagery play any role in two direct intellectual descendants of behaviorism: COGNITIVISM and THEORY. In COGNITIVISM, thought is information processing and the manipulation of symbols, a kind of abstract language of thought; there are no images anywhere in the process. In THEORY, all cognition is reduced to the encoding and decoding of signs, and to self-referential sign systems, of which language is the prototype; cognition is irreducibly logocentric and the world is textualized; there is no room for mental images. Both COGNITIVISM and THEORY take a reductionist approach to cognition and both privilege abstraction and language over contextualized reality and complexity. Both disregard the reality of perception as described by Gibson, Paivio, and others.

Ellen Esrock has studied the history of the role of mental imagery in literary theory in her book *The Reader's Eye* (1994, 1–78). Even when the author is declared dead and the reader supposedly reigns supreme, Esrock accurately reports a curious lack of recognition of mental imaging's role: "Whether readers are regarded psychoanalytically, as fragmented subjects, or politically, as members of competing interpretive communities, whether the actions of readers are better charted on a hermeneutic circle or on a semiotic grid, reader-oriented scholars still discuss only sound and meanings" (2). In Bartlett's 1932 study of memory, imagery plays a prominent role in human cognition, but behaviorism had no place at all for mental imaging—or any other kind of thought, for that matter. In cognitive science of recent decades, the "imagery debate" has been between the pictorialists or imagists, who argue that images involve depictive representations, and the descriptionalists or propositionalists, who argue that imagery is really propositional representation. After years of debate and experimentation, the overwhelming evidence came out in favor of the pictorial position (Tye 1991; Kosslyn 1994; Opdahl 2002, 33–56). Images do not depend on words, codes, signs, or symbols; they are similar to the images of perceptive systems, and they involve many of the same parts of the brain as actual perception.

The leader of the victorious pictorialist position has been Stephen Kosslyn, who defines a mental image as "a representation in the mind that gives rise to the experience of 'seeing' in the absence of the appropriate visual stimulation from the eyes" (1983, 29). His book *Image and Brain* (1994) is now the indispensable reference on the subject. Kosslyn meticulously and laboriously reviews the issues at hand, the history of research in the field, and his own painstaking studies on many different aspects of mental imaging over a period of decades. Kosslyn's conclusion is that mental imagery is "a basic form of cognition" which "plays a central role in many human activities—ranging from navigation to memory to creative problem solving" (1).

How do images function in the reading and understanding of literature? Readers regularly generate mental images based on what the text affords and as both enabled and constrained by their individual birth order, gender, memories, experiences, feelings and emotions, Theory of Mind, and schemata. Esrock (1994, 179–87) identifies three factors that influence the reader's creation of images: context, text, and reader. Mental imaging affects memory, clarifies spatial descriptions, and helps make a fictional world concrete. It also evokes the psychodynamics of vision and functions to position the reader within the text, and it serves as a means of establishing a formal contrast and producing semantic values (188–202). Esrock explicitly contrasts the assumptions of THEORY and the realities of embodied human cognition, arguing that

> we must discuss the imagist embodiment of meaning, where it occurs, and not simply the meaning or intentional description of the image. Imaging, as a conscious activity, can have functional consequences that are cognitive and affective and that can benefit the reading experience and the study of literature. Under this description, imagery is endowed with more powers than it possesses in most contemporary models, even if considerably less than in traditional ones. (203)

It is false to the reality of the human animal to ignore, let alone deny, the role of mental imagery in cognition, and, specifically in the case that interests us, in the reading and understanding of written texts. For an exhaustive review of "antiocularism" in THEORY, see Martin Jay's (1993) study of what he calls, in his subtitle, "the denigration of vision in twentieth-century French thought." See also Stafford's (1996) extended critique of what she calls the "entrenched antivisualism" of much contemporary theory, together with her defense of the "virtue" of images.

The most important study to date of mental imagery in the reading process is Elaine Scarry's *Dreaming by the Book* (1999). From the start, Scarry acknowledges that it is the nature of the mental image to be less vivid than the perceptual image, but that in no way reduces their importance. As Scarry notes, in no area is mental imaging more crucial than literature, "which is almost wholly devoid of *actual* sensory content" (5).

Grounding her approach to mental images largely in Gibson's theory of perception, Scarry studies the vivacity, solidity, and persistence of mental images. She insightfully investigates the ways we move and modify our images, taking advantage of the text's affordances to do so.

Mental images are also one of the elements at the heart of Keith Opdahl's book on the role of emotions in our imaginative activities of reading: *Emotion as Meaning* (2002). Specifically rejecting the mechanism and dehumanization of THEORY, Opdahl also turns to cognitive science in supporting his conviction that emotion is crucial to the act of reading. He discusses the two "codes" of understanding already recognized in cognitive science, the symbolic (language, discourse, propositions) and the imaginal (images), and then adds a third—the "affective code," a kind of emotional understanding. Like Scarry, Opdahl identifies imagination as the cognitive engine that drives the creation of mental images and what makes it possible for us to experience literature.

Things imagined need not be (and, indeed, rarely are) entirely unthought-of before. Rather, they are, as the earlier discussion of the work of Turner and Fauconnier suggests, most often blends, combinations, variations, or modifications of previous percepts and thoughts. Imagination can be creative, but it can also be re-creative (Currie and Ravenscroft 2002). Imagination can be rational (Byrne 2005) as well as fanciful as we wonder or speculate about "what if," "if only," and other kinds of counterfactual conditions. Imagination can also consist of role-playing and rehearsal, and it is essential to our frequent imaginary dialogues with ourselves and others (Watkins 1986). And the imagination includes the linguistic or symbolic as well as the visual or perceptual: we can imagine new narrative scenarios as well as we can picture little green men from Mars.

Imagination is not something extraordinary that some people do occasionally, but a constant, essential activity of the mind-brain. According to Mark Johnson, "Without imagination, nothing in the world could be meaningful" (1987, ix; see also 151). Frank Smith argues convincingly that "[i]magination is the dynamo of the brain, the source of all our intellectual energy and creativeness" (1990, 54; see 45–54). Ethel Person calls imagination "humankind's major adaptive tool," without which our consciousness "would be locked into the sensory present or the remembered past, denied the contemplation of infinite alternatives and possibilities, relegated forever to an animal-like existence" (1995, 32), a stance that recalls Edelman's remembered present and Donald's episodic culture.

In an important collaborative study between a neuroscientist and a cognitive linguist, Vittorio Gallese and George Lakoff (2005) discuss the implications of research that shows that the same areas of the brain involved in visual perception are also activated when we have visual mental images, and, further, the same is true for movements: "When one imagines seeing something, some of the same part of the brain is used as when one actually sees. When we imagine moving, some of the same part of the brain is used as when we actually move" (456). Therefore, Gallese and Lakoff propose the following hypothesis:

> The same neural substrate used in imagining is used in understanding.
> Consider a simple sentence, like "Harry picked up the glass." If you can't imagine picking up a glass or seeing someone picking up a glass, then you can't understand that sentence. Our hypothesis develops this fact one step further. It says that *understanding is imagination*, and that *what you understand of a sentence in a context is the meaning of that sentence in that context.*
> Our proposal is not an internalist theory of meaning. The reason is that imagination, like perceiving and doing, is *embodied*, that is, structured by our constant encounter and interaction with the world via our bodies and brains. The result is an *interactionist* theory of meaning. (456)

If, as Gallese and Lakoff suggest, "[*i*]*magination is mental simulation*" and it is "carried out by the same functional clusters used in acting and perceiving" (458), then imagination is as much a function of biology as seeing, moving about, or eating. All cognitive acts are carried out by our imaginative, creative, narrative mind-brains, which are fully integrated into our bodies, which in turn interact with the environment. Our biology, as much as our culture, makes us creatures of the imagination: *homo imaginans*.

Martha Nussbaum takes this point of view further as she makes the case for what she calls the "narrative imagination": "the ability to think what it might be like to be in the shoes of a person different from oneself, to be an intelligent reader of that person's story, and to understand the emotions and wishes and desires that someone so placed might have" (1997, 10–11; see also Mancing 2017 for a literary application of the concept). Few things increase imaginative powers as much as learning to read: literacy "shrank the perceived world in order to expand the imagined world" (Collins 1991, 14). A coevolutionary relationship is at work (Carroll 2004): the ability to understand and enjoy fiction depends on the imagination, while our imaginative powers are at the same time increased through the practice of fiction.

Uncoincidentally, Nussbaum uses the term so crucial in Bakhtin's moral philosophy: "sympathetic understanding." Taking her reasoning to a logical conclusion, Nussbaum argues that

> [i]f the literary imagination develops compassion, and if compassion is essential for civic responsibility, then we have good reason to teach works that promote the types of compassionate understanding we want and need. This means including works that give voice to the experiences of groups in our society that we urgently need to understand, such as members of other cultures, ethnic and racial minorities, women, and lesbians and gay men. (99–100)

True multicultural understanding comes from imaginatively interacting with those very different from yourself as well as with those who are more like you. Much research has been done on the notion of perspective-taking and its crucial role in facilitating empathy, particular with those who aren't in our "in-group" (see Galinsky and Moskowitz 2000; Husnu and Crisp 2015). The imaginative activity of reading literature promotes our understanding of and compassion with others as embodied, thinking, feeling individuals like ourselves, as it helps us understand that we, in different contexts, would be very different from what we are. To understand literary interpretation and textual meaning construction, we must consider imagination and conceptual blending as integral to those processes (as discussed in relation to Barbara Dancygier's pathbreaking work, e.g., in Chapter 14).

As this consideration of schema theory, categorization, blended spaces, mental imagery, and imagination makes clear: our normal, ongoing cognitive processes are embodied. Our cognitive processes are always grounded in bodily reality. Psychologist Raymond Gibbs explicitly stressed the mind-brain-body-environment, setting out what he calls the "embodiment premise":

> Cognition is what occurs when the body engages the physical, cultural world and must be studied in terms of the dynamical interactions between

> people and the environment. Human language and thought emerge from recurring patterns of embodied activity that constrain ongoing intelligent behavior. We must not assume cognition to be purely internal, symbolic, computational, and disembodied, but seek out the gross and detailed ways that language and thought are inextricably shaped by embodied action. (2006, 9)

The two sources for Gibbs's position are cognitive linguistics and dynamical systems theory. His arguments for an embodied cognition within the framework of dynamical systems theory reinforce the stance taken by prominent biologists, philosophers, linguists, and psychologists as discussed throughout this book.

The philosopher who has most consistently called attention to the role of the body in cognition is Mark Johnson, whose *Body in the Mind* (1987) is a landmark book in the field. In his later book, *The Meaning of the Body* (2007) he explores the full mind-body-environment continuum, this time specifically grounding his work in the autopoiesis of Maturana and Varela, the approach of the American pragmatists, especially James and Dewey, and the work of today's major neuroscientists like Damasio and LeDoux. Johnson states his position thus: "Change your brain, your body, or your environments in nontrivial ways, and you will change how you experience your world, what things are meaningful to you, and even who you are" (1–2). Johnson asserts, "Meaning is relational and instrumental" (268), and dependent on how we connect things to our bodily experience. Andy Clark's *Supersizing the Mind* (2008) also offers persuasive arguments for our embodied, autopoietic, extended, and active existence as it upholds that all human cognition is as much biological as it is social or cultural.

The embodied mind-brain is far more complex than what we have outlined briefly here. We have emphasized some aspects of this great cognitive system that seem most important to literary scholars, while leaving much without comment. In the next chapter, we consider how the mind-brain develops in individual humans. The way(s) in which infants and children come to know the world—how their minds begin to develop in the earliest years—is a field of knowledge that has been revolutionized in recent years, and that is our next chapter's starting point.

Notes

1. Even sociologist Erving Goffman, in his famous 1956 work *The Presentation of Self in Everyday Life*, is not arguing against the existence of a self, but rather asserts that the self engages in "impression management" during social interactions. The self for Goffman is a performer that "can be fully taken in by his own act; he can be sincerely convinced that the impression of reality which he stages is the real reality"—or, he can be cynical about it instead (10). Either way, situational context in self-presentation is important to Goffman, whose book is sometimes, but should not be, described as essentially denying the existence of self since all is only performance. There still needs to be a self that performs under this theory, and Goffman's language implies such.
2. Such studies continue to come out and with postmodernist undertones. Cocks (2016) for instance gives "a reading informed by Sigmund Freud and Judith Butler," analyzing "the processes of retrospection, reiteration, and repetition" within Lillian Hellman's play *The Children's Hour*, and asserts that "the handful of critical accounts thus far produced have underestimated the extent to which these structures disrupt appeals to a stable meaning" (363).
3. As bioengineer Rodrigo Quian Quiroga has stated: "In the story of Funes, Borges described very precisely the problems of distorted memory capacities well before neuroscience caught up. We now know that memory function is linked to a particular brain area, the hippocampus, which lies at the end of the neural pathway that processes sensory information" (Quiroga 2010).

References

Arbib, Michael A., and Mary B. Hesse. 1986. *The Construction of Reality*. Cambridge: Cambridge UP.

Armstrong, Paul B. 2013. *How Literature Plays with the Brain: The Neuroscience of Reading and Art*. Baltimore: Johns Hopkins UP.

Arnheim, Rudolf. 1969. *Visual Thinking*. Berkeley: U of California P.

Atran, Scott. 1990. *Cognitive Foundations of Natural History: Towards an Anthropology of Science*. Cambridge: Cambridge UP.

Barry, Ann Marie Seward. 1997. *Visual Intelligence: Perception, Image, and Manipulation in Visual Communication*. Albany: State U of New York P.

Barsalou, Lawrence W. 1983. "Ad-hoc Categories." *Memory and Cognition* 11: 211–27.

Bartlett, Frederic C. 1995 (1932). *Remembering: A Study in Experimental and Social Psychology*. Cambridge: Cambridge UP.

Belenky, Mary Field, et al. 1986. *Women's Ways of Knowing: The Development of Self, Voice and Mind*. New York: Basic Books.

Berthoz, Alain. 2000 (1997). *The Brain's Sense of Movement*. Trans. Giselle Weiss. Cambridge: Harvard UP.

Billig, Michael. 2006. "Lacan's Misuse of Psychology Evidence Rhetoric and the Mirror Stage." *Theory, Culture and Society* 23.4: 1–26.

Bloom, Paul. 2010. *How Pleasure Works: The New Science of Why We Like What We Like*. New York: W.W. Norton.

Bownds, M. Deric. 1999. *The Biology of Mind: Origins and Structures of Mind, Brain, and Consciousness*. Bethesda: Fitzgerald Science P.

Brown, Roger, and James Kulik. 1977. "Flashbulb Memories." *Cognition* 5.1: 73–99.

Bruner, Jerome. 1986. *Actual Minds, Possible Worlds*. Cambridge: Harvard UP.

———. 1991. "The Narrative Construction of Reality." *Critical Inquiry* 18: 1–21.

Byrne, Ruth M. J. 2005. *The Rational Imagination: How People Create Alternatives to Reality*. Cambridge: Bradford Book/MIT P.

Calvin, William H. 1990. *The Cerebral Symphony: Seashore Reflections on the Structure of Consciousness*. New York: Bantam.

Carroll, Noël. 2004. "Art and Human Nature." *Journal of Aesthetics and Art Criticism* 62.2: 95–107.

Carruthers, Mary. 1990. *The Book of Memory: A Study of Memory in Medieval Culture*. Cambridge: Cambridge UP.

Clark, Andy. 2008. *Supersizing the Mind: Embodiment, Action, and Cognitive Extension*. Oxford: Oxford UP.

Cocks, Neil. 2016. "Uncanny Repetitions in Lillian Hellman's *The Children's Hour*." *Modern Drama* 59.3: 363–79.

Cohen, Jonathan D., and Jonathan W. Schooler. 2014. *Scientific Approaches to Consciousness*. London: Psychology Press.

Cohut, Maria. 2019. "How the Brain Adapts to Hear Better After Vision Loss." *Medical News Today* April 25. Online.

Collins, Christopher. 1991. *A Poetics of the Mind's Eye: Literature and the Psychology of Imagination*. Philadelphia: U of Pennsylvania P.

Cook, Amy. 2018a. *Building Character: The Art and Science of Casting*. Ann Arbor: U of Michigan P.

Cozolino, Louis. 2006. *The Neuroscience of Human Relationships: Attachment and the Developing Social Brain*. New York: W. W. Norton.

Cunningham, Tomaz. 2017. "When Ourika Discovered She Wasn't White: A Lacanian Analysis of Romantic Ennui in Claire de Duras's *Ourika*." *The French Review: Journal of the American Association of Teachers of French* 90.4: 17–27.

Currie, Gregory, and Ian Ravenscroft. 2002. *Recreative Minds: Imagination in Philosophy and Psychology*. Oxford: Clarendon P.

Damasio, Antonio R. 1994. *Descartes' Error: Emotion, Reason, and the Human Brain*. New York: Grosset/Putnam.

———. 1999. *The Feeling of What Happens: Body and Emotion in the Making of Consciousness*. New York: Harcourt Brace.

de Bont, Leslie. 2017. "Portrait of the Female Character as a Psychoanalytical Case: The Ambiguous Influence of Freud on May Sinclair's Novels." *May Sinclair: Re-thinking Bodies and Minds*. Edinburgh: Edinburgh UP.

Dennett, Daniel C. 1991. *Consciousness Explained*. Boston: Little, Brown.

Dissanayake, Ellen. 1992. *Homo Aestheticus: Where Art Comes from and Why*. New York: Free P.

———. 2000. *Art and Intimacy: How the Arts Began*. Seattle: U of Washington P.

Domhoff, G. William. 2003. *The Scientific Study of Dreams: Neural Networks, Cognitive Development, and Content Analysis*. Washington, DC: American Psychological Association.

———. 2005. "Refocusing the Neurocognitive Approach to Dreams: A Critique of the Hobson Versus Solms Debate." *Dreaming* 15.1: 3–20.

Donald, Merlin. 1991. *The Origins of the Modern Mind: Three Stages in the Evolution of Culture and Cognition*. Cambridge: Harvard UP.

———. 2001. *A Mind so Rare: The Evolution of Human Consciousness*. New York: W. W. Norton.

Dufresne, Todd. 2003. *Killing Freud: Twentieth-Century Culture and the Death of Psychoanalysis*. London: Continuum.

Dunbar, Robin. 1996. *Grooming, Gossip, and the Evolution of Language*. Cambridge: Harvard UP.

Dutton, Denis. 2009. *The Art Instinct: Beauty, Pleasure, and Human Evolution*. New York: Bloomsbury.

Edelman, Gerald M. 1992. *Bright Air, Brilliant Fire: On the Matter of the Mind*. New York: Basic Books.

Engel, Susan. 1999. *Context Is Everything: The Nature of Memory*. New York: W. H. Freeman.

Esrock, Ellen J. 1994. *The Reader's Eye: Visual Imaging as Reader Response*. Baltimore: Johns Hopkins UP.

Evans, Dylan. 2005. "From Lacan to Darwin." In Gottschall and Wilson, 38–55.

Fauconnier, Gilles. 1994 (1985). *Mental Spaces*. New York: Cambridge UP.

———, and Mark Turner. 2002. *The Way We Think: Conceptual Blending and the Mind's Hidden Complexities*. New York: Basic Books.

Fivush, Robyn. 1991. "The Social Construction of Personal Narratives." *Merril-Palmer Quarterly* 37: 59–81.

———. 1994. "Constructing Narrative, Emotion, and Self in Parent-Child Conversations About the Past." In *The Remembering Self: Construction and Accuracy in the Self-Narrative*. Eds. Ulric Neisser and Robyn Fivush. Cambridge: Cambridge UP. 136–57.

Flanagan, Owen. 1992. *Consciousness Reconsidered*. Cambridge: Bradford Book/MIT P.

———. 1996. *Self Expressions. Mind, Morals, and the Meaning of Life*. New York: Oxford UP.

———. 2000. *Dreaming Souls: Sleep, Dreams, and the Evolution of the Conscious Mind*. Oxford: Oxford UP.

Fodor, Jerry A., and Z. W. Pylyshyn. 1981. "How Direct Is Visual Perception?: Some Reflections on Gibson's 'Ecological Approach.'" *Cognition* 9: 139–96.

Foulkes, David. 1999. *Children's Dreaming and the Development of Consciousness*. Cambridge: Harvard UP.

Frith, Chris. 2007. *Making Up the Mind: How the Brain Creates Our Mental World*. Oxford: Blackwell.

Frye, Joanne S. 1986. *Living Stories, Telling Lives: Women and the Novel in Contemporary Experience*. Ann Arbor: U of Michigan P.

Galinsky, Adam D., and Gordon B. Moskowitz. 2000. "Perspective-Taking: Decreasing Stereotype Expression, Stereotype Accessibility, and In-Group Favoritism." *Journal of Personality and Social Psychology* 78.4: 708–24.

Gallagher, Shaun. 2005. *How the Body Shapes the Mind*. Oxford: Clarendon P.

Gallese, Vittorio, and George Lakoff. 2005. "The Brain's Concepts: The Role of the Sensory-Motor System in Conceptual Knowledge." *Cognitive Neuropsychology* 23.3–4: 455–79.

Gazzaniga, Michael S. 1985. *The Social Brain: Discovering the Networks of the Mind*. New York: Basic Books.

———. 2005. *The Ethical Brain: The Science of Our Moral Dilemmas*. New York: Ecco.

Gelman, Susan A. 2003. *The Essential Child: Origins of Essentialism in Everyday Thought*. Oxford: Oxford UP.

Gibbs, Raymond W., Jr. 2006. *Embodiment and Cognitive Science*. Cambridge: Cambridge UP.

Gibson, James J. 1950. *The Perception of the Visual World*. Boston: Houghton Mifflin.

———. 1966. *The Senses Considered as Perceptual Systems*. New York: Houghton Mifflin.

———. 1979. *The Ecological Approach to Visual Perception*. Boston: Houghton Mifflin.

Gigerenzer, Gerd. 2007. *Gut Feelings: The Intelligence of the Unconscious*. New York: Viking.

Gillespie, Diane. 1992. *The Mind's We: Contextualism in Cognitive Psychology*. Carbondale: Southern Illinois UP.

Gilligan, Carol. 1993 (1982). *In a Different Voice: Psychological Theory and Women's Development*. Cambridge: Harvard UP.

Goffman, Erving. 1956. *The Presentation of Self in Everyday Life*. Edinburgh: U of Edinburgh P.

Gottschall, Jonathan. 2012. *The Storytelling Animal: How Stories Make Us Human*. Boston: Houghton Mifflin Harcourt.

———, and David Sloan Wilson, eds. 2005. *The Literary Animal: Evolution and the Nature of Narrative*. Evanston: Northwestern UP.

Gould, Stephen J., and Richard C. Lewontin. 1979. "The Spandrels of San Marco and the Panglossian Program: A Critique of the Adaptationist Programme." *Proceedings of the Royal Society of London* 205: 581–98.

Greenfield, Susan A. 2000. *The Private Life of the Brain: Emotions, Consciousness, and the Secret of the Self*. New York: Wiley.

Han, Hye Joo, Richard Schweickert, Zhuangzhuang Xi, and Charles Viau-Quesnel. 2016. "The Cognitive Social Network in Dreams: Transitivity, Assortativity, and Giant Component Proportion Are Monotonic." *Cognitive Science* 40: 671–96.

Hauser, Marc D. 2006. *Moral Minds: How Nature Designed Our Universal Sense of Right and Wrong*. New York: HarperCollins.

Hermans, Hubert J. M., and Harry J. G. Kempen. 1993. *The Dialogical Self: Meaning as Movement*. San Diego: Academic P.

Hilts, Philip J. 1995. *Memory's Ghost: The Strange Tale of Mr. M. and the Nature of Memory*. New York: Simon and Schuster.

Hirst, William, et al. 2009. "Long-Term Memory for the Terrorist Attack of September 11: Flashbulb Memories, Event Memories, and the Factors That Influence Their Retention." *Journal of Experimental Psychology: General* 138.2: 161–76.

Hobson, J. Allan. 1988. *The Dreaming Brain*. New York: Basic Books.

———. 1994. *The Chemistry of Conscious States: Toward a Unified Model of the Brain and the Mind*. Boston: Back Bay Books.

———. 1999. *Consciousness*. New York: Scientific American Library.

Holland, Norman N. 1988. *The Brain of Robert Frost: A Cognitive Approach to Literature*. New York: Routledge.

———. 1992. *The Critical I*. New York: Columbia UP.

———. 2009. *Literature and the Brain*. Gainesville: The PsyArt Foundation.

Humphrey, Nicholas K. 1992. *A History of the Mind*. New York: Simon and Schuster.

Husnu, Shenel and Richard J. Crisp. 2015. "Perspective-Taking Mediates the Imagined Contact Effect." *International Journal of Intercultural Relations* 44: 29–34.

Jay, Martin. 1993. *Downcast Eyes: The Denigration of Vision in Twentieth-Century French Thought*. Berkeley: U of California P.

Johnson, Mark. 1987. *The Body in the Mind: The Bodily Basis of Meaning, Imagination, and Reason*. Chicago: U of Chicago P.

———. 2007. *The Meaning of the Body: Aesthetics of Human Understanding*. Chicago: U of Chicago P.

Kandel, Eric R. 2006. *In Search of Memory: The Emergence of a New Science of Mind*. New York: W. W. Norton.

Kaschak, Ellyn. 1992. *Engendered Lives: A New Psychology of Women's Experience*. New York: Basic Books.

Kihlstrom, John. 1987. "The Cognitive Unconscious." *Science* 237: 1445–52.

Kosko, Bart. 1993. *Fuzzy Thinking: The New Science of Fuzzy Logic*. New York: Hyperion.

Kosslyn Stephen Michael. 1983. *Ghosts in the Mind's Machine: Creating and Using Images in the Brain*. New York: W. W. Norton.

———. 1994. *Image and Brain: The Resolution of the Imagery Debate*. Cambridge: MIT P.

Kotre, John. 1995. *White Gloves: How We Create Ourselves Through Memory*. New York: Free P.

Lakoff, George. 1987. *Women, Fire, and Dangerous Things: What Categories Reveal About the Mind*. Chicago: U of Chicago P.

———, and Mark Johnson. 1999. *Philosophy in the Flesh: The Embodied Mind and Its Challenge to Western Thought*. New York: Basic Books.

LeDoux, Joseph. 1996. *The Emotional Brain: The Mysterious Underpinnings of Emotional Life*. New York: Simon and Schuster.

———. 2002. *Synaptic Self: How Our Brains Become Who We Are*. New York: Viking.

Leverage, Paula. 2010. *Reception and Memory: A Cognitive Approach to the Chansons de Geste*. Amsterdam: Rodopi.

Lewis, Tanya. 2017. "Brain Activity During Sleep Can Predict When Someone Is Dreaming." *Scientific American* 1 July. Online.

Llinás, Rodolfo R. 2001. *I of the Vortex: From Neurons to Self*. Cambridge: Bradford Books/MIT P.

Loftus, Elizabeth, and Katherine Ketcham. 1996. *The Myth of Repressed Memory: False Memories and Allegations of Sexual Abuse*. New York: St. Martin's Griffin.

Luminet, Oliver, and Antonietta Curci. 2009. *Flashbulb Memories: New Issues and New Perspectives*. New York: Psychology Press.

McAdams, Dan P. 1988. *Power, Intimacy, and the Life Story: Personological Inquiries into Identity*. New York: Guilford P.

———. 2006. *The Redemptive Self: Stories Americans Live By*. New York: Oxford University Press.

Maiese, Michelle. 2011. *Embodiment, Emotion, and Cognition*. London: Palgrave Macmillan.

Mancing, Howard. 2017. "Narrative Imagination in *La dama duende*." In *Making Sense of the Senses: Current Approaches in Spanish* Comedia *Criticism*. Ed. Yolanda Gamboa and Bonnie Gasior. Newark: Juan de la Cuesta. 143–58.

Mandler, George. 1997. *Human Nature Explored*. New York: Oxford UP.

Minsky, Marvin. 1986. *The Society of Mind*. New York: Simon and Schuster.

Mithen, Steven. 1996. *The Prehistory of the Mind: The Cognitive Origins of Art, Religion and Science*. London: Thames and Hudson.

Neisser, Ulric. 1967. *Cognitive Psychology*. New York: Appleton-Century-Crofts.

———. 1976. *Cognition and Reality: Principles and Implications of Cognitive Psychology*. San Francisco: W. H. Freeman.

———. 1982a. "Snapshots or Benchmarks?" In *Memory Observed: Remembering in Natural Contexts*. Ed. Ulric Neisser. San Francisco: W. H. Freeman. 43–48.

———. 1982b (1981). "John Dean's Memory: A Case Study." In *Memory Observed: Remembering in Natural Contexts*. Ed. Ulric Neisser. San Francisco: W. H. Freeman. 139–59.

Noë, Alva. 2004. *Action in Perception*. Cambridge: MIT P.

———. 2009. *Out of Our Heads: Why You Are Not Your Brain, and Other Lessons from the Biology of Consciousness*. New York: Hill and Wang.

Nussbaum, Martha C. 1997. *Cultivating Humanity: A Classical Defense of Reform in Liberal Education*. Cambridge: Harvard UP.

Opdahl, Keith M. 2002. *Emotion as Meaning: The Literary Case for How We Imagine*. Lewisburg: Bucknell UP.

Paivio, Allan. 1986. *Mental Representations: A Dual Coding Approach*. New York: Oxford UP.

Panksepp, Jaak. 1998. *Affective Neuroscience: The Foundations of Human and Animal Emotions*. Oxford: Oxford UP.

Person, Ethel S. 1995. *By Force of Fantasy: How We Make Our Lives*. New York: Basic Books.

Pillemer, David. B. 1998. *Momentous Events, Vivid Memories*. Cambridge: Harvard UP.

Pinker, Steven. 2002. *The Blank Slate: The Modern Denial of Human Nature*. New York: Viking.

Price, Jill, with Bart Davis. 2008. *The Woman Who Can't Forget: The Extraordinary Story of Living with the Most Remarkable Memory Known to Science.* New York: Free P.

Pylyshyn, Zenon W. 1984. *Computation and Cognition: Toward a Foundation for Cognitive Science.* Cambridge: MIT P.

Quiroga, Rodrigo Quian. 2010. "In Retrospect: *Funes the Memorious.*" *Nature* 463: 611.

Reed, Edward S. 1988. *James J. Gibson and the Psychology of Perception.* New Haven: Yale UP.

Rock, Andrea. 2004. *The Mind at Night: The New Science of How and Why We Dream.* New York: Basic Books.

Rosch, Eleanor H. 1973. "Natural Categories." *Cognitive Psychology* 4: 328–50.

———. 1977. "Human Categorization." In *Advances in Cross-Cultural Psychology, Vol. 1.* Ed. Neil Warren. London: Academic P. 1–49.

———. 1978. "Principles of Categorization." In *Cognition and Categorization.* Ed. Eleanor H. Rosch and Barbara B. Lloyd. Hillsdale: Lawrence Erlbaum Associates. 27–48.

———. 1981. "Prototype Classification and Logical Classification: The Two Systems." In *New Trends in Cognitive Representation: Challenges to Piaget's Theory.* Ed. Ellin F. Scholnick. Hillsdale: Lawrence Erlbaum Associates. 73–86.

———, and Carolyn B. Mervis. 1975. "Family Resemblances: Studies in the Internal Structure of Categories." *Cognitive Psychology* 7: 573–605.

Sábato, Ernesto. 1988 (1950). *The Tunnel.* Trans. Margaret Sayers Peden. New York: Ballentine Books.

Sarbin, Theodore R, ed. 1986a. *Narrative Psychology: The Storied Nature of Human Conduct.* New York: Praeger.

———. 1986b. "Introduction and Overview." In Sarbin, ix–xviii.

———. 1986c. "The Narrative as Root Metaphor for Psychology." In Sarbin, 3–21.

Scalise Sugiyama, Michelle. 1996a. "Darwinian Literary Science?" *Skeptic* 4.4: 94–96.

———. 1996b. "On the Origins of Narrative: Storyteller Bias as a Fitness Enhancing Strategy." *Human Nature* 7: 403–25.

———. 2001a. "Food, Foragers, and Folklore: The Role of Narrative in Human Subsistence." *Evolution and Human Behavior* 22: 221–40.

———. 2001b. "Narrative Theory and Function: Why Evolution Matters." *Philosophy and Literature* 25: 233–50.

———. 2001c. "New Science, Old Myth: An Evolutionary Critique of the Oedipal Paradigm." *Mosaic* 121–36.

———. 2003. "Cultural Variation Is Part of Human Nature: Literary Universals, Context-Sensitivity and 'Shakespeare in the Bush.'" *Human Nature* 14.4: 383–96.

———. 2004. "Predation, Narration, and Adaptation: 'Little Red Riding Hood' Revisited." *Interdisciplinary Literary Studies* 5: 108–27.

———. 2005. "Reverse-Engineering Narrative: Evidence of Special Design." In Gottschall and Wilson, 177–96.

———. 2008. "Narrative as Social Mapping—Case Study: The Trickster Genre and the Free Rider Problem." *Ometeca* 12: 24–42.

———. 2017. "Oral Storytelling as Evidence of Pedagogy in Forager Societies." *Frontiers in Psychology* 29 March. Online.

Scarry, Elaine. 1999. *Dreaming by the Book*. New York: Farrar, Straus and Giroux.

Schacter, Daniel L. 1996. *Searching for Memory: The Brain, the Mind, and the Past*. New York: Basic Books.

Schank, Roger C. 1999. *Dynamic Memory Revisited*. Cambridge: Cambridge UP.

Schweickert, Richard. 2007. "Properties of the Organization of Memory for People: Evidence from Dream Reports." *Psychonomic Bulletin & Review* 14: 270–76.

Shaughnessy, Nicola. 2013. *Affective Performance and Cognitive Science: Body, Brain and Being*. London: Methuen.

Siegel, Daniel J. 1999. *The Developing Mind: Toward a Neurobiology of Interpersonal Experience*. New York: Guilford P.

Singer, Jefferson A., and Peter Salovey. 1993. *The Remembered Self: Emotion and Memory in Personality*. New York: Free P.

Smith, Frank. 1990. *To Think*. New York: Teachers College P.

Solms, Mark, and Oliver Turnbull. 2002. *The Brain and the Inner World: An Introduction to the Neuroscience of Subjective Experience*. New York: Other P.

Stafford, Barbara Maria. 1996. *Good Looking: Essays on the Virtue of Images*. Cambridge: MIT P.

Thompson, Evan. 2007. *Mind in Life: Biology, Phenomenology, and the Sciences of Mind*. Cambridge: Belknap P of Harvard UP.

Turner, Mark. 1996. *The Literary Mind*. New York: Oxford UP.

———. 2003. "Double-Scope Stories." In *Narrative Theory and the Cognitive Sciences*. Ed. David Herman. Stanford: CSLI Publications. 117–42.

———. 2014. *The Origin of Ideas. Blending, Creativity, and the Human Spark*. New York: Oxford UP.

Tye, Michael. 1991. *The Imagery Debate*. Cambridge: MIT P.

University of Chicago Medical Center. 2017. "Neuroscientists Call for More Comprehensive View of How Brain Forms Memories." *ScienceDaily* 5 July. Online.

Vassielieva, Julia. 2016. *Narrative Psychology: Identity, Transformation and Ethics*. London: Palgrave Macmillan.

Velmans, Max, and Susan Schneider, eds. 2017 (2007). *The Blackwell Companion to Consciousness*. Oxford: Blackwell.

Watkins, Mary. 1986. *Invisible Guests: The Development of Imaginal Dialogues*. Hillsdale: Analytic P.

Watson, Lyall. 2000. *Jacobson's Organ and the Remarkable Nature of Smell*. New York: W. W. Norton.

William, Jennifer Marston. 2002. "'Gut zugedeckt?' Metaphors for Memory in Kafka's 'Das Urteil'and 'Die Verwandlung.'" *The Journal of the Kafka Society of America* 26.1–2: 58–66.

Winograd, Eugene, and Neisser, Ulric, eds. 1992. *Emory Symposia in Cognition, 4. Affect and Accuracy in Recall: Studies of "Flashbulb" Memories*. Cambridge: Cambridge UP.

Winson, Jonathan. 1985. *Brain and Psyche: The Biology of the Unconscious*. Garden City: Anchor P.

Yates, Frances A. 1966. *The Art of Memory*. Chicago: U of Chicago P.

Zadeh, Lotfi A. 1965. "Fuzzy Sets." *Information and Control* 8: 338–53.

Zeki, Semir. 1993. *A Vision of the Brain*. Oxford: Blackwell Scientific Publications.

CHAPTER 9

Development of the Mind

Development occurs in context; the developing organism cannot be divorced from the environment in which it develops.

—Steven Rose

More than ever before, cognitive researchers are observing and learning from real babies and children in real situations, rather than making unmotivated assumptions or imposing their abstract theories upon babies. The development of neuroscience as a viable discipline has contributed significantly to this effort. This chapter treats the myth of the human brain as a blank slate (*tabula rasa*), new advances in infant cognition, and the role of language in early child development.

1 The Blank Slate Myth

An enduring belief in many circles, including THEORY, is that the mind is a general-purpose, undifferentiated tablet, upon which anything may be inscribed, essentially what John Locke called a *tabula rasa* (i.e., a scraped tablet), or, more popularly, a *blank slate.* In this view, development takes place as society or experience, perhaps by a process of association (as Saussure, e.g., says), imposes itself on the passive mind; essentially, this is also the doctrine of behaviorism. In this section, we draw substantially

H. Mancing and J. Marston William, *Restoring the Human Context to Literary and Performance Studies*, Cognitive Studies in Literature and Performance,
https://doi.org/10.1007/978-3-030-89078-0_9

from an influential work by Stephen Pinker, *The Blank Slate: The Modern Denial of Human Nature* (2002). Pinker defines the blank slate concept as follows: "The idea that the human mind has no inherent structure and can be inscribed at will by society or ourselves" (2). The pervasiveness of the idea of the blank slate in the social sciences can be seen in the collection of quotations Pinker provides from some of the most influential sociologists and anthropologists twentieth-century thought: Franz Boas, Emile Durkheim, Ruth Benedict, Margaret Mead, Albert Kroeber, Ashley Montagu, Clifford Geertz, and others (23–26). But, in the words of political scientist Paul H. Rubin, "The notion that humans are born as blank slates (*tabula rasa*, to use Locke's Latin phrase) is no longer intellectually respectable among serious people" (2002, ix).

Standing in direct opposition to the idea of an infinitely malleable *tabula rasa*, modern cognitive science, grounded in evolution and biology, neuroscience, and cognitive psychology, stresses the biological reality of the universal human mind-brain. The convincing arguments by Lenneberg and Chomsky that language is largely a function of human biology, and, especially, the fact of a critical period for the acquisition of language, are among the most important and undeniable factors contributing to the destruction of blank slate ideology in the cognitive sciences. Pinker's demolition, for example, of the naive anthropological argument of idyllic societies (the Noble Savage metaphor) in which violence is unknown is thoroughly convincing, as he shows how a mere two deaths in a band of fifty people would be the equivalent of 10,000,000 in the United States (2002, 56–58). Modern industrialized nations, in general, are characterized by a lower incidence of homicide and other forms of violence than traditional hunter-gatherer societies.

Pinker identifies four fundamental anxieties about human nature that most often underlie the blank slate position:

- If people are innately different, oppression and discrimination would be justified.
- If people are innately immoral, hopes to improve the human condition would be futile.
- If people are products of biology, free will would be a myth and we could no longer hold people responsible for their actions.
- If people are products of biology, life would have no higher meaning and purpose. (139)

These anxieties, many blank slate proponents assume, will lead inevitably to discrimination, Social Darwinism, Nazism, eugenics, genocide, racism, and a host of other evils. Pinker then shows the faulty logic behind each of these assumptions.

Pinker's Chapter 13, "In Touch with Reality" (197–218), in which he shows the relationship between blank slate thinking and postmodernism and other brands of social constructionism, is particularly relevant to our concerns. Pinker concludes:

> The view that humans are passive receptacles of stereotypes, words, and images is condescending to ordinary people and gives unearned importance to the pretensions of cultural and academic elites. And exotic pronouncements about the limitations of our faculties, such as that there is nothing outside the text or that we inhabit a world of images rather than a real world, make it impossible even to identify lies and misrepresentations, let alone to understand how they are promulgated. (217–18)

The price we pay for blind adherence to the blank slate is that we wind up "denying the possibility of objectivity and truth, dumbing down issues into dichotomies, replacing facts and logic with political posturing" (422). To cling desperately to the concept of an uncluttered, undifferentiated, empty mind upon which language, culture, and ideology impose themselves, in the face of the overwhelming evidence to the contrary that suffuses contemporary biology and psychology is a pathetic example of Bakhtin's theoreticism. Or as Pinker puts it, it is a matter of "a few intellectuals who have let their theories get the better of them" (422). The blank slate argument cannot be legitimately held in the twenty-first century.

2 Infant Cognition

Until about the last quarter of a century, it was usually assumed that infancy consisted of William James's "buzzing, booming confusion," a state of symbiosis and absolute identification with the mother, a period of profound (if not always well understood) sexual turmoil, and a regular march through a series of specific stages. These assumptions were made largely because we as adults could not imagine anything else for the essentially foreign beings that are our newborn children. Today we know the mental world of infants and very young children is really quite different from what we had assumed. By the beginning of the twenty-first century, it had become clear to all psychologists that "human infants are very social creatures from the moment they are born, if not before" (Tomasello 1999, 58). Psychologist Alison Gopnik, one of the leaders in the new approach to infant cognition, elaborates on the field in her book *The Philosophical Baby* (2009):

> In the last thirty years, there's been a revolution in our scientific understanding of babies and young children. We used to think that babies and young children were irrational, egocentric, and amoral. Their thinking and experience were concrete, immediate, and limited. In fact, psychologists and neuroscientists have discovered that babies not only learn more, but imagine more, care more, and experience more than we would ever have thought possible. In some ways, young children are actually smarter, more imaginative, more caring, and even more conscious than adults are. (5)

Recent years have witnessed the development of reliable and significant new techniques and methods for assessing the mind of beings (babies and non-human animals) who cannot speak, respond to our questions, take language-based tests, or otherwise reveal directly what may be on their minds. Foremost among these for human infants are three indicators: (1) head-turning; (2) looking; and (3) sucking (Stern 1985, 38–42). In brief, research suggests that when babies move their heads or eyes with greater or lesser animation to track certain things, and when they suck stronger or faster at the nipple or breast, they reveal their degree of attention or interest.[1]

A number of investigators have used the evidence of their research to argue for innate abilities in members of the human species, while others argue strongly against any sort of innate cognitive abilities in children. The degree to which something is an innate structure or process of the mind-brain as opposed to a feature that develops or emerges naturally as a part of embodied cognition—or the degree to which the same thing is referred to in different ways, essentially a matter of semantics or emphasis as much as anything else—is an issue that has yet to be worked out satisfactorily in biology and psychology.

One of the most influential works in child development is Annette Karmiloff-Smith's *Beyond Modularity* (1992), in which the author attempts to find a middle path between Fodor's strong modularity thesis and the structuralist thesis of Piaget, under whom she studied and with whom she later collaborated. A major focus of Karmiloff-Smith's work is the way in which representations change over time throughout development. Essential to her thesis is the idea of "representational redescription" (RR), by which the mind can "exploit internally the information that it has already stored (both innate and acquired)" (15). The result, she argues, is that development is not "*either* domain specific *or* domain general. It is clearly the intricate interaction of *both*—more domain general than is presupposed by most nativist/modularity views of development, but more domain specific than Piagetian theory envisages" (168).

This process of representational redescription, Karmiloff-Smith suggests, moves the mind beyond the domain-specific knowledge (modularity) that characterizes other species. Here she anticipates what Mithen calls cognitive fluidity, the ability of human beings creatively to call upon a variety of domains of knowledge to combine aspects of cognition that would otherwise remain separate. This gives the human species a kind of all-purpose mind-brain unique in the animal kingdom. Note that the mind-brain is not anything like a blank slate structure. Rather, during the course of development, the process of representational redescription makes Mithen's cognitive fluidity possible.

Language, for example, illustrates the process; Karmiloff-Smith argues that "language acquisition is both domain specific and domain general; i.e., that some initial domain-specific constraints channel the progressive building up of domain-specific linguistic representations but that, once redescribed, these representations become available to domain-general processes" (32). Similar processes are at work, she suggests, in other areas claimed for innate knowledge: physics, mathematics, psychology, and "notation"—a term Karmiloff-Smith (139) uses for external descriptions of internal representations; it is similar to Donald's external memory storage. In other words, just as we have seen the process of emergent qualities throughout evolutionary history, there is an emergence of cognitive generality, or cognitive fluidity, from domain specificity. Karmiloff-Smith's contribution is important, because it clearly lays out a middle way of cognition between two mutually exclusive systems. Karmiloff-Smith attempts to save the best of her mentor's system within the realities of modern biology and neuroscience: "I believe that it is possible to retain the essence of Piagetian theory while doing away with stage and structure" (173). If her work is to be faulted, it is only in not going far enough.

A view of infant cognition and child development that takes a very different approach from the ones described by Karmiloff-Smith has been elaborated in the brilliant work carried out by Esther Thelen and Linda B. Smith, who, in *A Dynamic Systems Approach to the Development of Cognition and Action* (1994), ground development in dynamical systems theory. Thelen and Smith propose a process of nonlinear, chaotic—or, at least, complex—emerging and self-defining. They fill their book, which is not an easy read, with diagrams of topographic maps and "attractors" in the brain. Further, they explicitly ground their work in

- Edelman's Theory of Neuronal Group Selection (TNGS),
- the approach to categorization initiated by Rosch and best exemplified by Lakoff (see our Sect. 5 in Chapter 8),
- Gibson's ecological approach to perception and theory of affordances (see our Sect. 2 in Chapter 8),
- Karmiloff-Smith's representational redescription (see above),
- the autopoiesis and enaction of Humberto Maturana and Francisco Varela (see our Sect. 2 in Chapter 12),
- the constructive interactionism of Susan Oyama (see our Sect. 2 in Chapter 12), and
- the experientialism of Lakoff and Johnson (see our Sect. 3 in Chapter 12).

Thelen and Smith reject mind–body dualism, the nature-nurture debate, the COGNITIVIST concept of decontextualized information, the mechanistic metaphor in general and in particular the computer metaphor for the mind (with its representation and information processing), connectionist models (although they allow that such models may hold some promise for future relevance), simplistic concepts of brain modularity (à la Fodor), any approach that relies on innate capabilities or competencies (such as the one proposed by Chomsky for language), and any sort of lock-step structural development of the logical mind (as proposed by Piaget, for example). Since they commit themselves to a "biologically consistent theory" of development, Thelen and Smith substitute for the mechanism metaphor one of "a fluid, organic system" (xix), but one that is in essence contextual (xix). Grounding cognition in perception and bodily experience, Thelen and Smith examine at length the most basic activities of infants, such as kicking, crawling, and reaching.

Slowly, they build their case that the mind-brain constructs itself in the contexts of everyday experiences. Learning to walk, for example, becomes "less a prescribed, logically inevitable process than a confluence of available states within particular contextual opportunities.... Walking self-organizes under these constraints because nonlinear, complex dynamic systems occupy preferred behavior states" (72). This approach leads them to conclude that "development is the outcome of the self-organizing processes of continually active *living* systems" (44) and not something that is predestined.

Devoting a whole chapter to "The Context-Specific Origin of Knowledge" (215–45), Thelen and Smith affirm their commitment to CONTEXTUALISM as they argue that intelligence is found or exhibited "not in context-free rules and abstract representations. Intelligence lies in our activity in authentic reality.... Intelligence means the ability to adapt, to fit behavior and cognition to the changing context" (244). Recall Plotkin's argument from Sect. 1 in Chapter 5 that knowledge is the product of evolutionary adaptations, that doing is knowing, a position explicitly restated in Maturana and Varela's theory of autopoiesis.

In Sect. 2 in Chapter 8, we discussed James J. Gibson's ecological approach to perception and to psychology, which inspired a movement of ecological psychology, but now we turn to more recent work in infant cognition also grounded in Gibson's theories. Eleanor J. Gibson (wife and colleague of James Gibson until his death in 1979) and Anne D. Pick have described this work in their book *An Ecological Approach to Perceptual Learning and Development* (2000). A distinguishing characteristic of the ecological approach is that it "takes as its unit of study the animal in its environment, considered as an interactive system. The relations within this system are reciprocal, with the reciprocity including a species evolving in an environment to which it becomes adapted, and an individual acting in its own niche, developing and learning" (14). Since learning is something that begins even before birth (as babies in the womb become accustomed to their mother's voice, for instance) and is characteristic of neonates from the first moments after birth, perceptual learning—defined by the authors as "a change in the relation between an active organism and some affordance of the environment" (50)—is crucial to physical and mental development, since it precedes by years the acquisition of language. Gibson and Pick identify three "modes of behavior" that affect human cognition and performatory activities: "communication with others; reaching for, manipulating, and using objects;

and locomotion—getting around in the environment" (50). All three modes of behavior are grounded in perception. In the development of all of them, infants themselves are active initiators of their learning. The idea of infants as active, creative, controlling, exploring, intentional agents is not one we have been used to in traditional approaches to psychology, but Gibson and Pick make a strong case for it (e.g., 160). This conceptualization of infants is absent from Freudianism and behaviorism. Research on the cognitive development of infants points again clearly to the fact that human beings are embodied agents who function pragmatically in concert with and as part of the world. So far in this chapter, attention has been directed primarily to the preverbal infant, but when language enters the picture, everything begins to change.

3 The Emergence of Language

According to Daniel N. Stern, when the verbal self begins to emerge in the second year of a child's life, "[a] new organizing subjective perspective emerges and opens a new domain of relatedness" (1985, 162), one in which it is possible to "be with" others in new ways, to share one's own experiences and create new experiences with others, to begin the construction of a narrative of one's life. Language is the key to the entry into that symbolic realm of existence known only to human beings, a way of life mediated by a conceptual and communicative tool that makes possible the higher-level consciousness that is uniquely characteristic of our species.

Katherine Nelson has extensively studied the significance of language in the developmental process, as detailed for instance in her superb book *Language in Cognitive Development* (1996). Nelson describes her project in this way:

> The basic premise of this book is that language is a catalyst of cognitive change during early to middle childhood. The multifunctional rules of language in cognition and communication during this important period of developmental change require deeper analysis than they have been given in contemporary theoretical writing. It is my conviction that to fulfill this goal the biological underpinnings and the sociocultural conditions of human development must be jointly analyzed in their contributions to the development of human cognition and language as a dynamic process. (3)

Her recognition of the interrelated and mutually defining roles of biology and culture, together with her evolutionary framework, give Nelson's work a contextual validity normally absent from, as she puts it, "contemporary theoretical writing." The usual approach to development in psychology and education (the classic example is that of Piaget) tends to be carried out within a paradigm in which evolution and biology virtually play no part. But Nelson explicitly relates her work to the ecological approach of James Gibson in an effort to avoid the nature-nurture binary; at the same time, she, like Thelen and Smith, rejects both the concept of a *tabula rasa* mind and the idea of an innate sense of biology or physics as described above. Nelson conceives of the child as having to construct knowledge by means of a gradual developmental process that "is *mediated* by virtue of living in a culturally arranged world in which caretakers direct most of the action" (8). Nelson relates the ideas of emergence and mediation to the concept of *representation*, not as conceived by COGNITIVISTS, but in a biological context that "implies neither symbols nor an interpreter, but it does imply an 'enactive mind'" (11). Describing Karmiloff-Smith's theory of RR as "powerful" (14), Nelson nevertheless notes that Karmiloff-Smith's

> stress on internal analysis and neglect of the influence of the external representations of others leaves the system analysis unfinished. In particular, the child's analysis of language cannot be independent of the context of the language in use. Thus the model does not take fully into account that language (and logic) are public constructions with private ramifications. (15)

Nelson's aim is to trace in much more detail, and with much more sophisticated and extensive theoretical grounding, the contexts—both internal and external, both biological and social—and the role of language in development.

Basic to Nelson's concept of development is a kind of building block (our term, not hers) of cognition that she calls "mental event representations" (MERs), the experienced-based nonlinguistic knowledge of the world that humans share with other mammals, although "each world model is different from every other in subtle and sometimes startling ways" (17). As should be clear, Nelson's MERs are similar to (but not exactly the same thing as) schemata as described by Neisser.

Nelson situates her approach within the sociocultural tradition of Vygotsky and Bakhtin, but combines this tradition's social construction of knowledge with a solid base in evolution and biology. Elegantly integrating evolution, biology, and the universality of the human mind-brain on one hand, and the collaborative, social, and cultural context within which cognition takes place on the other, Nelson offers a paradigm case of the program that we describe throughout this book. Nelson notes that the idea that ontogeny recapitulates phylogeny (individual development follows the same path as species development) is, while not a universal or hard-and-fast rule, often the case. And she makes the case for the process in this instance by taking as a model or metaphor (not an exact blueprint) Donald's (1991) theory of the evolution of the modern mind through four phases.

Nelson discusses different aspects of language development and learning throughout childhood, but for her most important point, she draws on the theory of relevance proposed by Dan Sperber and Deirdre Wilson (1995), discussed in Sect. 4 of Chapter 4. Nelson stresses that "communication between people always rests on *inference* and *interpretation*" (138) and that "the listener must *actively interpret* what the speaker's *intention* may be" (138–39). Both children and adults learn words by using *discourse context*, by making inferences about relevance via *cognitive context*, a model that is experientially based and situationally updated (139–40).

The remainder of Nelson's book is a detailed working out of her scheme based on Donald's model, with chapters devoted to the emergence of what she calls mediating language, the historical self, the storied mind, the paradigmatic mind, the temporal mind, and the projective mind. Nelson makes a compelling case that cognitive development is simultaneously a matter of an individual's unique sociocultural circumstances and experiences, as well as her biological heritage and instantiation.

Paul Bloom, in his *How Children Learn the Meanings of Words* (2000), argues against the commonsense notion that children learn new words automatically, perhaps by a (Saussurean or Skinnerian) process of association or some sort of stimulus–response process. He points out: (1) it is possible for children to learn words without there being a close spatiotemporal example between a word and its meaning; (2) full development of and access to all sensory systems (especially hearing) are not necessary for word learning to take place; (3) feedback is not always required for the acquisition of word meanings; and (4) there is no need for ostensive naming or child-directed discourse in order for the learning of words to occur (6–9). In effect, word learning illustrates Chomsky's poverty-of-stimulus thesis: there is not enough contextual usage or explicit teaching for children to learn from these processes all the new words they learn—and this is especially true for words without correspondence to material things.

Human beings' learning of new words is a prodigious feat. The typical high-school graduate knows (or at least recognizes in use) some 60,000 words—a conservative estimate according to Bloom: "Since word learning starts at about 12 months of age, this averages to learning 3,750 new words a year, or 10 words a day—a word every waking 90 minutes" (25). The process is a complex one, as children do not learn all there is to know about a word in a single instance, and early hypotheses about word meanings are often modified, expanded, or revised according to the way words are used by others in different contexts; rather than learning ten words a day, a child might learn "one-hundredth of each of a thousand different words" (25). And, further, the pace of word learning picks up as a child matures; a two-year old is not likely to learn her ten daily words as easily as an older child, so the latter are likely to be acquiring words at a much faster rate than very young children just learning to speak and understand language. Once literacy is achieved, most new word learning comes from written texts, so that a literate adult who reads regularly may well know over 100,000 words (192).

If children learn so many words so rapidly, and they are not systematically taught these words, how do they learn them? Bloom makes the argument that it is primarily by means of their developing Theory of Mind (55–87). Unlike philosophers (e.g., Putnam) who tend to ground word meaning solely in external reference, and unlike linguists (e.g., Chomsky) who tend to consider it exclusively in terms of individual psychology, Bloom logically assumes that "there are two aspects (or determinants) of the meaning of a word—an internal psychological aspect, sometimes called *narrow content*, and an external social and contextual aspect, sometimes called *broad content*. These work together to determine what words refer to" (21).

Just as the development of the mind-brain is a socio-biological process, so is the process of learning words: "When children learn the meaning of a word, they are—whether they know it or not—learning something about the thoughts of other people" (55). Here is where Theory of Mind (see our next chapter) comes in; it "underlies how children learn the entities to which words refer, intuit how words relate to one another, and understand how words can serve as communicative signs" (55). If the early stages of the development of a Theory of Mind—including the way in which babies attend to what others are saying, doing, and calling attention to by means of their gaze and pointing gestures—are accurate descriptions of infant cognition, then the very first aspects of language comprehension should be included as a part of the overall process.

It is important, also, to remind ourselves with Bloom, and with others cited above, such as Nelson, that not all thought is linguistic: "rich abstract thought is possible without words, and much of what goes on in word learning is establishing a correspondence between the symbols of a natural language and concepts that exist prior to, and independently of, the acquisition of that language" (242). Developmental psychology and post-Chomskyan linguistics provide ample evidence that the linguicentric convictions that underlie all of THEORY are not tenable. As indicated above, Bloom anchors his study of word learning in children in the concept of Theory of Mind, the subject of the next chapter.

Note

1. Even with new advances in eye-tracking technology, its potential to help scientists gain insight into human cognition is not always reached, due to human error and potential software or hardware limitations. With these devices, a large amount of data can be collected but it may not be conclusive, particularly if research questions aren't carefully formulated in advance, as Aslin (2012) warns: "Given access to an eye-tracker, infancy researchers are like the proverbial hammer in search of a nail. There is a tendency to gather data first and ask questions later" (130).

References

Aslin, Richard N. 2012. "Infant Eyes: A Window on Cognitive Development." *Infancy: The Official Journal of the International Society on Infant Studies* 17.1: 126–40.

Bloom, Paul. 2000. *How Children Learn the Meanings of Words*. Cambridge: Bradford Books/MIT P.

Donald, Merlin. 1991. *The Origins of the Modern Mind: Three Stages in the Evolution of Culture and Cognition*. Cambridge: Harvard UP.

Gibson, Eleanor J., and Anne D. Pick. 2000. *An Ecological Approach to Perceptual Learning and Development*. Oxford: Oxford UP.

Gopnik, Alison. 2009. *The Philosophical Baby: What Children's Minds Tell Us About Truth, Live, and the Meaning of Life*. New York: Farrar, Straus and Giroux.

Karmiloff-Smith, Annette. 1992. *Beyond Modularity: A Development Perspective on Cognitive Science*. Cambridge: Bradford Books/MIT P.

Nelson, Katherine. 1996. *Language in Cognitive Development: Emergence of the Mediated Mind*. Cambridge: Cambridge UP.

Pinker, Steven. 2002. *The Blank Slate: The Modern Denial of Human Nature*. New York: Viking.

Rubin, Paul H. 2002. *Darwinian Politics: The Evolutionary Origin of Freedom*. New Brunswick: Rutgers UP.

Sperber, Dan, and Deirdre Wilson. 1995 (1986). *Relevance: Communication and Cognition*. Cambridge: Harvard UP.

Stern, Daniel N. 1985. *The Interpersonal World of the Infant: A View from Psychoanalysis and Developmental Psychology*. New York: Basic Books.

Thelen, Esther, and Linda Smith. 1994. *A Dynamic Systems Approach to the Development of Cognition and Action*. Cambridge: MIT P.

Tomasello, Michael. 1999. *The Cultural Origins of Human Cognition*. Cambridge: Harvard UP.

CHAPTER 10

Theory of Mind (ToM)

Context is everything.
—Susan Engel

An area in which cognitive psychology and literary theory meet most readily is that of *Theory of Mind* (ToM). In all aspects of our daily life, we are thinking, and we think about (or infer) what other people must be thinking. This is a cognitive attribute that most separates human beings from other animals. In the past couple decades, the idea of ToM, sometimes referred to as "mindreading," has also been employed to understand what and how literary characters must be thinking. We treat literary characters as though they were human beings of flesh and bone, since this is their basis. This chapter consists of: an overview of the concept of Theory of Mind in non-human animals, children, and adults; ToM in relation to autism studies today; how and when a ToM might have come about during human evolution and different theories about what ToM might consist of; and, most importantly, the implications of ToM for understanding language and literature.

H. Mancing and J. Marston William, *Restoring the Human Context to Literary and Performance Studies*, Cognitive Studies in Literature and Performance,
https://doi.org/10.1007/978-3-030-89078-0_10

1 Primatology, Children and Adults, Folk Psychology

In 1978, primatologists David Premack and Guy Woodruff asked an intriguing question as part of their ongoing work with chimpanzees: "Does a chimpanzee have a Theory of Mind?" They described the problem thus:

> In saying that an individual has a Theory of Mind, we mean that the individual imputes mental states to himself and others (either to conspecifics or to other species as well). A system of inferences of this kind is properly viewed as a theory, first, because such states are not directly observable, and second, because the system can be used to make predictions, specifically about the behavior of other organisms. (515)

In other words, they wanted to know whether chimpanzees could attribute a belief or state of mind to other animals and act according to their belief about what that animal was thinking. This question should have been obvious enough to ponder before Premack and Woodruff, but it was not the sort of thing anyone would have even considered asking before the mid-1970s. Perhaps one reason for this was that sophisticated cognition was assumed to be related to such human characteristics as hunting and tool use and was not an agenda item outside of theories of human evolution and psychology. But in 1976, Nicholas Humphrey published an essay titled "The Social Function of Intellect." Calling into question the traditional assumption that intelligence functioned primarily to lead early humans to technological invention, Humphrey suggests that the principal function of imaginative intelligence in humans and other primates was, and is, primarily "to hold society together" (307). That is, primates, as social animals par excellence, found an evolutionary advantage in an increased ability to calculate and anticipate the behavior of others of their species. Social *inter*action is, above all, a *trans*action, a negotiation involving two or more individuals, so the development of a sophisticated social intellect is logical (311). It is difficult to overemphasize the importance or influence of Humphrey's essay. In the wake of his hypothesis that intelligence had a social origin, Dunbar and others began to place the evolution of language in the context of social relationships, and his work is also considered a founding document in the field of evolutionary psychology.

In the four decades since Premack and Woodruff asked their famous question, the issue of ToM in primates has been studied both in laboratories through ingenious experiments and in natural settings by observation. Chimpanzees and bonobos, monkeys and other primates, and various birds and mammals, have been the objects of such study, and the results have been, not surprisingly, inconclusive. Rigorous but artificial lab experiments involving self-contemplation in mirrors (Keenan 2003) or choosing options that result in more desired food for the subjects have, on the whole, suggested that non-human apes (except the gorilla) have at best only rudimentary aspects of a ToM, but that monkeys and other mammals do not display any such ability (Cheney and Seyfarth 1990; Sterelny 2003). More recent studies, such as one by Hayashi et al. (2020), point at least to the possibility of a simian ToM that anticipates other animals' behavior and false-belief-driven actions. Many more studies will be needed to determine the nature of this type of mindreading in comparison to that of humans.

Field observations, more natural but without experimental controls and subject to anthropomorphism and distortion in interpretation and reporting, have tended to be more positive, as reports of apes deceiving others to obtain food or sex do indeed suggest a much more fully developed ToM. Richard Byrne and Andrew Whiten (1988) coined the terms *tactical deception* and *Machiavellian intelligence* to describe how primates and other animals can deceive or mislead others to further their own goals, and they have collected convincing reports from ethologists around the world in support of their position that many animals have the ability to take advantage of others' mental states. Overall, however, the consensus seems comparable to that involving ape language: the rudiments of a ToM are present, but apes are far from having a full-fledged theory as do human beings.

But if the question of ToM originated in primatology, the greatest fruits of its contemplation come from cognitive and developmental psychology. The first psychologists to apply the concept to human beings were Heinz Wimmer and Josef Perner (1983), who began to explore the idea with respect to human children. Clearly human adults have a ToM, but what about children? When does a person's ability to have ideas about other people's thoughts develop? How can we tell? A key to determining when a child has a ToM is the "false belief test."

Here is how one version of the test works: A child observes two puppets playing together. Puppet Mary puts some candy in a drawer and then leaves the room. While she is gone, puppet John takes the candy out of the drawer and places it in a nearby box. Mary returns. The experimenter then asks the child where Mary will look for the candy. Younger children, those up to about the age of four, will inevitably say that Mary will look in the box. The child knows where the candy is and cannot conceive that her own knowledge is not the (only) knowledge that exists in the world. That is, she cannot understand that another person could believe something she herself knows not to be the case; this child does not yet have a fully developed Theory of Mind. But a slightly older child can tell the experimenter that Mary will look in the drawer, because that is where she left the candy and logically she assumes that it is still there. That is, by about the age of four or five years the child can understand that someone else can have a belief that is different from her own, that someone can have a false belief. This child has, at least to some degree, a ToM. The matter is not nearly that simple, but it illustrates the basic point: at first the child has no concept of other peoples' mental states, but at some time during development such a concept is acquired. It appears, from evidence gathered from other cultures, that this is a universal human phenomenon (see Whiten 1999, 188; Avis and Harris 1991).

Human infants begin to develop the most rudimentary aspects of understanding there are other people and that these people also have ideas and thoughts from their earliest days. A ToM is not something that springs suddenly into existence within the mind-brain of each individual when she realizes that John has hidden Mary's candy and Mary doesn't know it. The acquisition of a ToM is a slow, gradual, multi-phase process that takes place throughout the early years of childhood. One of the first indications that infants are beginning to think in terms of what others are thinking might be seen in their ability, already evident in the first year of life, to know that when someone is looking at something it may have some interest for you also, and to gaze in the same direction, a process known as *gaze monitoring* (Johnson 2004, 30–31; Goldman 2006, 193–95). Children, perhaps much like non-human animals, react first to emotional cues and bodily movements long before they develop a full-fledged ToM (Gibbs 2006, 235–36). This theory develops at different rates and to different degrees in each person.

James Russell (1996, especially 180–209) makes a good case that the development of a ToM is part and parcel of the development of general executive cognitive abilities and necessarily, and centrally, also involves the individual's sense of agency (see also Gibson and Pick 2000). Russell's definition of *agency* is the following: "*the ability to alter at will one's perceptual inputs*—motorically or atentionally" (1996, 3). Clearly, this sort of ability is available only to self-aware, embodied organisms. In a careful examination of a very large sample of recorded conversations of a number of children over a period of some years, Karen Bartsch and Henry M. Wellman (1995) provide further support for the gradual development of a ToM in children. Their findings suggest a consistent pattern of children talking about desires (wants and likes), when they first start talking around age two. At this age (or even somewhat earlier), children also begin to take part in pretend play, an important step in cognitive development (Leslie 1987; Gopnik 2009, 27–31). About a year later, some talk about beliefs begins to be detected along with the continued talk of desires. And then, toward the child's fourth year, the talk about beliefs becomes even more prominent. As a result, Bartsch and Wellman (1995, 143–73) propose a three-step development of a ToM that reflects the pattern suggested by their linguistic sampling: (1) an early-desire psychology; (2) an intermediate desire-belief psychology; and, finally, (3) a belief-desire psychology. By the time a belief-desire psychology (see the discussion of folk psychology below) is characteristic of a child's speech, they suggest, the child has acquired a ToM comparable to that of an adult. This timetable is consistent with the one outlined above in the discussion of the false-belief test and comes not from a contrived laboratory experiment, but from evidence suggested by children's normal speech in ordinary social contexts.

Human communication can never be understood if one remains at the superficial level of words or signs, as in the Saussurean tradition. Human communication necessarily involves understanding—inferring as best we can—the state of mind of the unique, embodied speaker, her thoughts and motivations, in the specific context. An important aspect of ToM is what is known as *metarepresentation*, the representation of representation (Sperber 2000; Zunshine 2006). Essentially, it is the understanding of something as understood by others and involves attributing source or context to something that is known. For example, an affirmation by a respected literary scholar tends to be given more weight in our minds than one made by a naïve novice reader; an utterance made in the context of a baseball game might be understood very differently from the same words uttered in the context of a marriage ceremony.

As with nearly everything else considered in this book, ToM is not an either-or proposition. Some people are superior mindreaders, taking note of the subtlest implications of intonation, direction of the gaze, gesture, body movement, previous experience with someone, sensitivity to cultural factors, and much more, to be experts at inferring others' states of mind. Some excel in certain circumstances, such as in their close family or in professional relationships, but are less sensitive in others. Some have become more sensitive to aspects of mind states, noting especially the implications of gender, birth order, race, disability, or social status. Some are good only on certain days and in certain contexts, but not others. And some may even be much better at knowing what others may be thinking than they are their own states of mind.

If we have a ToM in general, we also have a specific theory of the mind of everyone we deal with. And if our general theory may be pretty much a constant, our theories of individuals' minds can be very open and flexible. Biologist Stuart Kauffman sees it as finding ourselves "in a state of coevolving theories about one another, where our theories in turn are coupled with our actions" (qtd. in Frenay 2006, 358–59). We are often surprised by what people do—sudden, unexpected acts and events that seem out of character for a person you thought you knew triggers a massively revised theory of that person's mind. But our perception of another individual evolves gradually, and with it our theory of that person's mind. People change, you and your perceptions change, and thereby your theory of others' minds changes. The whole process can be an intricate coevolutionary dance of ebb and flow, give and take, action and reaction. A ToM is a constantly active and evolving matter, much like the schemata described by Ulric Neisser (see Sect. 5 in Chapter 8).

ToM is not infallible. Often we misread the intentions of others' words, deeds, gestures, expressions, hints, suggestions, or eye movements. Sometimes we infer something not intended at all, attributing to others motives or desires that they do not actually have. Many of our most serious problems in dealing with other people result from such misunderstandings. ToM is not—like Saussure's telementation model—to be understood as a perfect means of understanding. It is, rather, a heuristic—a quick-and-dirty, often uncertain, best guess: an inference. Here we recall Sperber and Wilson's (1986, 1995) relevance theory, in which implicit inferences are crucial to communication. Although misunderstandings abound in the world, our ToM is good enough in most cases that we seem to get by. A ToM has always been (since even before the term came into use) the bedrock foundation of folk psychology (Davies and Stone 1995).

Folk psychology is, as described by Janet Astington, "another name for belief-desire psychology. Folk psychology assumes that beliefs and desires exist, and that people's minds are the sum of their beliefs, desires, emotions, and intentions. Then it uses this assumption to explain why people act the way they do and to predict what they will do" (1993, 3). That is, folk psychology is the position that other people are (largely) autonomous agents, who have mental states commonly called beliefs and desires, and who are motivated by these mental states. When we rely on our folk psychology, we understand, define, and describe people from our perceived (or understood) beliefs, desires, feelings, values, experiences, and intentions. Because we understand peoples' actions in terms of these mental states, we can explain why people have done certain things and predict what they might do in certain contexts. In effect, the core of belief-desire, or folk, psychology is a ToM.

When we talk of folk psychology, we use the term *theory* in an informal rather than a rigorous sense. A ToM has no scientific status as does, for example, the theory of relativity or the theory of evolution. It is, rather, a theory in the popular sense of having an idea or a concept of something or an explanation of why something is the way it is: we may have a theory about why a friend is currently angry, a theory about what motivates our department head, or a theory about why so many English professors fail to acknowledge *Don Quixote* as the greatest novel of all time! Folk psychology includes much more than a ToM and, largely because of the often pernicious effect of some of these other common characteristics—such as the belief in an autonomous self, mind–body dualism, or the Cartesian theater of the mind—it is frequently (and often snidely) dismissed by serious psychologists and philosophers (e.g., Stich 1983; Pylyshyn 1984; Radcliffe 2007). Although we have argued throughout this book against several aspects of folk psychology, one does not have to reject everything; as always, a simple binary approach is not useful. Human beings are embodied agents, we have complex states of mind, and we understand others in terms of agency and intentionality.

We opened Chapter 5 with these words: "Human beings are animals." True, but, to be more specific and accurate, we should recognize with Sanjida O'Connell that we are "social animals" who use ToM to inform our relationships and predict behavior (1997, 3). Our predictions of what others are thinking or feeling, or our understandings of their desires, hopes, wishes, motives, needs, state of knowledge, or understandings, are not always correct. A ToM does not give us access to "the truth," but is a social construction in the sense that it is a function of our relationships with other people, an emergent product of our biology and our culture, as are all our understandings of the world. Literature, and perhaps especially fiction, depends on—according to Alan Palmer (2004, 244), it "embodies"—folk psychology.

A term often used by philosophers to describe mental states is *intentionality*, defined by John R. Searle as "that property of many mental states and events by which they are directed at or about or of objects and states of affairs in the world" (1983, 1). As elaborated by philosopher Daniel Dennett (1987, 1988), intentionality is the idea that all thought is about something, that we cannot think in the abstract or think of nothing, but must always be thinking of some specific thing. Intentionality can be expressed at various levels. I think that I can do it. That is one level of intentionality. You believe that I think I can do it. That is two levels. He wonders if you believe that I think I can do it. That is three. And so on. How many levels can an animal keep track of? Evidence suggests that a chimpanzee can, at best, barely get beyond the first level (Dunbar 2004, 57). Humans cannot normally get past about five levels. Dennett illustrates the concept of intentionality thus: "I suspect [1] that you wonder [2] whether I realize [3] how hard it is for you to be sure [4] that you understand [5] whether I mean to be saying [6] that you can recognize [7] that I can believe [8] you want me to explain [9] that most of us can keep track of only about five or six orders [of intentionality], under the best of circumstances" (1988, 185–86). Such a sentence, with nine levels of intentionality, would confuse most people most of the time, but it makes Dennett's point. We take for granted the ability to track multiple levels of intentionality, which we do constantly, and thus fail to recognize just how rare and wonderful it is. A ToM is "miraculous," in the words of V. S. Ramachandran: "Our species' highly sophisticated Theory of Mind is one of the most unique and powerful faculties of the human brain" (2011, 138).

2 ToM and Autism

Autism was first labeled in the 1940s when Leo Kanner and Hans Asperger, working independently at about the same time, both described versions of the syndrome and both called it "autism," from the Greek, based on *autos*, or self. Before that time, autistics were often labeled morons, imbeciles, idiots, or other atrocious popular terms. Today, milder forms of autism are usually called *Asperger syndrome* (or disorder), or simply Asperger's, and because autism can exist over a range of forms, the term *autism spectrum disorders* is often used. While the causes of autism are still being investigated, it is becoming increasingly clear that those on the autism spectrum are often well aware of and able to discuss their own feelings and thought processes, even when they have varying degrees of difficulty doing the same in regard to others' thoughts.

Given this difficulty in reading and understanding the minds of others, what can be said about autistics and ToM? First we must address what should not be said. Although the well-known psychologist Simon Baron-Cohen may have made some substantial scientific contributions in his studies of autistic people, his problematic notion of "mindblindness" (promoted in his 1995 book of the same name) has been dismissed in the meantime and must no longer be used. Autistic people, while often having great trouble with social interactions due perhaps to a ToM that functions differently or not as well as that of neurotypical people, cannot be assumed to possess no ToM at all (and see above comments regarding how even the neurotypical are often not stellar at ascertaining the intentions and feelings of others, and get into their share of social misunderstandings because of it). More commonly acceptable today in autism studies is to speak of "deficits" or "challenges" in using ToM, but not of a complete lack thereof.[1] Regardless of what the research ultimately indicates regarding ToM and autism, any terminology or assertions that imply autistic people are any less than human must be avoided at all costs.[2]

Temple Grandin, a highly successful professor of animal science and innovative entrepreneur, has written about her experiences as a person on the autism spectrum, providing us with some insights regarding differences between her ToM and that of someone considered to be neurotypical. As quoted by neurologist Oliver Sacks, Grandin explains that she can comprehend "simple, strong, universal" feelings or emotions, but not complex ones, or the sort of games that people play: "Much of the time … I feel like an anthropologist on Mars" (Sacks 1995, 259). Grandin does not understand the intentions behind figurative language, metaphor, irony, allusions, jokes, and so forth, but has less trouble with the more conventional and precise language of science. She says she does not understand what literary characters are thinking or what their motives and intentions are. Grandin compares her mind to a computer's CD-ROM, with rapid access to a large number of videotapes, which she plays through to recall how people are supposed to act in specific situations. Most of Grandin's thoughts are visual, rather than linguistic, and she understands the sub-symbolic cognition of animals in a personal way (Grandin and Johnson 2004). Much of her professional success is based on this visual cognitive ability, as she has designed slaughterhouses, feedlots, corrals, and other aspects of farm layout. The fact that she can do so because she thinks about the living situations from the animals' points of view is evidence that she certainly possesses a ToM, only different in its manifestation from the majority.

3 Evolutionary History and Theories of ToM

How, when, and why would an ability to think about the thoughts of other people come about in evolutionary history? Michael Tomasello has proposed that there has really only been one significant adaptation in the evolution of human beings, which, because of its overarching importance and inclusive ability to be manifest in numerous ways, forms the basis for all other human cognitive skills: "understanding others as intentional (or mental) agents (like the self)" (1999, 15). Tomasello proposes that this intentional understanding of others as agents makes language, tool manufacture and use, and other cognitive skills possible. While most authorities on human evolution would not go so far as subordinating everything to ToM, the ability to have ideas about what others are thinking, to anticipate their actions, and to influence what they do based on what you think they believe, is unquestionably one of the most powerful adaptations any species has ever developed.

The evolutionary advantage bestowed upon an animal's ability to understand something about the minds of conspecifics is obvious. If I see Lok pick up a rock and look menacingly at me, it very much increases my prospects of surviving and passing on my genes to the next generation if I understand that Lok probably wants to kill or maim me (perhaps because I think he may believe that I have secretly been mating with his partner). So, I take defensive action to eliminate Lok as a threat and continue my own survival and that of my descendants, who continue to develop such a skill. An increasing ability to understand the probable thoughts of others must have been one of the decisive steps in hominid evolution.

As the earliest hominids, the australopithecines, moved out of the forest and onto the African savannah, developed bipedalism, learned to control fire and use it to their advantage, modified their diet to include the high-protein source of meat, developed ever more complex social structures, began to make and use tools, and began to communicate mimetically, they must have also begun to develop an increasingly sophisticated ToM. The ability to communicate symbolically (i.e., linguistically), ToM, and the social structures and relationships uniquely characteristic of our species must all have developed together. By the time of early *Homo sapiens*, for example, the Cro-Magnons who made the beautiful cave paintings of Altamira and Lascaux some 15,000 years ago, or by the time agriculture was first practiced some 10,000 years ago, ToM was fully established as a defining characteristic of our species. In their book titled *Denial* (2013), Ajit Varki and Danny Brower have explored at length the evolution of ToM, its characteristics, how it makes possible such things as denial and self-denial, and other fascinating subjects.

There are four main streams in psychological and philosophical theorizing about theories of mind: the *innateness* theory, the *theory* theory, the *simulation* theory, and the *interaction* theory. The innateness (or nativist, or modularity) theory, whose major proponent is Alan Leslie (1987, 1992), is related to similar theories about infant cognition described in Sect. 2 of Chapter 9. It is assumed that human beings have, as a part of their genetic endowment, a structure and function of their mind-brain, a hard-wired module that gives them the ability to understand the thoughts and mind states of conspecifics. We find this explanation too insufficient to be adequately explanatory, at least when stated in its strong form. When invoked as part of a more complex biosocial understanding, then it is much more acceptable.

Those who prefer the theory theory compare, often very literally, the child with a scientist. Alison Gopnik and Andrew N. Meltzoff, in *Words, Thoughts, and Theories* (1997), for example, take the child-as-scientist as more than just a metaphor, making the case that "the processes of cognitive development in children are similar to, indeed perhaps even identical with, the processes of cognitive development in scientists" (3); see also Gopnik, Meltzoff and Kuhl (1999). It's not that children work consciously as do adult scientists by means of a hypothesis-test-revise algorithm, but that "the cognitive processes that underlie science are similar to, or indeed identical with, the cognitive processes that underlie much of cognitive development. It is not that children are little scientists but that scientists are big children" (32). The theory a child develops is a folk theory, not a logico-scientific theory; it is informal, nonconscious, flexible, often revised, and more pragmatic than abstract.

In contrast, an increasing number of others (e.g., Gordon 1986, 2001; Goldman 2006; Stueber 2006) prefer some version of the simulation theory. According to this view, we do not develop a (more or less) formal theory about what is going on in other people's minds, but we do imaginatively simulate in our own minds what we think must be taking place in the minds of others. The imagination is crucial to the simulation theory. Just as imagination and ToM facilitate the understanding of art in general, and narrative fiction in particular, imagination and ToM, in turn, may be developed to greater heights by the reading of fiction.

It is because we can infer others might (must, could) perceive situations similar to the way we know (think, feel) that we do ourselves, we can imagine being in their situation and understand their thoughts and actions. For Alvin Goldman, an important facet of the simulation ToM is what he calls "enactment imagination" (E-imagination), an act that involves "deliberate construction of a mental state with (quasi) visual character. The immediate output of the imaginative process is intended to resemble a *counterpart* state" (2006, 149). An act of E-imagination involves the creation of a mental surrogate of an act, a visualizing of something that might be done by another person or of something that you might do yourself.

The interaction approach is argued most strongly by Shaun Gallagher (2005), Matthew Radcliffe (2007), and Daniel Hutto (2008). This most radical position tends to reject the whole concept of any sort of "theory" of mind to argue that our nonconscious, embodied, sociocultural knowledge is all that is required. Gallagher bases much of his argument on the foundational work in development psychology by Colwyn Trevarthen on the themes of primary and secondary intersubjectivity, which develop literally from birth. Such intersubjectivities, Gallagher submits, include "imitation, intentionality detection, eye-tracking, the perception of meaning and emotion in movement and posture, and the understanding of intentional or goal-related movements in pragmatic contexts" (2005, 230) and involve nothing resembling a theory. Thus, Gallagher argues that "[t]he understanding of the other person is *primarily* neither theoretical nor based on an internal simulation. It is a form of embodied practice" (208). Likewise, Radcliffe argues that knowing what someone might do in many circumstances is not a matter of folk psychology but is merely "an understanding of public norms of conduct" and that such understanding in social contexts "is inextricable from *interaction* with the social world" (2007, 74, 86). Hutto, meanwhile, introduces the idea that most of our nonconscious social knowledge comes from our "encounters with stories about people who act for reasons" (5), which he calls the "Narrative Practice Hypothesis (NPH)."

Of these four approaches, the innateness theory is the most mechanistic and deterministic, although, as indicated earlier, a soft version of the innateness theory is quite consistent with the view of cognition for which we argue throughout this book. The simulation theory is the most contextual and humanistic, while the theory occupies something of a middle ground. Surely Gallagher and Radcliffe are correct that basic embodiment underlies all cognition, and surely Hutto is correct that the narratives we hear and read throughout our lives provide models for understanding and action. But this does not mean that these fundamental aspects of embodied cognition function in place of theory and simulation; rather, they underlie and inform them. We are disturbed that Gallagher misrepresents the theory and simulation theorists as taking a stance characterized by Cartesian mind–body dualism: "Both theory theory and simulation theory conceive of communicative interaction between two people as a process that takes place between two Cartesian minds" (211). That simply is not the case. Similarly, Radcliffe is correct in affirming that social behavior is embodied and contextualized activity (the point of this book's argument), but falls into the binary trap by asserting (incorrectly) that folk psychology assumes detached mental states rather than embodied cognition (2007, 107–20). He further misrepresents the thinking behind folk psychology and ToM as "mechanistic" (184, 230, 234), when it is exactly the opposite. The Gallagher et al. interaction theory is less a wholesale replacement for standard theories of mind than an important refinement of and supplement to them.

The four approaches are not mutually exclusive, but each stresses an aspect of what must be involved in understanding the minds of others. Clearly, we could have no ToM without the biological equipment, unique to our species, that makes it possible. But we could not have a ToM if we did not learn from embodied experience, generalize, and make practical assumptions and inferences—construct our theory—within our sociocultural milieu, in interaction with other people and the world. And it seems self-evident that we do indeed often understand others in light of what we know of ourselves, simulating the mind states of others. The reality of sympathy and empathy on one hand, and that of esthetic, creative, and imaginative literary experience on the other, both argue strongly for the simulation approach.

Vittorio Gallese and Goldman (1998) suggest that the basis for the simulation theory is to be found in the function of *mirror neurons.* Working with macaque monkeys, Gallese et al. (1996) found that when one individual saw another do something, such a reaching for a piece of food, some of the same neurons in its brain were activated as were involved in the other monkey's actions. That is to say, there seems to be among these monkeys a neural basis for understanding the actions of others. Rizzolatti and Arbib (1998) went on to suggest that such mirror systems might be what provides a bridge between performance and communication. Subsequently, it was confirmed that humans have similar mirror neurons that function in the same way (Rizzolatti et al. 2002).

If we vicariously experience what another is experiencing we are in effect simulating that action. Mirror neurons have become one of the hottest topics in recent neuroscience and cognitive psychology (see e.g., Iacoboni 2008; Ramachandran 2011; Ferrari and Rizzolatti 2015; for a different perspective, on the "myth of mirror neurons," see Hickok 2014, in which he argues that the entire concept of mirror neurons is based on foundationally weak assumptions). Alvin I. Goldman worked with Gallese and incorporated a useful summary of mirror neuron research in his book *Simulating Minds: The Philosophy, Psychology, and Neuroscience of Mindreading* (2006, 132–40, 205–11). Damasio (2003, 115–18) proposes that his earlier concept of "as-if body loops" draws on a variant of mirror neurons. Damasio and Damasio reconsider the issue and conclude that they "regard the remarkable evidence for mirror neurons as support for the 'as-if body loop' mechanism of placing the body in the mind. In turn, we consider the 'as-if body loop' system within each organism as the precursor to the mirror-neuron system" (2006, 20). Gallagher (2005, 77) concludes that newborn babies' imitative abilities imply that mirror neurons are already functioning at birth. Today, neuroscientists are still seeking to figure out mirror neurons and their function in social interactions, and again it comes back to monkeys. A relatively recent study published in *Science* indicates, through brain imaging of monkey-to-monkey interaction, a specific part of the primate brain dedicated to social interaction, suggesting "that this function is an evolutionary forerunner of human mind-reading capabilities" (Sliwa and Freiwald 2017). Scientists continue to study both monkeys' and humans' mirror neuron systems (MNS) with various research questions, for example, Errante and Fogassi (2021), who are concerned with lateralization (see our Sect. 2 in Chapter 6) and the roles of the brain's hemispheres in performing certain tasks and understanding others' actions.

What is sometimes lacking in the discussion of ToM, so dominated as it is by scientific (or scientific-sounding) emphases on innateness, brain modules, scientific theories, and simulation is an explicit recognition of the cultural context within which any ToM must develop. As Jerome Bruner reminds us: "To infer the mental state of another requires more than a Theory of Mind: it also requires a theory of culture" (1996, 113). Katherine Nelson (1996) and Leslie Brothers (1997), among others, also argue for the need to call explicit attention to the cultural context within which the social construction of a ToM must take place. It is not that most ToM proponents, of whatever stripe (except perhaps the strongest of innateness theorists), would argue that our mindreading abilities are strictly a biological or mechanistic process and that the cultural context is totally irrelevant. Rather, it is a matter of emphasis, and thus Bruner and others are correct to stress that our better understanding of ToM, as well as its acceptance by humanists and the non-scientific-oriented public in general, would be made possible by a more explicit recognition of the culturally embedded nature of all psychology. A ToM is, like everything else involving the human mind-brain, a facet of our embodied reality, an emergent product of the constant and complex organism-environment cooperative processes.

4 ToM's Roles in Language and Literature

The pragmatics of language use necessarily involves a Theory of Mind. Without a sense of how others will understand what we say in any specific context, we would be unable to lie. We would have no idea of how to shape our discourse rhetorically to persuade or convince someone else. We could not tell a joke, play with words, imply with silence, or speak ironically or satirically. To communicate linguistically is to be an embodied human agent who uses a cognitive tool in a social context to express thoughts to other embodied human agents. Bakhtin implicitly recognized all this in his writings about dialogism, heteroglossia, answerability, context, and related concepts.

William F. Allman lays out the basic premise that ToM is as essential for understanding a literary work as it is for understanding a verbal utterance or another human being's motives:

> Without our ability to form a Theory of Mind, human culture would not be possible. Much of the world of literature, drama, and humor relies on the supreme ability of humans not only to create theories about each character's mind but also to imagine simultaneously how each of these imaginary minds might view the minds of other characters. The tragic nature of Shakespeare's *Romeo and Juliet*, for instance, comes from a series of misconceptions among the characters that only the audience is aware of. Romeo's suicide is the result of his thinking that Juliet has died, and the audience is aware that if Romeo knew what they knew, this suicide would not have to have happened. To an audience of monkeys, however, Romeo's actions would make no sense, because they wouldn't be able to distinguish between their own beliefs and his. (1994, 68–69)

Robin Dunbar, whose work posits a theory of the relationship between grooming and language (discussed e.g., in our Sect. 5 of Chapter 5), attributes to the human ToM the "crucial ability to step back from ourselves and look at the rest of the world with an element of disinterest" (1996, 101). Just as Dennett outlined levels or orders of intentionality, Dunbar describes three orders of Theory of Mind:

1. the ability to be aware of our own thoughts;
2. the ability to understand other people's feelings; and
3. the ability "to imagine how someone who does not actually exist might respond to particular situations." (101–02)

Elaborating on this third-order ToM, Dunbar states its obvious implication: "we can begin to create literature, to write stories that go beyond a simple description of events as they occurred to delve more and more deeply into why the hero should behave in the way he does, into the feelings that drive him ever onwards in his quest" (102).

Lisa Zunshine's *Why We Read Fiction: Theory of Mind and the Novel* (2006) was a pioneering extended study on the importance of Theory of Mind to the study of literature. At the outset, Zunshine makes clear that she believes that a Theory of Mind "makes literature as we know it possible" (10). Here, Zunshine is continuing a line of thought first expressed by Brian Boyd: "Because Theory of Mind is so essential to our fine-grained perception and prediction of the behavior of others, it is essential to storytelling" (2001, 207). Writing at the same time as Zunshine, Robin Dunbar agrees: "Without Theory of Mind—indeed, without the higher orders of Theory of Mind—literature and much of everyday social intercourse would be impossible" (2004, 121). In reading a literary work, as Palmer (2004), Keen (2007), and others argue, we imagine the characters as though they were real people and simulate their states of mind, just as we use our Theory of Mind to understand the author's intentions as the work was being created. As Richard J. Gerrig states in his important book *Experiencing Narrative Worlds* (1993, 2), the two metaphors readers most often use to describe their experiences of reading fictional narratives are (1) being transported to another place, and (2) performing the narrative (see also Holland 1988). It is as though we were there, actively involved in what is going on; we escape (in the dual sense of get away *from* and get away *to*) this world and inhabit vicariously another one (Radway 1991; Blackford 2004); we become lost in a book (Nell 1988). Reading consists of much more than decoding the signs of the text. Reading is, above all, an esthetic experience that then becomes part of who we are: "As a work of art, it [i.e., a work of literature] offers a special kind of experience. It is a mode of living. The poem, the play, the story is thus an extension, an amplification, of life itself. The reader's primary purpose is to add this kind of experience to the other kinds of desirable experience that life may offer" (Rosenblatt 1995, 264). Frank Smith agrees: "Reading is experience. Reading about a storm is not the same thing as being a storm, but both are experiences. We respond emotionally to both, and can learn from both" (1994, 62).

We experience the text; we imaginatively recreate it and live it; we participate in the sense of the world that it affords us: we create, or construct, meaning and understanding. As David Olson notes, "In a sense reading requires every reader to become an actor; the reader has to 'interpret' the lines he or she reads just as an actor has to 'interpret' the lines for an audience, by uttering the lines with the correct expression" (1994, 108–9). Reader-response scholars whose work is grounded in a sound understanding of the psychology of real readers (e.g., Rosenblatt 1994; Holland 1988), as well as the psychologists who have studied human reading (e.g., Gerrig 1993; Smith 1994) all discuss reading in terms of an inferential, constructive, and imaginative process—one made possible, we suggest, by a ToM. Our evolved mind-brains simulate scenarios and tell stories, and we live our lives in terms of an informal (and unwritten) narrative autobiography. We understand what happens in the world, as well as why it happens, in terms of narratives. In this, our embodied mind-brains are literary to their very core. No one has expressed this better than Mark Turner in his book *The Literary Mind* (previously cited in Sect. 6 of Chapter 8), where he discusses those

> principles of mind we mistakenly classify as "literary"—*story*, *projection*, and *parable*. We notice these principles so rarely in operation, when a literary style puts them on display, that we think of them as special and separate from everyday life. On the contrary, they make everyday life possible. The literary mind is not a separate kind of mind. It is our mind. The literary mind is the fundamental mind. Although cognitive science is associated with mechanical technologies like robots and computer instruments that seem unliterary, the central issues for cognitive science are in fact the issues of the literary mind. (1996, v)

Turner elaborates further on this principle: "Narrative imagining—story—is the fundamental instrument of thought. Rational capacities depend upon it. It is our chief means of looking into the future, of predicting, of planning, and of explaining. It is a literary capacity indispensable to human cognition generally" (4–5). Rather than something frivolous or decorative in human life, literature is at the core of humanity.

And so where does the line between the natural and the created, the physical and the virtual, the real and the literary lie? A theory of literature that includes as a central tenet a ToM is thoroughly justified by evolution, neuroscience, cognitive psychology, and esthetics. Such a theory of literature is one with the convergent trends of all contemporary physical and biological science, the social sciences, and the arts and humanities. Overall, the brilliant and original work in infant cognition in the past few decades has given us a new view of humanity and the human condition, vistas that are in every way consistent with the evolved, biological, embodied, and socially contextualized human animal described throughout this book. The idea of a ToM is a particularly powerful concept, one that implies a great deal for literary and performance critics and theorists.[3] Accepting the validity of this new view necessarily means abandoning some long-held assumptions and theories about development, and how people conceive of minds (both fictional and real). It provides the basis for our development of a theory of the human mind-brain in context. But before examining the embodied and contextualist approach to cognition, it is necessary to take a close look at the disembodied and decontextualized concept of cognition that has characterized work in computation and artificial intelligence: COGNITIVISM.

Notes

1. For experts' perspectives on the ToM-related challenges that people with autism may experience, see, for example, a post from 2016 from the Seattle Autism Blog: http://theautismblog.seattlechildrens.org/autism-theory-mind/.
2. Humanities scholars working in cognitive studies have fallen into this trap and have realized their error in the meantime. See, for example, pioneering cognitive literary scholar Lisa Zunshine's paper she gave at the Modern Language Association in January 2013, "Real Mindblindness; or, I Was Wrong," in which she argued that the term "mindblindness" should no longer be used to describe those with autism.
3. For recent humanities and social science scholarship that uses ToM as an analytical focal point, see e.g., Bockting (2011), Leverage (2011), Mancing (2011, 2013, 2015) and William (2012b, 2017). For an empirical approach studying ToM and reading, see Kidd and Castano (2013).

References

Allman, William F. 1994. *The Stone Age Present: How Evolution Has Shaped Modern Life—From Sex, Violence, and Language to Emotions, Morals, and Communities*. New York: Simon and Schuster.

Astington, Janet Wilde. 1993. *The Child's Discovery of the Mind*. Cambridge: Harvard UP.

Avis, Jeremy, and Paul L. Harris. 1991. "Belief-Desire Reasoning among Baka Children: Evidence for a Universal Conception of Mind." *Child Development* 62.3: 460–67.

Bartsch, Karen, and Henry M. Wellman. 1995. *Children Talk About the Mind*. New York: Oxford UP.

Blackford, Holly Virginia. 2004. *Out of This World: Why Literature Matters to Girls*. New York: Teachers College P.

Bockting, Ineke. 2011. "The Importance of Deixis and Attributive Style for the Study of Theory of Mind: The Example of William Faulkner's Disturbed Characters." In Leverage et al. 175–86.

Boyd, Brian. 2001. "The Origin of Stories: *Horton Hears a Who*." *Philosophy and Literature* 25: 197–214.

Brothers, Leslie. 1997. *Friday's Footprint: How Society Shapes the Human Mind*. New York: Oxford UP.

Bruner, Jerome. 1996. *The Culture of Education*. Cambridge: Harvard UP.

Byrne, Richard, and Andrew Whiten, eds. 1988. *Machiavellian Intelligence: Social Expertise and the Evolution of Intellect in Monkeys, Apes, and Humans*. Oxford: Clarendon P.

Cheney, Dorothy L., and Robert M. Seyfarth. 1990. *How Monkeys See the World: Inside the Mind of Another Species*. Chicago: U of Chicago P.

Damasio, Antonio R. 2003. *Looking for Spinoza: Joy, Sorrow, and the Feeling Brain*. Orlando: Harcourt.

———, and Hanna Damasio. 2006. "Minding the Body." *Daedalus* 135.3: 15–22.

Davies, Martin, and Tony Stone, eds. 1995. *Folk Psychology: The Theory of Mind Debate*. Oxford: Blackwell.

Dennett, Daniel C. 1987. *The Intentional Stance*. Cambridge: MIT P.

———. 1988. "The Intentional Stance in Theory and Practice." In Byrne and Whiten, 180–202.

Dunbar, Robin. 1996. *Grooming, Gossip, and the Evolution of Language*. Cambridge: Harvard UP.

———. 2004. *The Human Story: A New History of Mankind's Evolution*. London: Faber and Faber.

Errante, Antonino, and Leonardo Fogassi. 2021. "Functional Lateralization of the Mirror Neuron System in Monkey and Humans." *Symmetry* 13.1: 77. https://doi.org/10.3390/sym13010077

Ferrari, Pier Francesco, and Giacomo Rizzolatti. 2015. *New Frontiers in Mirror Neurons Research*. Oxford: Oxford UP.

Frenay, Robert. 2006. *Pulse: The Coming Age of Systems and Machines Inspired by Living Things*. New York: Farrar, Straus and Giroux.

Gallagher, Shaun. 2005. *How the Body Shapes the Mind*. Oxford: Clarendon P.

Gallese, Vittorio, and Alvin Goldman. 1998. "Mirror Neurons and the Simulation Theory of Mind." *Trends in Cognitive Sciences* 12: 493–501.

———, et al. 1996. "Action Recognition in the Premotor Cortex." *Brain* 119.2: 593–609.

Gerrig, Richard J. 1993. *Experiencing Narrative Worlds: On the Psychological Activities of Reading*. New Haven: Yale UP.

Gibbs, Raymond W., Jr. 2006. *Embodiment and Cognitive Science*. Cambridge: Cambridge UP.

Gibson, Eleanor J., and Anne D. Pick. 2000. *An Ecological Approach to Perceptual Learning and Development*. Oxford: Oxford UP.

Goldman, Alvin I. 2006. *Simulating Minds: The Philosophy, Psychology, and Neuroscience of Mindreading*. Oxford: Oxford UP.

Gopnik, Alison. 2009. *The Philosophical Baby: What Children's Minds Tell Us About Truth, Live, and the Meaning of Life*. New York: Farrar, Straus and Giroux.

———, and Andrew N. Meltzoff. 1997. *Words, Thoughts, and Theories*. Cambridge: MIT P.

———, Andrew N. Meltzoff, and Patricia K. Kuhl. 1999. *The Scientist in the Crib: Minds, Brains and How Children Learn*. New York: William Morrow.

Gordon, Robert M. 1986. "Folk Psychology as Simulation." *Mind and Language* 1: 158–71.

———. 2001. "Simulation and Reason Explanation: The Radical View." *Philosophical Topics* 29: 175–92.

Grandin, Temple, and Catherine Johnson. 2004. *Animals in Translation: Using the Mysteries of Autism to Decode Animal Behavior*. New York: Scribner.

Hayashi, Taketsugu, et al. 2020. "Macaques Exhibit Implicit Gaze Bias Anticipating Others' False-Belief-Driven Actions via Medial Prefrontal Cortex." *Cell Reports* 30.13 (31 March): 4433–44.

Hickok, Gregory. 2014. *The Myth of Mirror Neurons: The Real Neuroscience of Communication and Cognition*. New York: W.W. Norton.

Holland, Norman N. 1988. *The Brain of Robert Frost: A Cognitive Approach to Literature*. New York: Routledge.

Humphrey, Nicholas K. 1976. "The Social Function of Intellect." In *Growing Points in Ethology: Based on a Conference Sponsored by St. John's College and King's College, Cambridge*. Ed. P. P. G. Bateson and Robert A. Hinde. Cambridge: Cambridge UP. 303–17.

Hutto, Daniel D. 2008. *Folk Psychological Narratives: The Sociocultural Basis of Understanding Reasons*. Cambridge: Bradford Books/MIT P.

Iacoboni, Marco. 2008. *Mirroring People: The New Science of How We Connect with Others*. New York: Ferrar, Straus and Giroux.

Johnson, Steven. 2004. *Mind Wide Open: Your Brain and the Neuroscience of Everyday Life*. New York: Scribner.

Keen, Suzanne. 2007. *Empathy and the Novel*. Oxford: Oxford UP.

Keenan, Julian Paul. 2003. *The Face in the Mirror: How We Know Who We Are*. New York: Ecco.

Kidd, David Comer, and Emanuele Castano. 2013. "Reading Literary Fiction Improves Theory of Mind." *Science* 342: 377–80.

Leslie, Alan M. 1987. "Pretence and Representation: The Origins of 'Theory of Mind'." *Psychological Review* 94: 412–26.

———. 1992. "Pretense, Autism and the Theory-of-Mind Module." *Current Directions in Psychological Science* 1: 18–21.

Leverage, Paula. "Is Perceval Autistic? Theory of Mind in the Conte del Graal." In Leverage et al. 133–52.

———, and Howard Mancing, Richard Schweickert, and Jennifer Marston William, eds. 2011. *Theory of Mind and Literature*. West Lafayette: Purdue UP.

Mancing, Howard. 2011. "Sancho Panza's Theory of Mind." In Leverage et al. 123–32.

———. 2013. "The Mind of a Pícaro: Lázaro de Tormes." In *Cognition, Literature, and History*. Ed. Mark J. Bruhn and Donald R. Wehrs. New York: Routledge. 262–88.

———. 2015. "Applying Theory of Mind to *Don Quixote*." In *Approaches to Teaching Cervantes's "Don Quixote*." 2nd ed. Ed. James A. Parr and Lisa Vollendorf. New York: Modern Language Association of America. 147–52.

Nell, Victor. 1988. *Lost in a Book: The Psychology of Reading for Pleasure*. New Haven: Yale UP.

Nelson, Katherine. 1996. *Language in Cognitive Development: Emergence of the Mediated Mind*. Cambridge: Cambridge UP.

O'Connell, Sanjida. 1997. *Mindreading: An Investigation into How We Learn to Love and Lie*. New York: Doubleday.

Olson, David R. 1994. *The World on Paper: The Conceptual and Cognitive Implications of Writing and Reading*. Cambridge: Cambridge UP.

Palmer, Alan. 2004. *Fictional Minds*. Lincoln: U of Nebraska P.

Premack, David, and Guy Woodruff. 1978. "Does the Chimpanzee Have a Theory of Mind?" *Behavioral and Brain Sciences* 1: 515–26.

Pylyshyn, Zenon W. 1984. *Computation and Cognition: Toward a Foundation for Cognitive Science*. Cambridge: MIT P.

Radcliffe, Matthew. 2007. *Rethinking Commonsense Psychology: A Critique of Folk Psychology, Theory of Mind and Simulation*. Basingstoke: Palgrave Macmillan.

Radway, Janice A. 1991 (1984). *Reading the Romance: Women, Patriarchy, and Popular Literature*. Chapel Hill: U of North Carolina P.

Ramachandran, V.S. 2011. *The Tell-Tale Brain: A Neuroscientist's Quest for What Makes Us Human*. New York: W. W. Norton.

Rizzolatti, Giacomo, and Michael A. Arbib. 1998. "Language within Our Grasp." *Trends in Neuroscience* 21: 188–94.

———, Laila Craighero, and Luciano Fadiga. 2002. "The Mirror System in Humans." In *Mirror Neurons and the Evolution of Brain and Language*. Ed. Maxim I. Stamenov and Vittorio Gallese. Amsterdam: John Benjamins. 37–59.

Rosenblatt, Louise M. 1994 (1978). *The Reader, the Text, the Poem: The Transactional Theory of the Literary Work*. Carbondale: Southern Illinois UP.

———. 1995 (1938). *Literature as Exploration*. New York: Modern Language Association.

Russell, James. 1996. *Agency: Its Role in Mental Development*. Hove: Erlbaum (UK) Taylor and Francis.

Sacks, Oliver. 1995. *An Anthropologist on Mars: Seven Paradoxical Tales*. New York: Alfred A. Knopf.

Searle, John R. 1983. *Intentionality: An Essay in the Philosophy of Mind*. Cambridge: Cambridge UP.

Sliwa, Julia, and Winrich Freiwald. 2017. "A Dedicated Network for Social Interaction Processing in the Primate Brain." *Science* 356.6339 (19 May): 745–49.

Smith, Frank. 1994. *Understanding Reading: A Psycholinguistic Analysis of Reading and Learning to Read*. Hillsdale: Lawrence Erlbaum Associates.

Sperber, Dan, ed. 2000. *Metarepresentations: A Multidisciplinary Perspective*. Oxford: Oxford UP.

———, and Deirdre Wilson. 1995 (1986). *Relevance: Communication and Cognition*. Cambridge: Harvard UP.

Sterelny, Kim. 2003. *Thought in a Hostile World: The Evolution of Human Cognition*. Oxford: Blackwell.

Stich, Stephen P. 1983. *From Folk Psychology to Cognitive Science: The Case Against Belief*. Cambridge: MIT P.

Stueber, Karsten. 2006. *Rediscovering Empathy: Agency, Folk Psychology, and the Human Sciences*. Cambridge: Bradford Books/MIT P.

Tomasello, Michael. 1999. *The Cultural Origins of Human Cognition*. Cambridge: Harvard UP.

Turner, Mark. 1996. *The Literary Mind*. New York: Oxford UP.

Varki, Ajit, and Danny Brower. 2013. *Denial: Self-Deception, False Beliefs, and the Origins of the Human Mind*. New York: Twelve.

Whiten, Andrew. 1999. "The Evolution of Deep Social Mind in Humans." In *The Descent of Mind: Psychological Perspectives on Hominid Evolution*. Ed. Michael C. Corballis and Stephen E. G. Lea. Oxford: Oxford UP. 173–93.

William, Jennifer Marston. 2012. "Against the Rhetoric of Sadness: Theory of Mind and the Writing Process in *The Curious Incident of the Dog in the Night-Time*." *PsyArt: A Journal for the Psychological Study of the Arts*, Summer 2012 (Online).

———. 2017. *Cognitive Approaches to German Historical Film: Seeing is Not Believing*. New York: Palgrave Macmillan.

Wimmer, Heinz, and Josef Perner. 1983. "Beliefs about Beliefs: Representation and Constraining Function of Wrong Beliefs in Young Children's Understanding of Deception." *Cognition* 13: 103–28.

Zunshine, Lisa. 2006. *Why We Read Fiction: Theory of Mind and the Novel.* Columbus: Ohio State UP.

PART IV

Context in Science and the Humanities

Context is one of the most central notions in cognition.

—Benny Shanon

There is no cognition independent of a social and normatively guided context.

—Marcus Jacobson

CHAPTER 11

Cognitivism

I don't think you can measure the function or even the existence of a computer without a cultural context for it.

—Jaron Lanier

Cognitive science is not a unitary effort, but one that includes a wide spectrum of approaches. The most accurate way to describe cognitive science is to trace a continuum between two very different poles. This chapter deals with the first of these poles, the one that provided the original impetus for much of what we now call cognitive science—computation and artificial intelligence (AI). Along with others, we use the term COGNITIVISM to describe this approach. In the next chapter, we will describe the second pole in cognitive science, which we term CONTEXTUALISM, and which posits embodied cognition. We assert that it is important for humanists to have a basic understanding of the concepts and theories behind COGNITIVISM because they are so ubiquitous in our contemporary society. Furthermore, many of them also underlie a great deal of theoretical assumptions in the humanities. For this reason, occasional references are made to certain linguistic and literary theories throughout this chapter and the next.

H. Mancing and J. Marston William, *Restoring the Human Context to Literary and Performance Studies*, Cognitive Studies in Literature and Performance,
https://doi.org/10.1007/978-3-030-89078-0_11

The point of departure in this chapter is the origins of the new discipline with emphasis on the computer model of the brain. Specifically, this chapter will in turn deal with: the cognitive revolution of the 1950s and the differences between cognitive science and behaviorism; the COGNITIVIST model of AI, where the human brain is conceived in the image of the computer; the nature of intelligence; robotics, the attempt to replicate human beings by non-biological means; and connectionism, the attempt to make computers in the image of the human brain.

1 Cognitive Science

Howard W. Gardner defines cognitive science—a discipline he first heard talk of in the 1970s—as "a contemporary, empirically based effort to answer long-standing epistemological questions—particularly those concerned with the nature of knowledge, its components, its sources, its development, and its deployment" (1985, 6). Somewhat more broadly, cognitive science can be considered as the study of all things related to human cognition, that is, to all possible kinds of thinking, knowing, feeling, and understanding. As such, cognitive science is not one of the traditional disciplines in the sense that mathematics, history, and philosophy have been for centuries. It is, rather, a vast inter- or multi-disciplinary field that only has a little more than a half-century of existence and in which several other disciplines cooperate in various models of convergent research and inquiry. But from its inception, and always at its core, has loomed the computer.

The standard history of the early decades of cognitive science is Gardner's *The Mind's New Science* (1985), in which he presents the discipline as an outgrowth of World War II work in *cybernetics* (the control relationship between human and machine; it typically involved feedback loops and on-site adjustments)—led by Norbert Wiener, who virtually defined the field—and the new enterprise of computer science. According to Gardner, research activities explicitly begin in the mid-1950s with emphasis on the following "key theoretical inputs": mathematics and computation, the neuronal model, the cybernetic synthesis, information theory, and neuropsychological syndromes (16–23).

More precisely, Gardner, like others, locates the "birthday" of the Cognitive Revolution on September 11, 1956. On that day at an MIT Symposium on Information Theory, there were three papers that, together, seemed to unite a set of new independent disciplinary orientations in such a way as to create something new. The first of these papers, by the computer scientists Allen Newell and Herbert Simon, titled "Logic Theory Machine," was the first example of the proof of a logical theorem worked out on a computer. The second was by a young and relatively unknown linguist named Noam Chomsky, whose "Three Models of Language" was the first public presentation of his revolutionary new approach to linguistics, one that would be published the next year as *Syntactic Structures* (1957). And the third paper was by psychologist George A. Miller, "The Magic Number Seven, Plus or Minus Two," on the capabilities of human short-term memory (1956). Miller and others who were there left the meeting excited and convinced that they had been present at the beginning of something new and important (Gardner 1985, 28–29). The convergence of computer science, linguistics, and psychology established the framework for a new approach to both machines and human beings, to what it means to be human, to the centrality of cognition in human affairs. The revolution that was initiated at that time had its roots in what Gardner describes as its foundational disciplines: psychology, linguistics, neuroscience, anthropology, artificial intelligence, and philosophy. It should be stressed that this new approach (and here the clear leader was Chomsky) was reacting to, criticizing, and attempting to replace the then-dominant paradigm of behaviorism.

Behaviorism was the major early-twentieth-century attempt in psychology to assume the mantle of an empirical science, on a par with the physical sciences. Led at various times by John B. Watson (see his 1925 book), Ivan Pavlov, B. F. Skinner, and others, this orientation was founded on two premises:

(1) that psychology must always adhere to the standards of empirical research, with replicable experimental results; and
(2) that the only thing that could be measured was behavior, and not any thought that may be behind it.

Basically, behaviorists reasoned somewhat like this:

(1) mental activities are private;
(2) private matters cannot be observed;

(3) what cannot be observed or measured cannot be studied scientifically; therefore
(4) cognition (mental activities, thoughts) cannot be the object of scientific inquiry.

Thus, behaviorists never spoke of mind, thought processes, imagination, emotion, feeling, intention, desire, interest, or any other non-observable and non-measurable quality. In the mechanistic worldview of behaviorism, the mind is conceived as a black box, impenetrable and outside the bounds of scientific inquiry. Strong versions of behaviorism went so far as to deny that there was such a thing as mind.

The attempt to penetrate the mind by assuming that it functions like a computer, in a way that is amenable to scientific observation, experimentation, and research, truly was a revolution in thinking. After Chomsky's (1959) demolition of Skinner's behaviorist approach to language in *Verbal Behavior* (1957), and the spectacular rise of computer science in the 1960s, behaviorism virtually collapsed and lost most of its legitimacy as a research paradigm. It must be noted, however, that if behaviorism died publicly, it has carried out a long afterlife as a *sub rosa* informing structure in much research and theory in the humanities and social sciences. Behaviorist assumptions have continued to inform much of the work in artificial intelligence, especially to the extent that this research effort claims empirical scientific status comparable to that of physics or chemistry.

2 Artificial Intelligence (AI)

Artificial intelligence has been defined as a research program that "seeks to produce, on a computer, a pattern of output that would be considered intelligent if displayed by human beings" (Gardner 1985, 140). AI is the research and development program according to which the computer stands as a model—and a way to understand the structure and function—of the human mind. AI is frequently divided into two general positions: weak and strong. Weak AI is merely the idea that computation may serve as an analog to, and show some truths about, the mind. It does not claim that the computer is a mind or that the mind is a computer. It recognizes that the computer model of the mind is an analogy, and a useful one, within a broad concept of cognition. It is hard to take exception with the weak version, as models often provide a useful, if limited, way of understanding something. As long as that which is being talked about is a certain basic-level similarity, and not a claim for identity, AI can only be a positive contribution to a more profound understanding of human cognition.

Artificial intelligence, and computation in general, is for many the only area in which cognitive science can claim legitimacy. Those who occupy this end of the spectrum tend, in a stance clearly reminiscent of behaviorism, to:

- assume a realist scientific model for research;
- deny the validity of anything not considered real science;
- rely on empirical evidence;
- assume a Cartesian mind-body dualism;
- take the sequential computer as a model for the human brain;
- align themselves with Anglo-American analytical philosophy;
- assume that language always has a literal meaning;
- equate cognition with information processing;
- de-emphasize emotion, feeling, context, culture, and history;
- lean toward functionalism, reductionism, or eliminative materialism; and
- assume that all thought consists of mental representations involving the manipulation of symbols.

This orientation is often called "strong AI," "COGNITIVISM," "Good Old-Fashioned Artificial Intelligence (GOFAI)" (a term whose origin is attributed to John Haugeland), or "High Church Computationalism" (coined by Daniel Dennett). The valuable products of this research orientation have been industrial robots and expert systems such as IBM's chess-playing Deep Blue and MYCIN, a system designed to diagnose bacterial infections. Exemplifying this tradition, a massive (more than 1500 pages long) two-volume history of cognitive science is titled *Mind as Machine* (Boden 2006) and takes the mind-as-computer model as its informing idea.[1]

Francisco Varela has drawn a "polar map" of cognitive science, consisting of a series of three concentric circles, within which he places the names of some of the most important researchers in the fields of cognitive psychology, neuroscience, AI, linguistics, and philosophy. Those closest to the center represent, for Varela, the hard-core computer-oriented, information-processing, approach that he calls COGNITIVISM (1992, 236–41; Varela et al. 1993, 37–57), the term we have adopted for this chapter (the small caps are ours, not Varela's). We will discuss Varela's next level, "emergence" (or connectionism) below and his third level, "enaction," in the next chapter.

The two figures that stand at the headwaters of twentieth-century computing are Alan Turing and John von Neumann. As early as 1936 Turing described an abstract mechanism, known after him as a "universal Turing machine," a device theoretically capable of carrying out any act of computation. The machine he described was never built, for it involved an infinitely long tape of ones and zeros that moved through a central mechanism capable of processing any task that could be expressed in binary code. Shortly before he died, in 1950, Turing proposed a test, which he called an "imitation game," for machine intelligence: if a machine could convince an observer that its answers had been written by a human being, it would be deemed to have achieved human intelligence. Turing confidently predicted a computer would pass this test within 50 years (i.e., by the year 2000). Now known as the *Turing test*, this idea is one of the subjects most discussed in computer science and AI.

Working in the 1940s, von Neumann began to design digital computers with an architecture (or structure) that has a number of essential characteristics:

- a central processing unit (CPU), where all the work is carried out;
- a separate memory unit (RAM, or random-access memory), to hold the data being worked on at a given time;
- disk storage, where the accumulated steps previously taken are stored for later retrieval;
- serial (one-thing-at-a-time) processing; and
- a digital structure (each step involves an either/or choice between two and only two options, coded as 1 or 0).

Von Neumann's format has become the one used in all digital and serial computers, which are often called von Neumann machines: all computers in the standard sense of the term—your desktop or laptop Mac or PC—has a von Neumann architecture.

Strong AI does not just conceive of the computer as a model or metaphor for the mind, it holds that the computer *is* a mind and that the mind *is* a computer. The only difference between a computer and a mind is the physical form in which it exists. In one case, it is a carbon-based biological entity that is the result of an evolutionary process. In the other, it is a silicone-based mechanical entity that is the result of manufacture. In both cases, you have a hardware-software distinction: the brain is the exact equivalent of the computer (hardware) and the mind is the exact equivalent of the program (software). The two may differ in certain details, but they both always work by processing information symbolically via serial moves, and they are functionally identical. This stance is, in fact, often called *functionalism*.

Zenon W. Pylyshyn became a prototypical representative of the strong AI position with his classic book *Computation and Cognition* (1984). Pylyshyn's basic claim is that "cognition *is* a type of computation" (xiii). The strong AI position assumes the validity of two long-held views in the Western tradition. The first is the Cartesian assumption of mind–body dualism, the idea that the body and the mind are two separate and distinct entities and that the mind functions in isolation from the body.

The second assumption, common at least since the age of Plato and Aristotle, is that that abstract reason and logical thought are the hallmarks of human cognition and human consciousness. For Pylyshyn, all thought becomes computation: "Thus the mind is depicted as continually engaged in rapid, largely unconscious searching, remembering and reasoning and generally in manipulating knowledge—that is 'cognizing'" (193). If cognition (mind) can be studied in isolation from the biological body, and if reason is the pinnacle of thought, then a machine that can reason is by definition the same thing as a mind.

Today, we do have computers much more powerful than Turing could have imagined, but as of yet computers do not regularly seem as human as human beings, not even in the simplistic manner described by the Turing test. At this writing, we have at-your-service voice-activated search engines on our phones and in our homes and cars—often with female human voices and names like Siri and Alexa—to give us driving directions, tell us the weather or time, and to provide us with seemingly infinite facts if we feel too lazy to google them. Yet as amazing and fun as these toys can be, they must be programmed to do these tasks and possess no creative, conscious minds of their own.

Perhaps it was largely due to the extravagant claims made for short-term success that helped insure enormous amounts of funding for any reasonably articulate research proposal in the field of AI. From the very beginning, strong AI received many millions of dollars in federal (especially Defense Department) and private research funding with the promise (never fulfilled) that machines capable of performing all human cognitive functions would be made. For an excellent introduction to the history and nature of AI research see Franklin (1995); also important is the collection of classic essays in Haugeland (1997).

Information is a special concept in the COGNITIVIST orientation where it has acquired a sense quite different from the more common one used outside the field of computation. The term is used in the way defined by mathematician Claude Shannon in his seminal 1938 master's thesis titled "A Symbolic Analysis of Relay and Switching Circuits" on the manipulation of informational signals. Information theory, as it developed in the wake of Shannon's thesis, is based on the notion that information is something separate from its content and is, rather, a matter of simple binary decisions. The basic unit of information in this sense is the *bit* (or *bi*nary uni*t*), and all that is needed to select from one of two possibilities. Everything can be encoded in such a way that it can be communicated by following a series of prescribed steps (an algorithm) in binary code. Information in this sense is a probability function that consists of context-free data, a pattern that can be encoded, transmitted, and decoded in such a way as to preserve the original data in exactly its original form. A perfect instantiation of this process is the telegraph: a set of words is translated into Morse Code, sent by means of telegraph wires, and decoded at the end back into the original words. The words on the sending end may be typed out and on the receiving end written longhand; that is, the material manifestation is different in each case, but they consist of the same information. It is difficult to overstate the importance of this concept of information for cognitive science, at least cognitive science of the COGNITIVIST variety. It ranks with Cartesian mind–body dualism and the assumption that thinking (and consciousness itself) is essentially a matter of rational thought as a fundamental tenet of strong AI. The strong AI line of thought goes something like this:

(1) the mind is separate from the body;
(2) the mind works rationally and logically;
(3) information is processed rationally and logically;
(4) information-processing is what the mind does;
(5) computer programs process information;
(6) computer programs are the same as mind; and therefore

(7) computers are the same as brains.

Mind–body dualism makes possible mind-information dualism, and the body becomes completely unnecessary. Body is mechanized, turned into a computer; brain is mechanized, turned into an information-processor. Again, Pylyshyn: "what the brain is doing is exactly what computers do when they compute numerical functions; namely, their behavior is caused by the physically instantiated properties of classes of substates that correspond to *symbolic codes*" (1984, 39).

Also essential to AI (and most other forms of cognitive science) is the concept of *representation*. This is the idea that to think, write, or talk about human cognitive functions, it is necessary to posit some level of activity that is separate from both biology and culture. That is, functions must be represented in the brain (or in the computer) on some level that is not explicitly limited to biological functions or sociocultural concepts. According to this approach, human cognitive activities must be considered in terms of symbols, images, schemas, or something comparable—Saussurean signs, perhaps. Pylyshyn (1999, 5) affirmed that one hypothesis "has become so deeply entrenched that it is simply taken for granted" within cognitive science and AI. This hypothesis is the following: What makes it possible for systems—computers or intelligent organisms—to behave in a way that is correctly characterized in terms of what they represent (say, beliefs and goals) is that the representations are *encoded* in a system of physically instantiated symbolic codes. The system behaves the way it does, through the unfolding of natural laws over the physical codes, because of the physical form these codes take on each occasion. So cognitive science *is* strong AI for Pylyshyn and many others.

One of the ironies of the early cognitive revolution, supposedly a major paradigm shift and an overthrow of the behaviorist agenda, is that—at least within the COGNITIVIST wing—behaviorism didn't really go away. Diane Gillespie (1992), whose important work will be discussed in the next chapter, has characterized the major tension between two orientations in cognitive science not so much in terms of COGNITIVISM vs. enaction, like Varela, but as *mechanism* vs. CONTEXTUALISM. The worldview that takes mechanism (the universe, the world, the body are all machines) as its guiding metaphor has been dominant in Western thought for centuries. As mentioned in earlier comments on behaviorism, a mechanistic assumption underlies all its premises: the image of the brain as black box, the mind–body separation and the rejection (or denial) of the former, the input–output/stimulus–response technique, the reductionism, the dependence on measurements, the presumed objectivism. The shift to cognitive science theoretically jettisoned all of this and opened things up again to discussions of thoughts, feelings, and intentions; consideration of mind, consciousness, and self. But did it, really?

Although mental matters were explicitly discussed in early cognitive science, it was almost always in terms of mechanical processes: encoding and decoding, information processing, representations, algorithms, and symbol manipulation. The mind was a machine—a computer. The unfathomable black box of behaviorism was opened up, but only to reveal electronic circuits and mechanical operations. No flesh, no blood; no feeling, no emotion; no body, no context. Cognition, and with it (human) intelligence, is reduced to a process, an abstraction, one governed by logical step-by-step procedures, cold and efficient—in short, machines all the way down. Scientists in various fields have taken issue with this approach; see for instance Oatley (2019), who points out deficiencies in psychological theories that equate embodied cognition to machine computation by outlining how much more complex the human brain is than a computer in its processes involving perception and representation. And as Amy Cook (2018) has pointed out, "When most cognitive scientists stopped thinking of the brain as a computer, it became easier for artists to think about science" (882). As we'll explore in later chapters, this shift has led to major developments in theater and performance studies as well as other humanities fields.

3 Intelligence

The mechanistic concept of representations is one of the most important factors in divorcing cognition from biology—the debiologization of cognition and intelligence—for if what is involved in thought is not specifically dependent on the messy work of neurons, synapses, and the ebb and flow of neurochemicals, but rather represented symbolically at another level, then in principle it does not matter whether this representation takes place in a biological organism or a mechanical device. Cartesian dualism is an essential tenet of strong AI. So are *reductionism* (reducing all cognition to symbolic representation) and *decontextualization* (removing information from any real-world context). It is convenient to make things simple to study them abstractly and theoretically; the problem comes in moving from theory to reality, from the disembodied to the embodied.

The so-called victory of IBM's Deep Blue chess-playing computer (capable of analyzing over 200 million moves per second; that is, some fifty billion moves for each three-minute period allowed by the rules) over world champion Garry Kasparov in 1997 is a perfect illustration of the pointlessness of attributing intelligence to a machine. Deep Blue did not defeat Kasparov in a chess match. Rather, a team of chess-playing engineers used Deep Blue's astonishing powers of calculation to beat Kasparov: human beings used a machine in a chess match with one unassisted man. Charles Jonscher's analysis of what happened during this famous match is perfectly accurate:

> A machine capable of analysing 200 million moves a second, developed by programmers with every advance in AI available to them, finally defeated a single man at what machines are best suited for: a game of logical combinations. So what? Short of the sort of ludicrously simple arithmetic task ("please add the following three numbers") that would have insulted the programmers engaged on the project, a more perfectly confined, constrained "digital" challenge pitting machine against person could hardly have been devised.
>
> And if, during one of the games in the New York match, the room had started filling with smoke from a raging fire, every adult and every child—even a bee with a pinprick-sided brain containing just 7,500 neurons—would have known to leave, but the computer would have gone on playing. Where in the room was the intelligence and where the dumbness? (1999, 144)

A machine, even the most powerful and sophisticated computer, is only a tool—like a hammer or a pencil, a microwave oven or an airplane—that knows nothing and does nothing. It is the intelligent person who *uses* the tool who *knows* and *does* things.

Alan Wolfe argues that the basis of human (biological) intelligence lies in the fact that human beings are "dumber" than computers, that we need not depend on having all the knowledge stored internally in our "programs," which gives our species the ability to get by with *not* knowing a great amount: "We can define the situation because the situation is not defined. We can construct meaning because the meaning is not known" (1993, 67). This allows us to deal with ambiguity and context: "Not knowing everything there is to know in advance, they [i.e., human beings] have to rely on social practices, the cues of others, experience, definitions of the situation, encounters, norms, and other ways of dealing with uncertainty that enable mind to develop"; thus, Wolfe continues, "[i]t is because humans have minds that we can speak of artificial intelligence, but almost never of artificial wisdom or judgment" (1993, 68).

In Sect. 6 of Chapter 8, we explored the idea of narrative as one of the most fundamental of all cognitive processes. The history of approaches to narrative within an AI context is best represented by Roger Schank and his associates, who undertook one of the most ambitious of all AI projects when they attempted to program computers to know how to act in certain prototypical situations: in a restaurant, at the barbershop, in a classroom. By writing complex programs to analyze and reproduce activities according to "scripts," following certain understood plans and goals, the hope was that a computer could do (not actually *do*, physically, but print out language about what would be done) what a human being would do in these typical situations (Schank and Ableson 1977). Ultimately, Schank and Ableson admit that there was "at present, no program that understands complex stories using all the mechanisms we have described here" (176).

With his book *Tell Me a Story* (1990), Schank finally came to realize the hopelessness of his task:

> The interesting subject underlying any discussion about the potential for artificial intelligence should be the nature of intelligence in general… . *Intelligence can only mean intelligent behavior.* Any entity capable of behavior could, at least in principle, be capable of intelligent behavior. The issue for anyone interested in intelligence, then, is what behaviors signify intelligence. (x–xi; emphasis added)

Schank concluded that until you can make a computer live, gain contextual and implicit knowledge over time, have experiences in the real world, and talk about it, you cannot have a computer that can be intelligent in human terms. Key to doing all this, Schank suggests convincingly, is the ability to understand and tell stories; real-world intelligence consists in having narrative, rather than logical, skills. Computers can't tell or understand original and truly humanly interesting narratives, but intelligent human beings can, and therein lies the enormous difference between them.

4 ROBOTS

The field of robotics illustrates more graphically than any other the impracticality of the COGNITIVIST dream. Universal robots—true mechanical beings—represent the ultimate phase in computation: supremely intelligent, superhumanly strong, freely moving and acting, superior to *Homo sapiens* in every way. *Homo mechanicus*, it is said, is the inevitable way of the future. If Turing and other early AI scientists and engineers were absurdly optimistic about how intelligent computers would be by the end of the twentieth century, their equivalents in the subfield of robotics took things to an even greater extreme. Most notoriously, roboticist and futurist Hans Moravec (1988, 1999) has repeatedly claimed that within fifty years it would be common to see robots with intelligence comparable to that of human beings and that within a century, AI computer systems would "mature into entities as complex as ourselves, and eventually into something transcending everything we know—in whom we can take pride when they refer to themselves as our descendants" (1). This was inevitable because early successes in information processing had already demonstrated the unlimited potential of strong AI, and once the genie was out of the bottle, it would be impossible to put it back in it again.

In his second book, Moravec admits that he was overly optimistic a decade earlier, but then makes the same sort of predictions, only a bit more extravagant, and adds: "But, this time *for sure*!" (1999, viii). Our "mind children" will succeed us as the planet's primary intelligent species before the end of the twenty-first century; *Homo sapiens* will suffer at the hands of our mind children the fate we may have inflicted on the Neanderthals some 30,000 years ago: extinction. We had better live fast (*carpe diem* in a very literal sense), for some of us alive today will, in Moravec's scenario, be no more than the detritus shoved aside by our glorious successors.

It would be nice to think of Moravec as an aberration, blind to the cognitive reality of embodied agents. But he is not alone. Inventor and futurist Ray Kurzweil, in *The Age of Spiritual Machines* (1999), matches Moravec's exuberant claims of what computers are about to do, claiming that by the end of the third decade of the twenty-first century-computers will be reading and understanding

> all of the world's literature—books, magazines, scientific journals, and other available material. Ultimately, the machines will gather knowledge on their own by venturing into the physical world, drawing from the full spectrum of media and information services, and sharing knowledge with each other (which machines can do far more easily than their human creators). (3)

Alas, we find ourselves in the third decade of the twenty-first century and there is nothing remotely like computers that can do what Kurzweil was confident would take place.

In the past few decades, Kurzweil has been predicting the coming of the "Singularity," which he defines as "a future period during which the pace of technological change will be so rapid, its impact so deep, that human life will be irreversibly transformed" (2005, 7). Based on the sort of thinking described above, Kurzweil predicted when this will occur: "I set the date for the Singularity—representing a profound and disruptive transformation in human capability—as 2045. The non-biological intelligence created in that year will be one billion times more powerful than all human intelligence today" (136). Kurzweil has maintained this prediction and has designated 2029 as the year that "an AI will pass a valid Turing test and therefore achieve human levels of intelligence" (Reedy 2017). We aren't holding our breath.

The elementary logic of modeling is useful here: you can build a simple model based on a complex reality (e.g., a tinker-toy model of a real skyscraper), but you cannot build a complex reality based on a simple model (e.g., a skyscraper based on the tinker-toy model). You cannot (not now and, some would argue, never) make a computer as complex as a human brain, but you can use the human brain as a model to make a (simpler) computer with some similar traits. It is a stunning reversal of the elementary principles of logic (one of the COGNITIVISTS' buzzwords) to say that since we can do something easy in a special set of circumstances, we can also do something extremely complex in a set of circumstances we cannot control. That simply does not compute.

Although the real game, as at least some members of the strong AI community have finally begun to learn, is not merely a simplistic one of processing information, but one of embodied, purposeful intelligence, it is worth our while looking briefly at the fondest dream of many COGNITIVISTS: the replacement of human beings by our "mind children," super-intelligent robots. The name *robot* comes from Karel

Čapek's 1920 play entitled *R.U.R.*, where the initials stand for "Rossum's Universal Robots." The idea of human-like machines is an old one that has always inspired flights of fantasy. Čapek's robots were creatures who are made increasingly more competent and who eventually rebel against their human makers and replace them—Hans Moravec's dream. It was in the 1950s, as Alan Turing was predicting intelligent machines by the end of the twentieth century, that the prodigy of science fiction, Isaac Asimov, published his enormously influential book composed of the robot stories he had written in the previous decade titled *I, Robot* (1950). Asimov brilliantly has his robots programmed with his "three laws of robotics":

(1) A robot may not injure a human being, or, through inaction, allow a human being to come to harm.
(2) A robot must obey the orders given it by human beings except where such orders would conflict with the First Law.
(3) A robot must protect its own existence as long as such protection does not conflict with the First or Second Law. (As stated in the *Handbook of Robotics*, 56th edition, 2058 A.D., and included at the beginning of all editions of *I, Robot*)

The stories are characteristically about robots that apparently violate one or more of these laws, but, upon closer inspection (most often by the great robopsychologist Susan Calvin), are shown to have conformed to all of them. Sequels and elaborations by Asimov and other writers have kept this vision of robots alive and well for decades. In 1976, as part of the U. S. bicentennial celebration, Asimov published "The Bicentennial Man," in which a robot overcomes anthropocentric prejudice and is conceded full status as a human being—only to achieve the one human quality denied his "race": death.

More recently, films have brought robot dreams even more graphically to the popular imagination. Robby the Robot, the clumsy but loyal and admirable figure from the 1956 film *Forbidden Planet*, is arguably the founding father of all the mechanical men of film. But R2D2 and CP3O from *Star Wars* (1977)—the Don Quixote and Sancho of robotdom—are the figures everyone recognizes. More recently the *cyborg* (a creature made of a hybrid of human being and machine) Terminator of Arnold Schwarzenegger (more technologically and conceptually sophisticated than his predecessors in the television series "The Six Million Dollar Man" and "The Bionic Woman") in the films *The Terminator* (1984), *Terminator 2: Judgment Day* (1991), and *Terminator 3: Rise of the Machines* (2003) has further blended the human and the machine. The remakes and additional sequels to earlier cyborg films, like *Robocop* (2014) and *Terminator Genisys* (2014), indicates the continuing popularity of this topic—but still as fiction![2]

Perhaps the most important contribution to AI lore in film is not exactly a robot, but rather a "conscious, computer-bodied intelligent agent" (Dennett 1997, 356): HAL, the computer who runs the ship on the mission to Jupiter at the end of the 1968 film *2001, A Space Odyssey*. HAL demonstrates more emotion than any of the ship's human crewmembers, becomes paranoid, and kills every member of the crew but one, who finally manages to disconnect him. Able to play chess, read lips, infer thoughts, and act independently and intelligently, HAL represents a supreme vision of COGNITIVISM. The fact that we are already well past the date when HAL was supposed to have been created (1997) and nothing even remotely like it has been constructed is yet another illustration of the way the dreams of science fiction writers and predictions of over-confident COGNITIVISTS have not been born out in real life. We are still waiting for our first robot miners, babysitters, housekeepers, and soldiers. It turns out that what is easiest for us to *do* is by far most difficult for us to *build*. There is no logical or computational procedure that our machines cannot do rapidly and accurately for us. What the roboticists have taught us above all is how very wonderful the blind process of evolution, functioning over a period of millions of years, really is. True intelligence consists of real world, embodied, contextual capabilities, not wires or microchips.

Foremost among the critics of Moravec, Kurzweil, and others is Rodney A. Brooks, director of the MIT Artificial Intelligence Laboratory and builder of some relatively sophisticated robots. Brooks has proposed a radically new bottom-up, rather than top-down, approach to robotics. He wants to construct robots capable of moving, seeing, communicating, and otherwise functioning in real-life environments—as they would if they had an organic body—and then add on the easy things of computation, logic, and so forth only at the end. The key concepts in his "behavior-based robotics" are situatedness, embodiment, intelligence, and emergence (2002, 416). So far, Brooks's "Creatures" (as he calls them) have begun to move about independently in the AI labs and offices at MIT (not quite the real world), learning for themselves as they go.

Brooks's conclusion is that "[w]ithout an ongoing participation in and perception of the world, there is no meaning for an agent—everything is empty symbols referring only to other symbols" (417). Literary theorists, as well as COGNITIVISTS, have much to learn from Rodney Brooks. For Brooks, "*Intelligence is determined by the dynamics of interaction with the world*" (418; emphasis in the original). *Pace* Derrida, one can only be intelligent in the very "hors texte"—context—he claims does not exist.

5 CONNECTIONISM

In Varela's (1992, 242–48; Varela et al. 1993, 85–103) polar map of cognitive science, the second circle of activity deals with those who have moved beyond the sequential computer model of the mind and into connectionism, or, as Varela calls it, "emergence." Here, the core metaphor is not that the mind is structured and functions like a (sequential) computer, but that the computer is structured and functions like a (self-organizing, parallel, distributed, massively connected) human brain. This represents a radical redirection of thinking, in which the emphasis shifts from the assumption that the human mind processes information like a machine, to the assumption that we must build computers that function more like the human mind. Rather than trying to conceive of biological cognition in machine terms, emergence entails modeling computation on a biological model. Meaning, in this paradigm, is not the result of logical calculations, but something that emerges in ways not readily defined and programmed, as a result of experience and in specific contexts. The idea, in effect, is that the machine "learns" from experience (Bechtel and Abrahamsen 1991, 2). Not dependent on the traditional COGNITIVIST symbol-processing mode, this architecture is sometimes referred to as *subsymbolic*. There must be some initial training (by a human programmer) but at a certain point the machine becomes its own judge of what is or is not correct. Early examples of connectionism in practice have often involved shape detection and identification. The potential of a computer in which several things are happening at the same time in different places (thus it is *parallel and distributed processing*, PDP), which does not depend on a pre-programmed algorithm, and which learns from experience is obviously great. The connectionist paradigm is derived from the example of the human brain and is not a simple one imposed on the brain. Furthermore, the process can resemble the emergent, chaotic, non-linear activities of complex systems in general, not just the brain. Thus a new optimism re-energizes AI.

One problem with connectionism is that it is awfully hard to build a machine with more than a few levels and a few hundred nodes; thousands are almost unthinkable. Since the human brain has many billions of such nodes, or synapses, it will likely be quite a while before connectionism comes anywhere near replicating the human brain. Interestingly, while the excitement about connectionism grows in certain areas, few actual connectionist machines are being constructed. Rather, connectionist models are *simulated*—on a standard serial computer! It is therefore premature to proclaim the accomplishments of this approach.

The main problem with AI is that what is easy and natural for human beings—coordinated physical movement, stereoscopic color vision, understanding language in context—has turned out to be hard for computers, and what is hard for human beings—logic, mathematical calculation, perfect memory—has turned out to be easy for computers. And this simple formulation of the matter is exactly the case. Ironically, then, the most important thing we have learned from COGNITIVISM is that evolution and biology far outpace computation and information processing. The next chapter presents the major alternative to COGNITIVISM–CONTEXTUALISM–in a discussion of some of the most interesting approaches to embodied cognition.

Notes

1. For a comprehensive, relatively recent review of views and scholarship in this area, see the *Stanford Encyclopedia of Philosophy*'s entry "The Computational Theory of Mind" (2015).
2. In addition, an intriguing blog details successes, challenges, and mishaps with neural networks "trained" to generate original ideas, such as new names for action figures or new designer paint colors (among which "Copper Panty," "Farty Red," or "Dorky Brown" probably won't be a hit anytime soon … cultural context is indeed a hard thing to teach). The blog replete with fun examples of "the sometimes hilarious, sometimes unsettling ways that machine learning algorithms get things wrong" is found here: aiweirdness.com.

References

Asimov, Isaac. 1991 (1950). *I, Robot*. New York: Bantam Books.

Bechtel, William, and Adele Abrahamsen. 1991. *Connectionism and the Mind: An Introduction to Parallel Processing in Networks*. Cambridge: Basil Blackwell.

Boden, Margaret A. 2006. *Mind as Machine: A History of Cognitive Science*. 2 vols. Oxford: Clarendon P.

Brooks, Rodney A. 2002. *Flesh and Machines: How Robots Will Change Us*. New York: Vintage Books.

Čapek, Karel. 2004 (1920). *R.U.R. (Rossum's Universal Robots)*. London: Penguin.

Chomsky, Noam. 1957. *Syntactic Structures*. The Hague: Mouton.

———. 1959. Review of *Verbal Behavior* by B. F. Skinner. *Language* 35: 26–58.

"The Computational Theory of Mind." 2015. *Stanford Encyclopedia of Philosophy*. Online.

Cook, Amy. 2018. "4E Cognition and the Humanities." In Newen et al. 875–90.

Dennett, Daniel C. 1997. "When HAL Kills, Who's to Blame? Computer Ethics." In *HAL's Legacy: 2001's Computer as Dream and Reality*. Ed. David G. Stork. Cambridge: MIT P. 351–65.

Franklin, Stan. 1995. *Artificial Minds*. Cambridge: Bradford Books/MIT P.

Gardner, Howard. 1985. *The Mind's New Science: A History of the Cognitive Revolution*. New York: Basic Books.

Gillespie, Diane. 1992. *The Mind's We: Contextualism in Cognitive Psychology*. Carbondale: Southern Illinois UP.

Haugeland, John, ed. 1997 (1981). *Mind Design II: Philosophy, Psychology, Artificial Intelligence*. Cambridge: Bradford Books/MIT P.

Jonscher, Charles. 1999. *The Evolution of Wired Life: From the Alphabet to the Soul-Catcher Chip—How Information Technologies Change Our World*. New York: Wiley.

Kurzweil, Ray. 1999. *The Age of Spiritual Machines: When Computers Exceed Human Intelligence*. New York: Viking.

———. 2005. *The Singularity Is Near: When Humans Transcend Biology*. New York: Viking.

Miller, George A. 1956. "The Magical Number Seven, Plus or Minus Two: Some Limits on Our Capacity for Processing Information." *Psychological Review* 63: 81–97.

Moravec, Hans. 1988. *Mind Children: The Future of Robot and Human Intelligence*. Cambridge: Harvard UP.

———. 1999. *Robot: Mere Machine to Transcendent Mind*. New York: Oxford UP.

Newen, Albert, Leon De Bruin, and Shaun Gallagher, eds. 2018. *The Oxford Handbook of 4E Cognition*. Oxford: Oxford UP.

Oatley, Keith. 2019 (1978). *Perceptions and Representations. The Theoretical Bases of Brain Research and Psychology*. New York: Routledge.

Pylyshyn, Zenon W. 1984. *Computation and Cognition: Toward a Foundation for Cognitive Science*. Cambridge: MIT P.

———. 1999. "What's in Your Mind?" In *What Is Cognitive Science?* Ed. Ernest Lepore and Zenon Pylyshyn. Oxford: Blackwell. 1–25.

Reedy, Christianna. 2017. "Kurzweil Claims That the Singularity Will Happen by 2045." *Futurism*, 5 October (Online).

Schank, Roger C. 1990. *Tell Me a Story: A New Look at Real and Artificial Memory*. New York: Charles Scribner's Sons.

———, and Robert P. Ableson. 1977. *Scripts, Plans, Goals and Understanding: An Inquiry into Human Knowledge Structures*. Hillsdale: Lawrence Erlbaum Associates.

Skinner, B. F. 1957. *Verbal Behavior*. New York: Appleton-Century Crofts.

Trefil, James. 1997. *Are We Unique? A Scientist Explores the Unparalleled Intelligence of the Human Mind*. New York: Wiley.

Varela, Francisco J. 1992. "Whence Perceptual Meaning? A Cartography of Current Ideas." In *Understanding Origins: Contemporary Views on the Origin of Life, Mind and Society*. Ed. Francisco J. Varela and Jean-Pierre Dupuy. Dordrecht: Kluwer Academic. 235–63.

———, Evan Thompson, and Eleanor Rosch. 1993. *The Embodied Mind: Cognitive Science and Human Experience*. Cambridge: MIT P.

Watson, John B. 1925. *Behaviorism*. New York: W. W. Norton.

Wolfe, Alan. 1993. *The Human Difference: Animals, Computers, and the Necessity of Social Science*. Berkeley: U of California P.

CHAPTER 12

Contextualism

In being open, contextualism often looks more chaotic, less disciplined, even (shudder) feminine.

—Diane Gillespie

Humanists sometimes recoil from any mention of cognitive *science*, perhaps because they either have no understanding of what it is all about, or they associate the enterprise with strong AI—COGNITIVISM as detailed in the previous chapter. The COGNITIVIST position is, obviously, not the one that we endorse. Instead, we are passionate about introducing more humanists to CONTEXTUALISM, an approach supported by the branches of science centering on what makes literature and the arts possible: the human body and mind. This chapter is divided into three sections: an overview of CONTEXTUALISM; a discussion of some major contextualist/interactive approaches to evolution and biology, including autopoiesis and systems theory; and brief presentations of some important contextualized approaches to psychology and philosophy.

H. Mancing and J. Marston William, *Restoring the Human Context to Literary and Performance Studies*, Cognitive Studies in Literature and Performance,
https://doi.org/10.1007/978-3-030-89078-0_12

1 The Contextual Metaphor

The most exciting work in cognitive science today is being carried out by those who are located toward the end of the disciplinary spectrum farthest from COGNITIVISM, the orientation we are calling CONTEXTUALISM. Those who occupy this end of the spectrum tend to:

- assume a constructivist model for research;
- accept a very broad definition of science;
- deny that empiricism is the only way to reliable knowledge;
- adopt a non-Cartesian stance toward mind–body dualism;
- claim that the human brain is radically unlike a sequential computer;
- align themselves with Continental existential, phenomenological, and hermeneutical—but *not* poststructural—philosophy;
- recognize that language is virtually always metaphorical or poetic, rather than literal;
- equate cognition with meaning-making;
- emphasize emotion, feeling, context, culture, and history;
- lean towards constructionism, enaction, and experientialism; and
- assume that thought is embodied, metaphorical, narrative, and inferential.

Thus, the full range of work within the cognitive sciences can be considered to fall at various points along the line represented by the two polar opposites. Much work may well be located more toward the center of the spectrum, but it seems more logical to conceive of any stance as basically leaning toward, if not fully in agreement with, one of the two poles.

We take the term CONTEXTUALISM from Diane Gillespie's *The Mind's We: Contextualism in Cognitive Psychology* (1992), still arguably the best single introduction to the field for humanist scholars and students. Drawing on the important (and too neglected) work of Stephen Pepper (1942), Gillespie identifies the major worldview that has informed psychology (and the social sciences in general) in our time: *mechanism*. Mechanism takes as its root metaphor the machine: "The machine metaphor entails parts that have specified, precisely defined locations. Each part can be expressed quantitatively, and the parts function according to laws which describe the machine's functioning. Truth resides in statements expressing cause and effect relationships between the parts" (Gillespie 1992, 16). No more obvious expression of mechanism exists than the COGNITIVIST assumption that the mind is like a computer, although a close second is the structuralist telementation or conduit model of linguistic communication. Indeed, structuralism—with its emphasis on individual pieces of the whole (phonemes, words, signs), its insistence on relationships (systems of differences) and causal processes (signifier-signified-signification), and its decontextualization—can be considered a paradigm case of mechanism.

In place of mechanism, Gillespie proposes the worldview Pepper calls CONTEXTUALISM. The root metaphor of CONTEXTUALISM, she notes, is the historical event:

> For the contextualist, experience consists of total events that are rich in features... . Because the event takes up the knower in the known, contextualism is an interactive, dynamic worldview. Moreover, nothing in the event is permanent or immutable because each particular changes with the flux of time. The contextualist focuses on the richness of experience and on shared meanings that arise out of interaction with others. Truth lies in the process of taking up the whole context of the event ... And so meaning is embodied in our experience of the world. (18)

In contrast to Gardner's (1985) definition of cognitive science, with its emphasis on empirical methods and its strong COGNITIVIST orientation, Gillespie defines the field as one which "poses fundamental questions about knowing and acting, about how we come to understand our experiences in the world" (1). Note the series of differences between this concept of cognition and one drawing on computation:

- knowing, not information processing;
- embodied acting or doing, not a radical mind–body dualism;
- understanding, not representing; and
- personal experiences in the real world, not disembodied computation.

Contextualized, embodied, cognition is as far from COGNITIVISM as one can get. It literally represents the opposite pole from COGNITIVISM on the spectrum of cognitive science. CONTEXTUALISM is always explicitly situated in the real world and in the actual activities of living organisms; as Pepper indicates (1942, 141), it is closely associated with pragmatism.

If COGNITIVISM ignores or removes context to isolate mind from body and body from environment, a contextualist approach does exactly the opposite. CONTEXTUALISM always maintains a sense of the past, both the historical past and the past of an individual's personal experiences. In Gillespie's terms, it "works to preserve the dynamic, integrative, and situated nature of past experience in cognition"; this "recognition of the multilayered nature of past experience in the present" leads the contextualist naturally toward a narrative epistemology: "Well-told stories contain the best traces we have of the creative process of cognition itself" (58).

CONTEXTUALISM is arguably the richest of all the frameworks for a cognitive science that can be used to inform a coherent and meaningful approach to literature, and thus it is the name we have chosen for the second defining pole on the cognitive science spectrum. The concept of CONTEXTUALISM characterizes (at least implicitly) nearly all the work we have most strongly endorsed throughout this book.

2 Evolution and Biology in Context

Philosopher Patricia S. Churchland is one of the more prominent scholars who, in the 1980s, perceived what she calls a shift from "the grand old paradigm" of a traditional representational epistemology to a "naturalized" (i.e., embodied, contextualized) approach to knowledge in an age when we understand more about the way the brain works than we did a mere two decades before—the "age of neuroscience":

> The fundamental epistemological question from Plato onward is this: How is it possible for us to represent reality? How is it that we can represent the external world of objects, of space and time, of motion and color? How do we represent our inner world of thoughts and desires, images and ideas, self and consciousness? Since it is, after all, the nervous system that achieves these things, the fundamental epistemological question can be reformulated thus: *How does the brain work?* Less cryptically and more accurately, the question is: *How, situated in its bodily configuration, within its surrounding physical environment, and within the social context it finds itself, does the brain work?* (1987, 546)

We continue investigating these questions today, more than three decades later, under the premise that human cognition—including everything that has to do with literature—takes place within the overall interrelated complex of mind-brain, body, and context, including both the physical surroundings and the social/historical situation. It is no longer sufficient to reduce everything to linguistic structures, the function of signs, abstract (debiologized) subjectivity, or endless deferral. We are thinking animals with bodies, and we function in, and in conjunction with, contexts.

In Sect. 1 of Chapter 5, we described briefly the standard contemporary neo-Darwinian synthesis and the Central Dogma in evolutionary theory. According to this, evolution is viewed largely in terms of genetics: DNA sequences, the human genome, inherited traits, adaptation, and genes. Organism and environment are usually seen as independent entities, and their relationship is sometimes described in terms of domination, or adaptation, rather than cooperation; dualisms—mind-body, to some extent, and body-environment, to a large extent—are the norm; mechanism is often the reigning metaphor (see Capra 1996). Ideally, the reality of evolution and biology could be reconciled with the contextualist paradigm. In fact, there have been some brilliant attempts to bridge the self-other divide and place evolution in a contextualist framework. Three of these projects are (1) the *autopoiesis* of Humberto Maturana and Francisco Varela; (2) the *constructivist interactionism* of Susan Oyama; and (3) Steven Rose's *lifelines*.

Pioneering proponents of our embodied reality were the neurobiologists Humberto R. Maturana and Francisco J. Varela. Following up on their earlier collaborative work (1980; originally published in Spanish in 1973), the two collaborated on *The Tree of Knowledge* (1992), an introduction to the biology of understanding built around the concept of autopoiesis. Maturana has described how he coined the term that became central to his and Varela's work:

> It was in these circumstances that one day, while talking with a friend (José Bulmes) about an essay of his on Don Quixote de la Mancha, in which he analyzed Don Quixote's dilemma of whether to follow the path of arms (*praxis*, action) or the path of letters (*poiesis*, creation, production), and his eventual choice of the path of *praxis* deferring any attempt at *poiesis*, I understood for the first time the power of the word "poiesis" and invented the word that we needed: *autopoiesis*. This was a word without a history, a word that could directly mean what takes place in the dynamics of the autonomy proper to living systems. (1980, xvii)

It is interesting that this CONTEXTUALIST biologist finds inspiration in literature, something that no bona fide COGNITIVIST would consider doing.

For Maturana and Varela, the paradigmatic model of an autopoietic system is the living cell, which constantly makes and remakes itself in conjunction with its surroundings. In contrast to most traditional work in biology, Maturana and Varela contextualize the organism within its environment. The result of this contextualization, they propose, is the need for the autopoietic (i.e., self-organizing, or self-making) organism to "bring forth" its cognitive world; that is, to create its own pragmatic understanding of its relation to external reality. Thus cognition becomes self-defining action: "*All doing is knowing, and all knowing is doing*" (1992, 26); "to live is to know (living is effective action in existence as a living being)" (174). As we will see below, these ideas echo throughout the recent work of numerous contemporary biologists, psychologists, and philosophers. Further, the similarity between this concept of biology and Bakhtin's contextualized, dialogic, emergent approach to language also becomes clear.

An animal, unlike a machine, is an "autopoietic system" in the sense that "it pulls itself up by its own bootstraps and becomes distinct from its environment through its own dynamics, in such a way that both things are inseparable" (46–47). This organism-environment inseparability comes about by means of a process Maturana and Varela call "structural coupling," the result of "recurrent interactions leading to the structural congruence between two (or more) systems" (75).[1] For human beings, everything we do is part of "a world brought forth in coexistence with other people" (239). Maturana and Varela insist—completely obviating the validity of any subject-object, mind-body, self-other, or nature-nurture dualism—that knowledge is "enactive," that "human cognition as effective action pertains to the biological domain, but it is always lived in a cultural tradition … for cognition is effective action; and as we know how we know, we bring forth ourselves" (244; see also Varela 1992).

Varela first used the term *enaction* in an essay published in 1987. He was inspired to take the term from a 1930 poem by Antonio Machado that includes the line "Al andar se hace camino" ("In walking you lay down a path") (63). Noteworthy is that both Maturana (with Don Quixote) and Varela (with Machado) found part of the inspiration for their biological theories in literature. Maturana and Varela conclude, again erasing the line between the biological and the social, as follows: "Whatever we do in every domain, whether concrete (walking) or abstract (philosophical reflection), involves us totally in the body, for it takes place through our structural dynamics and through our structural interactions. Everything we do is a structural dance in the choreography of coexistence" (1992, 248).

Maturana and Varela also stress that human activity (almost) always involves the use of language, and they coin the term *languaging* to describe the ongoing process of using language: "In other words, we are in language or, better, we 'language' only when through a reflexive action we make a linguistic distinction of a linguistic distinction. Therefore, to operate in language is to operate in a domain of congruent, co-ontogenic structural coupling" (210). Languaging is at the heart of human autopoiesis as "[w]e find ourselves in this co-ontogenic coupling, not as a preexisting reference nor in reference to an origin, as an ongoing transformation in the becoming of the linguistic world that we build with other human beings" (234–35).

Some poststructuralists have taken this concept of languaging to be virtually synonymous with the view that it is language that determines our subjectivity. Nothing could be farther from Maturana and Varela's belief. For them, language is an ongoing activity of contextualized, autopoietic, human biological systems—unique persons—who bring forth their worlds. In literary discourse, the socially constructed subject has nothing to do with the embodied agent considered here.

Shortly after completing *The Tree of Knowledge*, Varela teamed with philosopher Evan Thompson and psychologist Eleanor Rosch (the latter of whom, as we saw in Sect. 5 of Chapter 8, did groundbreaking research on categorization) to write an extraordinary book that blends Western cognitive science with Buddhist meditative psychology: *The Embodied Mind* (1993). The way the authors define two key terms from the Buddhist tradition illustrates their program. "Mindfulness" is the idea that "the mind is present in embodied everyday experience; mindfulness techniques are designed to lead the mind back from its theories and preoccupations, back from the abstract attitude, to the situation of one's experience itself" (22); its purpose is "to experience what one's mind is doing as it does it, to be present with one's mind" (23). "Embodied" means "reflection in which the body and mind have been brought together ... reflection is not just *on* experience, but *is* a form of experience itself" (27).

Varela, Thompson, and Rosch specifically reject any COGNITIVIST concept of knowledge as representation in the sense of computer-like manipulation of symbols that represent real-world features (8). More in line with a concept of cognition that has its roots in European phenomenology (especially Merleau-Ponty), they propose a "philosophical hermeneutics" in which "knowledge depends on being in a world that is inseparable from our bodies, our language, and our social history—in short, from our *embodiment*" (149). Thus, they maintain, we must "negotiate a middle path between the Scylla of cognition as the recovery of a pregiven outer world (realism) and the Charybdis of cognition as the project of a pregiven inner world (idealism)" (172). More precisely, they define *embodied action* as follows:

> By using the term *embodied* we mean to highlight two points: first, that cognition depends upon the kinds of experience that come from having a body with various sensorimotor capacities, and second, that these individual sensorimotor capacities are themselves embedded in a more encompassing biological, psychological, and cultural context. By using the term *action* we mean to emphasize once again that sensory and motor processes, perception and action, are fundamentally inseparable in lived cognition. Indeed, the two are not merely contingently linked in individuals; they have also evolved together. (172–73)

Clearly, our critique of nonbiological intelligence in Sect. 3 of Chapter 11 is reinforced by this definition of embodied action.

This concept of embodied action also has, as the authors emphasize, important implications for evolutionary science (185–214). Whereas the norm in most writing on evolution stresses the idea of adaptation, Varela, Thompson, and Rosch call attention to the conceptual implications of this term. The adaptationist, or neo-Darwinian stance, they argue, implies that an organism is separate from its environment, which forms an external reality, and must adapt to that reality. Rather, they insist, an enactive view of evolution constantly foregrounds the reality that organism and environment interact through the process of structural coupling, i.e., "living beings and their environments stand in relation to each other through *mutual specification* or *codetermination*" (198). According to Maturana and Varela, this distinction is more than a mere refinement of mainstream evolutionary thought, for although it appears to be little more than a shift in emphasis, it is a radically different way of conceiving of the process of natural selection. It is the authors' claim that "organism and environment are mutually enfolded in multiple ways, and so what constitutes the world of a given organism is enacted by that organism's history of structural coupling" (1992, 202). But Maturana and Varela may somewhat overstate their case in their discussion of adaptation. It seems that the concept of structural coupling is, in effect, little more than a version of adaptation: when an organism couples with its environment it is adapting to that environment. We view structural coupling as complementing, rather than contradicting, the more traditional idea of adaptation.

Conceiving of the world as being made up of mutually self-defining processes, rather than of discrete entities interacting with other discrete entities, is an important reconceptualization of the world. The originality of the Maturana-Varela approach to embodied cognition can hardly be overemphasized. The fact that their work (like that of James Gibson discussed in Sect. 2 of Chapter 8) is increasingly gaining recognition suggests its foundational importance in contemporary, embodied, contextualized approaches to cognition. The concept of autopoiesis becomes a natural lens through which to view literary characters and narrative structures (see e.g., the analyses of Dillon 2011 and Mancing 2016). In literary theory, the stance closest to Maturana and Varela's autopoiesis is the transactional concept of reading developed by Louise Rosenblatt (1994a, b, 1995). In her earliest work, in the 1930s, Rosenblatt initially used the word "interaction," but soon replaced this with "transaction," specifically to downplay the idea of "two distinct entities acting on each other, like two billiard balls" and to underscore her "emphasis on the to-and-fro, spiraling, nonlinear, continuously reciprocal influence of reader and text in the making of meaning" (1995, xvi; see also 191–92). She specifically likens the transactive reading relationship to the autopoietic biological concept under consideration here: "The underlying metaphor is organic, as in the ecological view of human beings in a reciprocal relation with the natural environment" (26; see also 131–32; see also 1994b, 1058–59). The pragmatic idea of a transactive reading relationship developed by Rosenblatt and Bakhtin's comparable stress on dialogic relationships in both speaking and reading link their work of nearly a century ago with today's embodied cognitive science and radically distinguish it from intervening poststructuralist literary theory.

Susan Oyama (whose work is cited by Varela et al. 1993, 199–200) is the primary exponent of an important related view of evolution that also does away with the mechanistic and dualistic underpinnings of standard evolutionary theory in her two widely respected and highly praised books *The Ontogeny of Information* (1985) and *Evolution's Eye* (2000). The concept of "interaction," as Oyama uses it, involves "the necessity of viewing transactions between an entity and its surround [sic] as aspects of a single system. When we speak of organisms altering their environments, both animate and inanimate, and being altered by their interactions, it is this mobile interchange that is being highlighted" (1985, 6). She specifically equates this concept with CONTEXTUALISM in psychology (19), thus linking her work to Gillespie's and to that of others described below. She also explicitly relates her work to Maturana and Varela's, dealing with "biology as a way of knowing, the relationship between knowers and knowledge, and the necessity of taking responsibility for our contributions to the knowledge (and therefore the Nature) we construct" (2000, 143–44).

Crucial to Oyama's position is the elimination of all forms of determinism, mechanism, and binary thought. Oyama specifically calls attention to the apparently enlightened approach that denies that we are determined either by biology or by the environment in favor of one that holds both may be present, each to a certain degree. Although this may be a step in the right direction, it still maintains the nature-nurture dualism, and she insists that any trace of such dualistic thinking should be completely abandoned: "What all this means is not that genes and environment are necessary for all characteristics, inherited or acquired (the usual enlightened position), but that there is no intelligible distinction between inherited (biological, genetically based) and acquired (environmentally mediated) characteristics" (122).

Maturana, Varela, Thompson, Rosch, and Oyama explore the full implications of the mind-brain-body-environment continuum, concentrating on the body-environment aspect. One of the strengths of this argument is the complete and seamless integration of the biological and the social. And this integration, in all its complexity, is what is lacking in the contemporary theories that attribute everything to social forces. By completely ignoring the evolutionary and biological reality that informs everything human, such theories distort their subject and ring hollow, if not vacuous, in the light of convergent modern thought.

Steven Rose is one of the biologists most adamant about the inadequacy of determinism, either social constructivist or biological (see below). In his elegant book *Lifelines* (1997), he argues that we are "the products of the constant dialectic between 'the biological' and 'the social' through which humans have evolved, history has been made and we as individuals have developed" (6). Rose's primary target is what he perceives as biological determinism, which he refers to as "ultra-Darwinism" (see especially 174–249), but much of his criticism of this stance is equally applicable to the sort of facile social determinism found at the core of many humanist and social science theories. Rather than buy into any sort of one-cause determinism, Rose offers a conceptualization of biological organisms' "radically indeterminate" trajectories, or "lifelines" (7). His book ends on this note: "Thus for humans, as for all other living organisms, the future is radically unpredictable. This means that we have the ability to construct our own futures, albeit in circumstances not of our own choosing. And it is therefore our biology that makes us free" (309). It has rarely been stated better.

Evolution makes it clear that the human mind-brain is part of the human body and inseparable from it in any way. As Maturana, Varela, their colleagues, Oyama, and Rose have elegantly argued, in modern contextualized biological science and allied fields there is no room for any sort of Cartesian mind–body dualism. The brain only functions as part of a larger system that is the body; it is not a self-contained unit located within the head and separable from everything else. What happens in the body affects the brain and vice versa. Further, the human mind-brain functions within a continuum that starts at the molecular level and extends through the environment. All the lines of division that we see so clear and hold so dear are fuzzy and negotiable. In the words of neuroscientist Ira B. Black, "Biology becomes behavior, and behavior becomes biology" (1990, 167). The point Black makes is the same Varela, Oyama, and the others have made: self, mind, brain, body, environment—all interact at every point along the unbroken continuum that makes up reality. To talk of mind as separate from body, or body as separate from context and environment, is meaningless.

The scientists cited above have not worked in a vacuum. The approach to existence and cognition for which they argue can itself be contextualized, as Oyama continually stresses, within the general movement known as *systems theory*, which is derived from Gestalt psychology, cybernetics, and information theory, and which includes, or is also related in various ways to, process theory, general system theory, dissipative structures, the mathematics of complexity, dynamical systems theory, chaos theory, the theory of fractals, nonlinearity, and even the Gaia hypothesis. As described by Fritjof Capra in *The Web of Life* (1996), "systems thinking is 'contextual' thinking; and since explaining things in terms of their context means explaining them in terms of their environment, we can also say that all systems thinking is environmental thinking" (37). Capra sees in the work of Maturana and Varela, which he calls the "Santiago theory" in recognition of the origins of their project at the University of Santiago de Chile, "the first coherent scientific framework that really overcomes the Cartesian split. Mind and matter no longer appear to belong to two separate categories but are seen as representing merely different aspects, or dimensions, of the same phenomenon of life" (175). Important is the concept that the mind is not a thing, but "a process—the process of cognition, which is identified with the process of life. The brain is a specific structure through which this process operates" (175). Mind is, in other words, what the brain does.

Some feminist biologists at the turn of the century also duly recognized the value of systems thinking. Two important books in this field are Lynda Birke's *Feminism and the Biological Body* (1999) and Anne Fausto-Sterling's *Sexing the Body* (2000). Birke argues that feminist theorists, especially those like Judith Butler who have incorporated their work entirely within the social constructionist framework, have tended to deal with the body only on the surface (the body as text cliché) such that the body of this theory "seems to be disembodied—or at the very least disemboweled. Theory, it seems, is only skin deep" (1999, 2). Moving beyond the nature-nurture dichotomy and the idea of a socially constructed subject, Birke emphasizes "an understanding of agency that emerges out of the engagement of the organism with its surrounding; it is thus an agency in relation, not an essential property of the individual" (152). She makes a plea "for a view of the biological body that embraces transformation and change, and prioritises the body in relation—thus emphasising the connections of the 'body's world'" (158).

In Fausto-Sterling's earlier book (1992, 7–8), she presented a relatively balanced view of the relationship between biology and social factors. Later she admitted that, from her position as a feminist, she has "good reason to be scared of bringing biology into the picture" largely because she is only too familiar with "centuries of arguments in which the body has been used to justify power inequities" (2000, 258, n14). And it would seem that at times in the past she let a political agenda interfere with her scientific judgment; see Anne Campbell's (2002, 16) brief exposé of a gaffe by Fausto-Sterling who let her simplistic, but politically correct, social constructionist stance lead her into a mistake about (and thus ironically confirm) an important principle of evolutionary psychology. See also the blunt criticism of Fausto-Sterling by Griet Vandermassen (2005, 107–9).

However, Fausto-Sterling eventually moved away from her earlier, more hardline, social constructionist stance to argue, like Birke, that the dichotomous positions many feminists and others adopt "conspire to make a sociocultural analysis of the body seem impossible" (2000, 22). In place of this either/or biophobic position, Fausto-Sterling proposes that

> [s]uccessful investigations of the process of gender embodiment must use three basic principles. First, nature/nurture is indivisible. Second, organisms—human and otherwise—are active processes, moving targets, from fertilization until death. Third, no single academic or clinical discipline provides us with the true or best way to understand human sexuality. The insights of many, from feminist critical theorists to molecular biologists, are essential to the understanding of the social nature of physiological function. (235)

The way to move beyond such dualisms, Fausto-Sterling suggests, is by thinking in terms of "developmental systems theory, or DST" (25). Drawing much of her theoretical understanding of the function of dynamic systems from the previously cited work of Susan Oyama and others, Fausto-Sterling concludes that "development within a social system is the sine qua non of human sexual complexity. Form and behavior emerge only via a dynamic system of development. Our psyches connect the outside to the inside (and vice versa) because our multi-year development occurs integrated within a social system" (243). Her conclusion, which is antithetical to social constructionist thought, is the following:

> The cell, the individual, groups of individuals organized in families, peer groups, cultures, and nations and their histories all provide sources of knowledge about sexuality. We cannot understand it well unless we consider all of these components. To accomplish such a task, scholars would do well to work in interdisciplinary groups. And while it is not reasonable, for example, to ask all biologists to become proficient in feminist theory or all feminist theorists to be proficient in cell biology, it *is* reasonable to ask each group of scholars to understand the limitations of knowledge obtained from working within a single discipline. (254–55)

Fausto-Sterling, in effect, calls here for the sort of disciplinary convergence (or consilience) that is discussed elsewhere in this book. Feminists, literary scholars, and anthropologists (and the thankfully growing body of feminist literary scholars and feminist anthropologists!), for example, do not have to become cell biologists, primatologists, or evolutionary theorists. Nor is the reverse true. But each of these groups has, we suggest (and indeed implore), the intellectual obligation to become somewhat familiar with the general concepts and limits of the other disciplines. They further have the intellectual obligation to pursue their own research and scholarship in such a way that it does not fundamentally contradict the sound basic findings of the other fields.

A literary theory articulated in reductionist terms and that ignores biological reality suffers from its absence of empirical rigor. All systems approaches to cognition are rich with relationships to and implications for evolutionary theory, biology, psychology, cognitive science, the social sciences in general, and the human sciences, including literary theory. Conceiving of literature as an activity of embodied, contextualized, autopoietic agents is radically different from conceiving of literature as a vast, socially constructed, independent, structure of signs and signification. Systems thinking easily and comfortably couples with biological and psychological reality, and with contemporary scientific understanding, without surrendering or compromising anything of significance to traditional humanistic interests.

3 Psychology and Philosophy in Context

As an extension of the views of evolution and biology outlined above, we now focus attention on several trends in contemporary psychology and philosophy that can be profitably studied by literary scholars. The primary areas to be discussed in this section are ecological psychology, constructivism, experientialism, and cultural psychology. These trends all share a similar view of the human agent (not subject), who actively cocreates herself and her world (is not socially constructed), in rich, thick contexts (not *hors texte*). It is a view replete with potential for a pragmatic (not postmodern THEORETICAL) meaning and understanding (not signification).

Previously we discussed James J. Gibson's (1979) ecological approach to visual perception. The concept of ecology is crucial to Gibson because it places perception in the actual practices of embodied human beings who live and function in the real world. It is meant to stand in contrast to perception as studied in the artificial setting of the psychology laboratory or as conceived theoretically by philosophers. Within recent decades a movement has grown in psychological studies that places Gibson's fundamental ecological concept at the center of attention: take psychology out of the laboratory and place it fully within the actual world of living, breathing human beings; in short, it is an ecological psychology.

The most important early convert to Gibson's ecological position in psychology was Ulrich Neisser. Neisser's *Cognitive Psychology* (1967) was the first book ever published on the subject and one that largely defined the field in the early years. But after he joined Gibson on the faculty of Cornell University, Neisser was won over to an ecological position that was very different from the strong COGNITIVIST orientation of his earlier work. Neisser's significantly titled *Cognition and Reality* (1976) very substantially and explicitly incorporates Gibson's concepts of perception, ecology, affordances, and information pickup, among others. Revising his earlier position that cognition consisted primarily of information processing, Neisser prophetically foresees only trouble ahead if cognitive psychology stubbornly adheres to the computer model: "Lacking in ecological validity, indifferent to culture, even missing some of the main features of perception and memory as they occur in ordinary life, such a psychology could become a narrow and uninteresting specialized field" (7). As an alternative, Neisser proposes constructing a different kind of psychology in more realistic terms (see Neisser's Gibsonian approach to perception and schemata as discussed in Sect. 5 of Chapter 8).

Although Neisser is much more than a mere spokesperson for Gibson, and *Cognition and Reality* is much more than just a Gibsonian work, Neisser's change in position is substantial. Unfortunately, this extraordinarily important and original book did little to stem the COGNITIVIST tide, as the computer model of the mind, representations, and information processing continued to dominate the field for some time to come. The result is that, like Gibson's own work, this book by Neisser seems in many ways more contemporary today than it did more than four decades ago.

Another prominent figure within ecological psychology is Edward S. Reed, who describes Gibson's work as nothing less than "the first truly new theory in [the psychology of perception] in the last four hundred years" (1988, 2; see also 1996b, 30–31). In the 1990s, ecological psychologists like Neisser (1997) and Reed (1997) extended Gibson's ecological approach far beyond perception—which, however, remains the firm ground upon which all cognition is based—into all areas of psychological inquiry. Reed states his fundamental position as follows:

> That we are embodied, made up of cycling hormones and intricate networks of nerves, is a fact. But it is also a fact that we exist in a different way, at a different level: as explorers of our surroundings, as actors who strive to make a difference in the world, and as interactors who enter into both conflict and cooperation with our fellows. That we—sometimes—think in symbols is a fact. But it is also a fact that there are other ways to think, and that it is we who use the symbols and not the other way around. That human beings have made and remade themselves throughout the course of history is a fact. But it is also a fact that the process of self-making (and the conditions that constrain it) is as important to what we are as the resulting product. That our actions and experiences are heavily laden with the symbols, practices, and norms of our cultures is a fact. But it is also a fact that it is because we are in touch with our surroundings that symbols, practices, and norms emerge as useful ways of organizing our mental lives—and without our connectedness to the world, all symbols, practices, and norms would vanish. (1996a, 5)

The themes of this sort of ecological approach are those we have seen before: embodied reality, constructivism, contextualization, self-making. Ecological psychology is a commonsense, realistic, direct, enterprise that strives for relevance in a complex and contingent world. It is, in other words, a meaningful psychology. Like Maturana and Varela, Oyama, and others cited above, ecological psychologists shun reductionism, determinism, and other simple cause-effect relationships, to look for ways to understand the complex interrelated effects of multiple factors. Although Reed does not talk in terms of systems theory, it is essentially a systems approach he and other ecological psychologists employ.

Reed prominently locates Gibson's concept of affordance at the heart of his ecological psychology, stating that one of his fundamental hypotheses is that "affordances and only the relative availability (or nonavailability) of affordances create selection pressure on the behavior of individual organisms; hence, behavior is regulated with respect to the affordances of the environment for a given animal" (1996a, 18). Affordances are not things that determine subjectivity, but they are those things employed by active agents, people who "make things happen, they *make their way in the world*" (19). Following Gibson (1979, 129), Reed asserts that an affordance is neither subjective nor objective in the usual sense of those words (1996a, 184). Psychology deals with our face-to-face, ongoing, direct, ecological encounter with the world that is primary; all secondhand "mediated" (Gibson's term) or "processed" information is secondary, both derivative of and dependent on our ecological cognition (Reed 1996b, 2–3). In these terms, no behaviorist structuralism, no self-referring semiotic system, and no decontextualized COGNITIVISM can be considered as any sort of psychology.

Language, for an ecological psychologist like Reed, is not a social system like a Saussurean *langue*, but "a repertoire of skills—cognitive and social as well as communicative—that enable [children] to become competent (junior) partners in their community" (153). In other words: "Language is found within communities of users. Language is a fact of the populated environment. The identical language is never found in two people" (165). Thus, Reed sees language as internal (an I-language), not external (an E-language), as described by Chomsky. Rejecting the telementation model of Saussure and his followers, Reed affirms that "[*l*]*anguage is not a means of transmitting ideas or representations; it is a means of making information available to others*" (155). With this view, linguistically provided information becomes an affordance that others may use (in any of a variety of ways); it is a pragmatic tool and not a representation or a duplicate copy of someone else's mental concept. I use my I-language to make available information, an affordance, that you, using your I-language, understand and use to the degree that it is relevant to you, that you understand as you can or will in your context. Communication—as Bakhtin and his colleagues saw some nine decades ago—is a contingent, emergent, constructive, contextually significant, social act. Ecological psychology is a contextual psychology in every respect and one which literary scholars should find useful in their own teaching and writing.

The construction of our selves, worlds, and realities is a theme that runs through the work of Gillespie, Oyama, Reed, and others. But what does this mean? In what way is it different from determinism? What is constructed? How? By whom—or by what? One important answer comes jointly from philosophy and psychology under the rubrics of *constructivism* and *internal realism*. Nelson Goodman is the primary proponent of constructivism in philosophy, particularly as articulated in his books *Ways of Worldmaking* (1978) and *Of Mind and Other Matters* (1984). Goodman begins from a pluralist perspective, one that asserts more than one way of knowing (or describing) any reality, depending on one's frame of reference. Thus there are multiple versions of anything, each valid within its own framework; there are, in other words, multiple "right" versions of the world (1978).

The versions of the world which each of us constructs are not made up of whole cloth, with nothing in common among them. Goodman's constructivism is not, as he states, solipsistic. Rather, our worldmaking "always starts from worlds already on hand; the making is a remaking" (6). As a result, once we recognize that "some supposed features of the world derive from—are made and imposed by—versions, 'the world' rapidly evaporates. For there is no version-independent feature, no true version compatible with all true versions" (Goodman 1984, 33). We literally construct our versions of the world: "The world of a true version is a construct; the features are not conferred upon something independent of the version but combined with one another to make the world of that version" (34).

One of the more interesting corollaries of Goodman's worldmaking enterprise is that "the arts must be taken no less seriously than the sciences as modes of discovery, creation, and enlargement of knowledge in the broad sense of advancement of the understanding" (1978, 102). Fictional versions of the world are as important as any other in our worldmaking processes: "our worlds are no more a heritage from scientists, biographers, and historians than from novelists, playwrights, and painters" (103). See also Hilary Putnam's (1981, 1987, 1988) internal realism, the fact "that what is (by commonsense standards) the same situation can be described in many different ways, depending on how we use the words…" (1988, 114).

The idea that we construct our realities is not merely a philosophical or psychological concept; many biologists and neuroscientists also endorse some version of constructionism. Typical is Antonio Damasio, who describes the process of self-construction and remodeling as occurring largely nonconsciously (1999, 224). Again, we see the convergence of significant lines of thought from various fields. The fact that a constructivist position has biological and neurological validity reinforces the significance of the philosophical stance. The world in which we live, with which we interact, within which we make (construct) our selves and our realities is, to some degree, unique to each of us. How could it be otherwise if our individual mind-brains are all unique? How could it be otherwise if we make and define ourselves within and mutually with our environments or contexts? This is no different from the fact that each of us has, in Chomsky's terms, an I-language that is the product of our unique biology and psychology and that is, to some degree, also unique to each of us. One might say that each of us lives in our own I-world.

The concept of *social construction* has its origin in the important book titled *The Social Construction of Reality* (1966) by Peter L. Berger and Thomas Luckmann, for whom the term meant that the reality of the worlds we individually construct takes place within and is very much influenced by social factors. This is a situated, pragmatic, individualized concept of reality: to understand what people know to be real ("the sociology of knowledge"), we must first understand the social construction of reality (15). Thus, although it is the individual who creates her reality, she must by definition do so within the context of society. Within this context, it is our relationships with others that most directly participate in such social construction, and "[t]he most important experience of others takes place in the face-to-face situation, which is the prototypical case of social interaction. All other cases are derivatives of it" (28). We orient what we say and do to others, just as they do to us, according to Berger and Luckmann, "and this continuous reciprocity of expressive acts is simultaneously available to both of us" (29). The social construction of our realities is thus, as Bakhtin says, dialogic.

Berger and Luckmann's description of the processes of social construction does not imply that we are passively determined or constructed *by* society or language or anything else. Furthermore, Berger and Luckmann do not make the social the sole determiner of human subjectivity. Rather, they suggest that biology (even though not a particularly sophisticated version of biology) is always a factor (183). The further, more sophisticated, version of this socially situated self-construction is Maturana and Varela's structural coupling and autopoiesis. As Jerome Barkow puts it, the soft (original) version of social constructionism "is not only compatible with evolutionary psychology, it is required by it" (2006a, 25).

An important contribution to the establishment of social constructionism in mainstream psychology is an article by Kenneth J. Gergen (1985). In it, Gergen states that "[s]ocial constructionist inquiry is principally concerned with explicating the processes by which people come to describe, explain, or otherwise account for the world (including themselves) in which they live" (266). Note how Gergen places the social construction of an individual's reality firmly within that person's socio-historical context, affirms the person's agency, and stresses the process of interchange and negotiation. The main point is that there is no objective social world, but a series of worlds individually constructed within specific contexts. There is no hint here, nor in Berger and Luckmann, that subjects are passive entities constructed by linguistic or ideological forces. It is clear, then, that the term social construction came into academic discourse in a perfectly reasonable way and with a perfectly reasonable meaning.

It is noteworthy how the original sense of the term is maintained and enriched by systems theorists, specifically by developmental psychologists Esther Thelen and Linda Smith:

> Language, logic, consciousness, imagination, and symbolic reasoning are not 'above' the processes of motivated perception, categorization, and action that we have been describing. Rather they are part and parcel of these processes, seamless in time and mechanism. Above all, we maintain, higher cognition is developmentally situated. It grows from and carries with it the history of its origins. In particular, cognition is *embodied* and *socially constructed.* (1994, 321)

In its logical, practical, pragmatic understanding of how human beings construct their realities (their versions of their worlds) within a social context, the original idea of social constructionism is virtually indistinguishable from Goodman's constructivism; at best, more emphasis rests on the social, but the process is essentially the same.

An enormous gulf exists between this concept of social construction and the strong form of linguistic, ideological, or cultural determinism explicit or implied in the same term when used by THEORISTS. The latter turn a valuable concept into a cliché, thus devaluing the currency of the original term. The term social construction has in many ways become a code or shorthand for identifying oneself as a radical deconstructor of the metaphysics of presence or one in the know about power-knowledge relationships; in short, a card-carrying member of the sort of THEORY we have seen so often in postmodern circles. For strong social constructionists, material reality in general or a biological body in particular are non-existent; everything is relative, created solely by language and other social forces such as ideology, power, or interpellation. Radical social constructionism provides a classic case of Bakhtin's *theoreticism*.

For us, the literature scholars who are the authors of this book, the cognitive revolution began in the 1990s (Mancing in 1992, and William five or so years later). Mancing and William were both influenced significantly in these beginning forays into cognitive studies by the work *Metaphors We Live By* (1980), by linguist George Lakoff and philosopher Mark Johnson, as discussed in Sect. 4 of Chapter 4. Lakoff and Johnson's new approach to metaphor, seeing it as an activity of mind rather than as a rhetorical embellishment, represents one of the most important shifts in cognitive science from COGNITIVISM toward CONTEXTUALISM, and it has informed much of our own work in one way or another over the years.[2] In addition to their radical revision of our concept of metaphor, Lakoff and Johnson developed a philosophy they call *experiential realism* or *experientialism*.

Criticizing the myths of objectivism (absolute objectivity) and subjectivism (subjective intuition), together with the facile assumption that these two extremes are the only possibilities (binarism), Lakoff and Johnson propose a middle ground. They start from the constructivist premise that "truth is relative to our conceptual system, which is grounded in, and constantly tested by, our experiences and those of other members of our culture in our daily interactions with other people and with our physical and cultural environments" (193). Whereas both traditional objectivism and subjectivism assume and incorporate essentially into their premises an absolute person-environment dualism, experientialism stresses embodied, contextualized cognition: "From the experientialist perspective, truth depends on understanding, which emerges from functioning in the world" (230). Clearly, experientialism's embodied, contextualized, interactive concept of philosophy and cognition are consistent with the evolutionary, biological, and psychological approaches discussed above.

Lakoff and Johnson followed up their earlier work with the massive *Philosophy in the Flesh* (1999), a brash and uncompromising criticism of analytical philosophy and other types of disembodied approaches to knowledge (including COGNITIVISM) that have dominated Western thought since the ancient Greeks. The authors describe what they call "first-" and "second-generation" cognitive science (74–93). First-generation cognitive science, as practiced in the 1950s and 1960s, prominently featured symbolic computation, a disembodied and literal concept of reason, and the premises of Anglo-American analytical philosophy. Its basic premises included "early artificial intelligence, information-processing psychology, formal logic, generative linguistics, and early cognitive anthropology" (75). Essentially, this is the COGNITIVISM described in the previous chapter.

By the late 1970s, Lakoff and Johnson suggest, a new cognitive paradigm was gaining ground, one that was based on two radically different kinds of evidence: "(1) a strong dependence of concepts and reason upon the body and (2) the centrality to conceptualization and reason of imaginative processes, especially metaphor, imagery, metonymy, prototypes, frames, mental spaces, and radial categories" (77). Not surprisingly, this is a concise description of precisely the work carried out by the authors themselves and their colleagues and students. That is, they limit the scope of experientialism and claim more uniqueness for their own emphases than may be warranted. Experientialism as a term has not been taken up by a significant number of other researchers, including many of those who work in areas of cognitive linguistics, schema theory, and categorization within the Lakoff-Johnson model.

Furthermore, the assertion by Lakoff and Johnson that the second, or contextual, model has *replaced* the first, or COGNITIVIST, model, is more a statement of wish than one of fact. As indicated previously, COGNITIVISM may have lost very much of the glamor it had in the earliest days of cognitive science, but it remains a strong research tradition and for many it is still the only thing cognitive science consists of. One can only hope that the Lakoff-Johnson statement proves to be more prophetic than it is accurate today. No approach described in this chapter is more specifically and intimately involved in so many basic aspects of cognition and language than this one. The facts that Lakoff and Johnson write with clarity and precision and that they deal at length with matters of language should endear their work particularly to literary scholars.

Psychologist Jerome Bruner, one of the founders of the new field of cognitive psychology in the 1950s, has lamented (1990, 1–32) that the original intent of the cognitive revolution was to stress the construction of knowledge and meaning in ways that were radically antithetical to the then-dominant behaviorism: "Its aim was to discover and to describe formally the meanings that human beings created out of their encounters with the world, and then to propose hypotheses about what meaning-making processes were implicated" (2). But, somehow, the "originating impulse … became fractionated and technicalized…. Very early on, for example, emphasis began shifting from 'meaning' to 'information,' from the *construction* of meaning to the *processing* of information" (4), a process in which the context-free computational model of the mind was dominant. Once "mind" became "program," insists Bruner, "[t]here could be no place for 'mind' in such a system—'mind' in the sense of intentional states like believing, desiring, intending, grasping a meaning" (8) and any concept of human agency is lost (9). (There may be some creative memory and revisionism in Bruner's nostalgia. All evidence indicates that the computer model of cognition was dominant in cognitive science from the very beginning.) Bruner pleads for a "culturally oriented psychology," that is, a psychology that "neither dismisses what people say about their mental states, nor treats their statements only as if they were predictive indices of overt behavior" (19). The idea of cultural psychology is one that seems to be true both to human biology and to what we know of the mind-brain. It is one more version of the sort of contextualized approach to cognition that we are convinced can be meaningful and useful to literary studies.

Overall, there is ample evidence, both in theory and practice, in favor of a sense of cognition that is totally at odds with both the decontextualized, dehumanized representation and information processing of COGNITIVISM on the one hand, and the socially constructed subjects brought into being by language or ideology of THEORY on the other. When cognition, thought, knowledge, intelligence, and communication are placed in evolved, embodied, ecological, contextualized, contingent, autopoietic agents, our view of all things human, specifically including literature and literary theory and performance, are seen in a radically new light. We see it as nothing less than a paradigm change.

NOTES

1. The structural coupling concept with its emphasis on interconnection is used in environmental and sustainability discourse, see e.g., Cull (2014).
2. Lakoff and Johnson's groundbreaking book continues to profoundly shape scholarly studies in embodied cognition as well as "crossover" nonfiction. For a recent example of the latter see architecture critic Sarah Williams Goldhagen's *Welcome to Your World: How the Built Environment Shapes Our Lives* (2017). Further, a recent review of a resource on Plato's similes criticizes the work for not taking *Metaphors We Live By*, in addition to some other cognitive-studies works, into account (Kokkiou 2017).

References

Barkow, Jerome H. 2006a. "Introduction: Sometimes the Bus Does Wait." In Barkow, 3–59.

———, ed. 2006b. *Missing the Revolution: Darwinism for Social Scientists*. Oxford: Oxford UP.

Berger, Peter L., and Thomas Luckmann. 1966. *The Social Construction of Reality: A Treatise in the Sociology of Knowledge*. Garden City: Doubleday.

Birke, Lynda. 1999. *Feminism and the Biological Body*. New Brunswick: Rutgers UP.

Black, Ira B. 1990. *Information in the Brain: A Molecular Perspective*. Cambridge: Bradford Books/MIT P.

Bruner, Jerome. 1990. *Acts of Meaning*. Cambridge: Harvard UP.

Campbell, Anne. 2002. *A Mind of Her Own: The Evolutionary Psychology of Women*. Oxford: Oxford UP.

Capra, Fritjof. 1996. *The Web of Life: A New Scientific Understanding of Living Systems*. New York: Anchor.

Churchland, Patricia Smith. 1987. "Epistemology in the Age of Neuroscience." *Journal of Philosophy* 84: 543–53.

Cull, Jane. 2014. "The Circularity of Life." *The Ecologist*, 4 November (Online).

Damasio, Antonio R. 1999. *The Feeling of What Happens: Body and Emotion in the Making of Consciousness*. New York: Harcourt Brace.

Dillon, Sarah. 2011. "Chaotic Narrative: Complexity, Causality, Time, and Autopoiesis in David Mitchell's *Ghostwritten*." *Critique: Studies in Contemporary Fiction* 52.2: 135–62.

Fausto-Sterling, Anne. 1992 (1985). *Myths of Gender: Biological Theories About Women and Men*. New York: Basic Books.

———. 2000. *Sexing the Body: Gender Politics and the Construction of Sexuality*. New York: Basic Books.

Gardner, Howard. 1985. *The Mind's New Science: A History of the Cognitive Revolution*. New York: Basic Books.

Gergen, Kenneth J. 1985. "The Social Constructionist Movement in Modern Psychology." *American Psychologist* 40: 266–75.

Gibson, James J. 1979. *The Ecological Approach to Visual Perception*. Boston: Houghton Mifflin.

Gillespie, Diane. 1992. *The Mind's We: Contextualism in Cognitive Psychology*. Carbondale: Southern Illinois UP.

Goldhagen, Sarah Williams. 2017. *Welcome to Your World: How the Built Environment Shapes Our Lives*. New York: Harper Collins.

Goodman, Nelson. 1978. *Ways of Worldmaking*. Indianapolis: Hackett.

———. 1984. *Of Mind and Other Matters*. Cambridge: Harvard UP.

Johnson, David Martel, and Christina E. Erneling, eds. 1997. *The Future of the Cognitive Revolution*. New York: Oxford UP.

Kokkiou, Chara. 2017. Rev. of *Plato's Similes: A Compendium of 500 Similes in 35 Dialogues* by John E. Ziolkowski (2014). *The Classical Journal* (CJ-Online), CAMW.

Lakoff, George, and Mark Johnson. 1980. *Metaphors We Live By*. Chicago: U of Chicago P.

———. 1999. *Philosophy in the Flesh: The Embodied Mind and Its Challenge to Western Thought*. New York: Basic Books.

Mancing, Howard. 2016. "Embodied Cognition and Autopoiesis in *Don Quijote*." In *Cognitive Approaches to Early Modern Spanish Literature*. Ed. Isabel Jaén-Portillo and Julien Simon. Oxford: Oxford UP. 37–52.

Maturana, Humberto R., and Francisco J. Varela. 1980 (1973). *Autopoiesis and Cognition: The Realization of the Living*. Dordrecht: D. Reidel.

———. 1992 (1984). *The Tree of Knowledge: The Biological Roots of Human Understanding*. Trans. Robert Paolucci. Boston: Shambhala.

Neisser, Ulric. 1967. *Cognitive Psychology*. New York: Appleton-Century-Crofts.

———. 1976. *Cognition and Reality: Principles and Implications of Cognitive Psychology*. San Francisco: W. H. Freeman.

———. 1997. "The Future of Cognitive Science." In Johnson and Erneling, 247–60.

Oyama, Susan. 1985. *The Ontogeny of Information: Developmental Systems and Evolution*. Cambridge: Cambridge UP.

———. 2000. *Evolution's Eye: A Systems View of the Biology-Culture Divide*. Durham: Duke UP.

Pepper, Stephen C. 1942. *World Hypotheses: A Study in Evidence*. Berkeley: U of California P.

Putnam, Hilary. 1981. *Reason, Truth and History*. Cambridge: Cambridge UP.

———. 1987. *The Many Faces of Realism: The Paul Carus Lectures*. LaSalle: Open Court.

———. 1988. *Representation and Reality*. Cambridge: Bradford Books/MIT P.

Reed, Edward S. 1988. *James J. Gibson and the Psychology of Perception*. New Haven: Yale UP.

———. 1996a. *Encountering the World: Toward an Ecological Psychology*. New York: Oxford UP.

———. 1996b. *The Necessity of Experience*. New Haven: Yale UP.

———. 1997. "The Cognitive Revolution from an Ecological Point of View." In Johnson and Erneling, 261–73.

Rose, Steven. 1997. *Lifelines: Biology beyond Determinism*. Oxford: Oxford UP.

Rosenblatt, Louise M. 1994a (1978). *The Reader, the Text, the Poem: The Transactional Theory of the Literary Work*. Carbondale: Southern Illinois UP.

———. 1994b. "The Transactional Theory of Reading and Writing." In *Theoretical Models and Processes of Reading*. Ed. Robert B. Ruddell, Martha

Rapp Ruddell, and Harry Singer. Newark: International Reading Association. 1057–92.
———. 1995 (1938). *Literature as Exploration*. New York: Modern Language Association.
Thelen, Esther, and Linda Smith. 1994. *A Dynamic Systems Approach to the Development of Cognition and Action*. Cambridge: MIT P.
Vandermassen, Griet. 2005. *Who's Afraid of Charles Darwin? Debating Feminism and Evolutionary Theory*. Lanham: Rowman and Littlefield.
Varela, Francisco J., Evan Thompson, and Eleanor Rosch. 1993. *The Embodied Mind: Cognitive Science and Human Experience*. Cambridge: MIT P.

CHAPTER 13

Evolutionary Psychology

Evolutionary psychology does not deny personal experience, but rather places such experiences in the context of our biology.

—Robert Jay Russell

We have left until last in this part of the book a consideration of the relatively new field of research known as evolutionary psychology. Its very name makes clear its intention to merge the two general areas that we have argued are most disregarded or misunderstood in THEORY: biology (evolution) and (cognitive) psychology. The basic premises behind this interdisciplinary movement are that human beings as they exist today are a product of a long evolutionary history, and that this history has important implications for the structure and function of our mind-brains and for the way we perceive the world and think. According to these researchers, a comprehensive study of the human mind must take into consideration the implications of our evolutionary history. This chapter provides an overview of evolutionary psychology, a look at the earlier sociobiology, and the surrounding debates.

H. Mancing and J. Marston William, *Restoring the Human Context to Literary and Performance Studies*, Cognitive Studies in Literature and Performance,
https://doi.org/10.1007/978-3-030-89078-0_13

1 The Evolved Mind

We begin with definitions of evolutionary psychology from leading figures in the field. The first is from the editors of "the bible of modern evolutionary psychology" (Miller and Kanazawa 2007, 11): *The Adapted Mind* (1992), a collection of seminal and foundational essays edited by three leaders in the field, anthropologist Jerome H. Barkow and the husband-and-wife team of anthropologist John Tooby and psychologist Leda Cosmides. Their goals as stated in their introduction are first, to increase the understanding of evolutionary psychology within the scientific world and, second, "to clarify how this new field, by focusing on the evolved information-processing mechanisms that comprise the human mind, supplies the necessary connection between evolutionary biology and the complex, irreducible social and cultural phenomena studied by anthropologists, sociologists, economists, and historians" (Barkow et al. 1992, 3). Thus, it is clear from the outset that the purpose of this endeavor is to combine the study of biology, especially evolutionary theory, and the social sciences into an integrated approach. The editors cite the concept of "conceptual integration," the idea that "the various disciplines within the behavioral and social sciences should make themselves mutually consistent, and consistent with what is known in the natural sciences as well" (4); see also Barkow, who uses the term "vertical integration" (1989, 3–5; 2006, 29–34), and E. O. Wilson who prefers "consilience" (1998). The natural sciences have already achieved this integration among themselves, as nothing in the study of astronomy, biology, chemistry, geology, or physics could contradict the principles of any of the others. But some social sciences and humanities are not often well integrated into this overarching conceptual scheme. On consilience, see Slingerland and Collard (2012), especially the editors' introduction, and also Stephen J. Gould's final book, published posthumously in 2003 with the hopeful subtitle "Mending the Gap between Science and the Humanities."

In their long opening of *The Adapted Mind*—an essay that has been called "the single most influential theoretical essay in evolutionary psychology" (Carroll 2004, 154)—Tooby and Cosmides again begin by stressing the conceptual integration of knowledge across the disciplines and then return to the theme of the interdisciplinary nature of evolutionary psychology (1990, 21). The authors' stance aligns with the arguments recurrent throughout this book and that are explicitly endorsed by the autopoiesis, interactionism, lifelines, experientialism, systems theory, and ecological psychology described in the previous chapter. The intellectual honesty and profound interdisciplinary integrational approach of evolutionary psychology are made clear in their elaboration of the field.

As further evidence of this stance we cite someone mentioned in earlier chapters, the peripatetic linguist Steven Pinker, whose work is fully integrated both within the paradigm of generative linguistics and that of evolutionary psychology. Basing his approach on his understanding of the embodied biology and psychology of language, Pinker insists that "we can discard the pre-scientific, magical model in which the issues are usually framed" (1994, 407). This "magical model" is a simplistic one in which it is held either that heredity *causes* behavior or environment *causes* behavior. Pinker then continues: "The 'controversy' over whether heredity, environment, or some interaction between the two causes behavior is just incoherent. The organism has vanished; there is an environment without someone to perceive it, behavior without a behaver, learning without a learner" (407–8). Then Pinker provides a model that he describes as "simplistic" but more realistic than the unidirectional, single-causal models he rejects. In it, "innate psychological mechanisms, including learning mechanisms" *cause* behavior. But these innate mechanisms are *built* by heredity, while the environment *provides input* to them, and at the same time a set of skills, knowledge, and values *develop and access* these mechanisms and are in turn *developed and accessed* by them. Commenting on the way this model makes it clear that behavior is always, and necessarily, the result of a complex interaction of multiple factors, Pinker argues that "[l]earning is not an alternative to innateness; without an innate mechanism to do the learning, it could not happen at all," adding that he also wants to "reassure the nervous: yes, there are important roles for both heredity and environment" (408).

Elsewhere, Pinker has written that "[i]n this scientific age, 'to understand' means to try to explain behavior as a complex interaction among (1) the genes, (2) the anatomy of the brain, (3) its biochemical state, (4) the person's family upbringing, (5) the way society has treated him or her, and (6) the stimuli that impinge upon the person" (1997, 53). Evolutionary psychologists agree on this: an individual's specific behavior in any context is brought about by a complex set of biological and social factors, and it adds immeasurably to our understanding of human psychology if we take the evolved architecture of the human mind-brain into consideration as we attempt to understand our socially embedded embodied cognitive processes. Everything is simultaneously biological (genetic) and social (cultural) all the way down.

Finally, we can see the same stance taken by the authors of the textbook *Human Evolutionary Psychology* (2002), Louise Barrett, Robin Dunbar, and John Lycett. Their opening statement on what evolutionary psychology is all about can be read as a direct response to the position assumed by hardline social constructionists:

> Unless you are a Creationist, you have to accept that humans have been subject to the same processes of evolutionary change as all other living things on earth. A full understanding of human nature therefore requires an understanding of biological as well as sociological processes. Indeed, it is actually impossible to separate the two. We are products of an interaction between biology and culture, or to put it in its more familiar guise, nature and nurture, genes and environment. To separate the two is a false dichotomy. Many would argue that human nature cannot be reduced to mere biological processes—and they would be right. But to infer from this, as many do, that biology is now completely irrelevant … is to commit an egregious logical error. (2)

The central premise of this version of evolutionary psychology is "that there is a universal human nature, but that this universality exists primarily at the level of evolved psychological mechanisms, not of expressed cultural behaviors" (5). The various manifestations of culture that are observed worldwide are not inconsistent with this central premise; nothing in the idea of a universal human nature even remotely implies that there should be a uniformity of cultural practices any more than it does that there should be but a single language worldwide. Donald Brown's *Human Universals* (1991) is the most authoritative presentation of those universal characteristics of humanity that underlie cultural variety.

The subjects most studied by evolutionary psychologists include adaptations, altruism, kin selection, inclusive fitness, and reciprocity; beauty or physical attractiveness; comparative (i.e., human and animal) psychology; decision-making, social exchange, and game theory; homicide, despotism, violence, jealousy, and rape; mate-selection, marriage, sexual choice, sexual selection, and parent– (or caretaker–) child relationships; morality and ethics; patriarchy and sex differences; and more.

Often, evolutionary psychologists approach their subjects by using a process of *reverse engineering*, the opposite of engineering (designing a machine to do something): figuring out what a machine was designed to do (see e.g., Pinker 1997, 21–23). The "machine" involved here, however, is the human mind. The question often asked is something like, "What evolutionary explanation might there be for our tendency to do X or Y or Z?" The tentative answers evolutionary psychologists come up with are often surprising, frequently thought-provoking, and sometimes controversial. Some critics reject such speculative evolutionary explanations out of hand as *just-so stories* (after Rudyard Kipling's fancifully charming explanations of why leopards have spots or why rhinoceroses have wrinkly skin), as fanciful fictions. But surely the proposals suggested by evolutionary psychologists are no more fanciful than Freudian, Marxist, or social-constructionist theories that many of these critics have always taken so seriously and debated or defended so vigorously. All evolutionary psychological explanations may not be exactly right the first time—some of them might be absolutely wrong—but they are serious proposals that deserve honest intellectual consideration and debate.

Tooby and Cosmides have written at some length contrasting two conceptual and theoretical models for the social sciences. The first is what they refer to as the *Standard Social Science Model* (SSSM): "The consensus view of the nature of social and cultural phenomena that has served for a century as the intellectual framework for the organization of psychology and the social sciences and the intellectual justification for their claims of autonomy from the rest of science" (1990, 23). The SSSM is grounded in the early sociological theories of Emile Durkheim and others and is best exemplified in the work of influential contemporary anthropologists such as Clifford Geertz. After tracing a long and complex ten-step series of motivating considerations of SSSM (25–30), they outline an eleven-point summary of its characteristics. This list is too lengthy to reproduce here, but some highlights are as follows:

> Particular human groups are properly characterized typologically as having "a" culture, which consists of widely distributed or nearly group-universal behavioral practices, beliefs, ideational systems, systems of significant symbols, or informational substance of some kind… . The individual is the more or less passive recipient of her culture and is the product of that culture… . What is organized and contentful in the minds of individuals comes from culture and is socially constructed… . In discussing culture, one can safely neglect a consideration of psychology as anything other than the nondescript "black box" of learning, which provides the capacity for

> culture… . Evolved, "biological," or "innate" aspects of human behavior or psychological organization are negligible, having been superseded by the capacity for culture. (31–32)

Note that in this scheme, "[t]he causal flow is overwhelmingly or entirely in one direction. The individual is the acted upon … and the socio-cultural world is the actor" (26). While noting some validity of certain aspects of the SSSM, Tooby and Cosmides fault this model on the grounds it is based on outmoded and erroneous developmental theories, a dubious nature-nurture dichotomy, and an impossible psychology (33–34). Basically, the SSSM is nothing but an elaborate description of the strong social-constructionist stance that dominated the social sciences and humanities for most of the twentieth century.

In contrast to the SSSM, Tooby and Cosmides propose what they call the *Integrated Causal Model* (ICM), which is characterized by a human mind with a set of evolved content-specific "information-processing mechanisms," adaptations that are specialized for specific cognitive functions, and that can generate cultural content (24).

Under this model, it is the set of evolved universal structures that make human diversity possible. Human social realities are not determined by any single factor, but are the result of a complex and context-specific combination of factors from biological structure, culture, the individual, and their interactions.

There has been some criticism of the SSSM and ICM models as outlined here. And, clearly, many social scientists would not endorse all elements in the first model nor reject all those of the second. But just as we have contrasted COGNITIVISM and CONTEXTUALISM by making opposing lists of descriptors, it is fair to take the contrasting models of Tooby and Cosmides as reasonable descriptions of two basically opposing worldviews.

One of the most frequent metaphors Tooby and Cosmides, along with other evolutionary psychologists, employ to describe the mind is that of the "Swiss Army knife." The mind-brain, they maintain, is massively modular; it consists of a relatively large number of domain-specific neural assemblies or "mental algorithms." This is not Fodorian modularity (see Sect. 2 in Chapter 6); rather, the concept of modules described in this context are brain circuits or neural networks, that might be distributed throughout various areas of the brain, that come together to perform certain kinds of specific tasks. Pinker, with his characteristic bent for metaphor and image, suggests, "[a] mental module probably looks more like roadkill, sprawling messily over the bulges and crevasses of the brain" (1997, 30).

Critics of evolutionary psychology sometimes take the Swiss Army knife metaphor quite literally and snidely criticize the concept of a brain filled with independent, or encapsulated, task-specific devices. One of the problems with metaphors is that they can provide critics with a bludgeon that can be wielded mercilessly. We are not especially fond of the Swiss Army knife metaphor either, but it has at least some degree of validity as opposed to, say, the blank slate assumed by social constructionists. Overall, however, given the degree to which evolutionary psychologists insist on massive modularity and are unwilling to nuance their discussions, they do invite criticism. Dunbar and Barrett lay out what is probably the best position when they write that

> an evolutionary perspective does not of itself necessarily commit us to the claim that the mind is entirely organized on modular principles, even though some have argued trenchantly for such a case. Full-scale modularity is just one of a continuum of possibilities. Indeed, we do not doubt that the mind has *some* degree of modular structure, but it is a purely empirical question as to exactly what form this modularity takes, and how many modules there are. (2007, 5)

Another source of misunderstanding is the way in which Tooby and Cosmides, Pinker and many other evolutionary psychologists prominently use the computer metaphor for the mind and frequently write of information processing. But no evolutionary psychologist is aligned with the Artificial Intelligence (AI) enterprise. In AI the brain is literally conceived as a computer, processing data in exactly the way your desktop Mac or PC does. Tooby and Cosmides, on the other hand, use the term "information processing" in the more general sense of "any psychological process: Reasoning, emotion, motivation, and motor control … In cognitive science, the term *mind* refers to an information-processing description of the functioning of an organism's brain and that is the sense in which we use it" (1990, 65). No COGNITIVIST would ever describe computation in any sense that would involve either emotion or motivation, as such factors are not part of machine computation. Here we have another case of a loose, or metaphorical, use of the term. Evolutionary psychologists would make their point more clearly and more effectively if they specifically avoided the computation metaphor, the mechanistic imagery of the body, brain, and mind, and other aspects of COGNITIVIST discourse, instead employing a vocabulary grounded in the contextualist stance they obviously assume. A healthy dose of autopoiesis and dynamical systems theory would go a long way to curing the biggest mistakes many evolutionary psychologists have consistently made.

Psychologist Catherine Salmon has proposed that, despite the great gulf separating evolutionary psychology and the study of literature, "there is no good reason why that should be the case" (2005, 244). Her proposal that this gulf be closed rests on the same grounds we explore throughout this book:

> Many mainstream humanists have had a tendency to assume that human nature is constructed, that everything is nurture and nothing is nature. Recent research in cross-cultural anthropology and psychology suggests that this is incorrect, that almost everything that is important about human behavior and psychology always develops through a combination of nature and environment. One can't truly understand and interpret text without a deep understanding of basic human nature, those human universals that transcend cultural differences. (244)

Salmon makes her point by means of a highly original comparative study of female-oriented romance novels and male-oriented pornography; see also Salmon and Symons (2003, 2004). This exemplary study, together with the proposition that evolutionary approaches can bring new and significant insights into the study of literature might be a good place to end this chapter and Part IV of this book.

But before doing that, we find it advantageous to review one of the ugliest controversies that has marred the intellectual scene in recent decades and the implications of the stances taken in the dispute. The subject is sociobiology.

2 Sociobiology

In many ways, evolutionary psychology develops a line of thought initiated by E. O. Wilson with his book *Sociobiology: The New Synthesis* (1975). He defines sociobiology as "the systematic study of the biological basis of all social behavior" (4). As the *Stanford Encyclopedia of Philosophy* documents, Wilson's book can be considered an ancestor to more contemporary projects that explore behavior from an evolutionary perspective and are often influenced by, but also divergent from, Wilson's theories ("Sociobiology," 2018). Wilson dealt primarily with a wide variety of social behaviors in animals, from insects to primates, and only in the last chapter (which occupies less than 5% of the book) addressed the way in which much human behavior might have a biological basis. The book attempts to undo the nature-nurture binary and see all behavior as an interaction between the social and the biological, thus the name: *socio-biology*. In this, Wilson believed he was contributing to a fuller understanding of the modern synthesis of neo-Darwinism as described in this book in Sect. 1 of chapter 5. Throughout the book he made it clear that genes and environment work together in determining any specific trait of an individual organism. Later, in his final chapter, he specifically affirms, "genes promoting flexibility in social behavior are strongly selected at the individual level" (548). Wilson's message is that biology (nature) works together with the environment (nurture) to produce behavior. Unfortunately for him, his book was not read as he had hoped.

3 Debates

Wilson's book had barely appeared when the *New York Review of Books* published, in May 1975, a scathing letter of denunciation (Allen et al. 1975), accusing Wilson of racism, promoting biological determinism, and raising the ugly specters of eugenics and Social Darwinism. The letter was from a group of Boston-Cambridge academicians calling itself the "Sociobiology Study Group," led by Wilson's Harvard colleagues Richard Lewontin and Stephen J. Gould, together with other scientists and students from the Boston-Cambridge area. It was the opening volley in what became known as the *sociobiology debate*, a fierce intellectual-moral fight that lasted for at least two decades and still lingers today. The attack on Wilson seems to have consisted of various degrees of genuine misunderstanding, intellectual dishonesty, willful misreading, hypocrisy, fear of potential consequences of certain kinds of research, the promotion of a (Marxist) political agenda, and other factors, in which participants on all sides saw themselves as "defenders of the truth" (Segerstråle 2000a). Wilson is not a biological determinist in anything like the sense his critics charge; sociobiology is not a reductionist discipline promoting the idea that our genes alone determine our behavior. Wilson does not believe, nor did he write, that instincts trump culture. It is not a necessary conclusion that if our genes influence us, we are helpless pawns of evil politicians; Wilson did not write to promote a conservative political agenda.

Although at first led by the Americans Lewontin and Gould, the British soon became involved in the debate after the publication of Richard Dawkins's 1976 book *The Selfish Gene* and the criticism of the Wilson-Dawkins position (as the two were quickly conflated) by Steven Rose. The vivid and detailed account of the debate by sociologist Ullica Segerstråle (2000a) reads like a soap opera, comic at times, pathetic at others, with dramatic confrontations, self-righteous moral posturing, gossip, simplistic thinking, accusations, and broken hearts. It represents one of the saddest and ugliest episodes in modern academia.

Once Wilson's colleagues opened the doors to a critique of sociobiology, hardline postmodernist social constructionists jumped in and expanded the fray into what became known as the *Science Wars*.[1] Not content to argue that work like Wilson's represents both bad science and bad morals, many postmodernists maintained that all of science is nothing more than another (largely fictional) discourse, that science is irrelevant to any social concern, or that "science today serves primarily regressive social tendencies" (Harding 1986, 9). The critique of sociobiology thus became just another weapon in the promotion of pantextuality ("*Il n'y a pas de hors-texte*"), all-inclusive intertextuality, infinite semiosis, endless deferral of meaning, and all the other clichés.

Wilson and Dawkins, together with other sociobiologists, have written in defense of their position and in rebuttal to the accusations of determinism and reductionism (see Segerstråle 2000a). They have made perfectly clear that they do not claim that genes *determine* anything. For example, Wilson, in his Pulitzer Prize-winning book *On Human Nature*, writes, "Each person is molded by an interaction of his environment, especially his cultural environment, with the genes that affect social behavior" (1978, 18). Logically, from his professional point of view and to combat what he calls "ultra-environmentalists," he places emphasis on the "decisive" importance of genetic inheritance (19), but it is a willful misreading to take this to mean genetic determinism. One of the best and most coherent of the responses to the accusation of genetic determinism can be found in the second chapter, titled "Genetic Determinism and Gene Selectionism," of Dawkins's *The Extended Phenotype* (1999, 9–29), originally published in 1982. In it, Dawkins addresses his critics' "wanton eagerness to misunderstand," and he describes the "myth" of genetic determinism as "pernicious rubbish on an almost astrological scale" (10).

We bring up the sociobiology debate in this context because evolutionary psychology evidently shares much with that earlier movement and is, to some extent, derived from it. The primary difference between the two is that sociobiology related *genes and behavior*, while evolutionary psychology relates *the evolved mind and behavior*; the former is grounded in genetics, the latter in evolution, neuroscience, and psychology. Evolutionary psychology is a subtler and more accurate way of framing the discussion. And evolutionary psychologists have gone out of their way to underscore this difference and demonstrate the significance of their new approach, all the while striving "to be as politically correct of possible" (Segerstråle 2006, 135) to deflect the unfair criticism leveled at sociobiology. It hasn't helped much. Time after time critics equate sociobiology and evolutionary psychology, continue to hurl the epithet of biological determinism, and talk in terms of reductionism, racism, eugenics, Social Darwinism, and right-wing politics. Deliberate distortions of the positions taken by evolutionary psychologists and dishonest summaries of their research are not uncommon.[2]

From the start, evolutionary psychologists have made it abundantly clear that their work does not concentrate on biology at the expense of environment or culture. In an early essay, David Buss made the point as explicitly as possible, noting that "few other perspectives within psychology place greater importance on a detailed and complex treatment of environmental, situational, and contextual factors. Contextual evolutionary analysis takes place at several levels in the causal sequence" (1995, 10–11). Buss then goes on to discuss the historical selective context, the ontogenetic context, and immediate situational inputs to further develop the point most decisively: "A central goal of evolutionary psychology is to explicate all these forms of contextual input—historical, ontogenetic, and experiential" (11).

An egregious example of this unfair criticism of evolutionary psychology by fellow scientists—Henry Plotkin accurately calls it "vituperative" (2007, 4)—is the book edited by Hilary Rose and Steven Rose titled *Alas, Poor Darwin* (2000a). The editors announce that their aim is to challenge what they call "one of the most pervasive of present-day intellectual myths" (2000b, 1), the excessive (and often improper) use of Darwinian and evolutionary terminology to explain virtually everything—but their specific target is evolutionary psychology. In the introduction, the editors pillory the field, misrepresenting it in precisely the terms mentioned above: they argue that evolutionary psychology is "little different from old-style sociobiology ... not merely mistaken, but culturally pernicious ... [with] a directly political dimension" (4); it gives rise to "new forms of biological determinism ... [claiming that] our biology is our destiny" (5); its political stance "is transparently part of a right-wing libertarian attack on collectivity, above all the welfare state" (9); and so forth. The 14 essays (including one by each of the editors) that follow are mostly in the same vein.

The reader may have noted already that Steven Rose is the very person cited earlier in the previous chapter for his book *Lifelines* (1997), explicitly written in the tradition of Maturana and Varela's autopoiesis. Much of that book, beyond the author's own perceptive and discerning thesis, is unfortunately also a critique of evolutionary psychology. Rose had been England's most prominent critic of sociobiology. It seems that his political and ideological solidarity with Lewontin and Gould led him to side with those who unfairly criticized Wilson, Dawkins, and others, as biological determinists. Rose's apparent assumption that anyone evoking a biological explanation, even when nuanced and when stressing the interaction between the evolved mind and cultural factors, must be a biological determinist results in the situation that even when others are saying almost *exactly* what he says, he accuses them of saying the *opposite* of what he says. This is a very strange and misguided example of Bakhtin's theoreticism: one's political ideology blinds one to the use of the English language. It is also an ugly example of Theory of Mind: X says Y, but I know he means Z. Books such as anthropologist Roger N. Lancaster's *The Trouble with Nature* (2003), which critiques sociobiology for its reductionism while maintaining a problematic commitment to binary oppositions, exemplify this stance.

One wonders why autopoiesis, interactionism, lifelines, ecological psychology, and other forms of contextualist thought are not criticized as biological determinism, reductionism, and naive realism the way evolutionary psychology and sociobiology are? It could be that postmodern THEORISTS and other social constructionists assume that they are implicitly covering all the bases by naming just these two fields. But whatever the reason(s), the knee-jerk accusation that anyone who isn't like you must be the polar opposite of you is apparently at work again, as it is explicitly in Lancaster's book. If someone is not a strong social constructionist, it (il)logically follows that this person must be a biological determinist. The case of evolutionary psychology illustrates very well the reasons why we are writing this book.

Notes

1. For more on the Science Wars and their aftermath from various disciplinary perspectives, see e.g., Segerstråle (2000b) and Parsons (2003).
2. For a more recent critique of sociobiology, see Bennett (2015), and for a recent balanced summary of the controversy, see Schnettler (2019).

References

Allen, Elizabeth, Jon Beckwith, Barbara Beckwith, Steven Chorover, David Culver, Margaret Duncan, et al. 1975. "Against 'Sociobiology.'" *New York Review of Books* 22.18 (13 November): 182, 184–86.

Barkow, Jerome H. 1989. *Darwin, Sex, and Status: Biological Approaches to Mind and Culture*. Toronto: U of Toronto P.

———, Leda Cosmides, and John Tooby, eds. 1992. *The Adapted Mind: Evolutionary Psychology and the Generation of Culture*. New York: Oxford UP.

———, ed. 2006. *Missing the Revolution: Darwinism for Social Scientists*. Oxford: Oxford UP.

Barrett, Louise, Robin Dunbar, and John Lycett. 2002. *Human Evolutionary Psychology*. Princeton: Princeton UP.

Bennett, Franklin Roy. 2015. *Evolution and Ethics. A Critique of Sociobiology*. New York: Palgrave Macmillan.

Brown, Donald E. 1991. *Human Universals*. New York: McGraw-Hill.

Buss, David M. 1995. "Evolutionary Psychology: A New Paradigm for Psychological Science." *Psychological Inquiry* 6.1: 1–30.

Carroll, Joseph. 2004. *Literary Darwinism: Evolution, Human Nature, and Literature*. New York: Routledge.

Dawkins, Richard. 1989 (1976). *The Selfish Gene*. Oxford: Oxford UP.

———. 1999 (1982). *The Extended Phenotype: The Gene as the Unit of Selection*. Oxford: Oxford UP.

Dunbar, Robin, and Louise Barrett. 2007. "Evolutionary Psychology in the Round." In *The Oxford Handbook of Evolutionary Psychology*. Ed. R. I. M. Dunbar and Louise Barrett. Oxford: Oxford UP. 3–9.

Gottschall, Jonathan, and David Sloan Wilson, eds. 2005. *The Literary Animal: Evolution and the Nature of Narrative*. Forewords by E. O. Wilson and Frederick Crews. Evanston: Northwestern UP.

Gould, Stephen J. 2011 (2003). *The Hedgehog, the Fox, and the Magister's Pox. Mending the Gap between Science and the Humanities*. Cambridge: Harvard UP.

Harding, Sandra. 1986. *The Science Question in Feminism*. Ithaca: Cornell UP.

Lancaster, Roger N. 2003. *The Trouble with Nature: Sex in Science and Popular Culture*. Berkeley: U of California P.

Miller, Alan S., and Satoshi Kanazawa. 2007. *Why Beautiful People Have More Daughters: From Dating, Shopping, and Praying to Going to War and Becoming a Billionaire—Two Evolutionary Psychologists Explain Why We Do What We Do*. New York: Perigee Book.

Parsons, Keith. 2003. *The Science Wars: Debating Scientific Knowledge and Technology*. Amherst: Prometheus Books.

Pinker, Steven. 1994. *The Language Instinct: How the Mind Creates Language*. New York: William Morrow.

———. 1997. *How the Mind Works*. New York: W. W. Norton.

Plotkin, Henry. 2007. *Necessary Knowledge*. Oxford: Oxford UP.

Rose, Hilary, and Steven Rose, eds. 2000a. *Alas, Poor Darwin: Arguments against Evolutionary Psychology*. New York: Harmony Books.

———. 2000b. "Introduction." In Rose and Rose, 1–16.

Rose, Steven. 1997. *Lifelines: Biology beyond Determinism*. Oxford: Oxford UP.

Salmon, Catherine. 2005. "Crossing the Abyss: Erotica and the Intersection of Evolutionary Psychology and Literary Studies." In Gottschall and Wilson, 244–57.

———, and Donald Symons. 2003 (2001). *Warrior Lovers: Erotic Fiction, Evolution and Female Sexuality*. New Haven: Yale UP.

———. 2004. "Slash Fiction and Human Mating Psychology." *Journal of Sex Research* 41: 94–100.

Schnettler, Sebastian. 2019. "The Sociobiology Wars." *Encyclopedia of Evolutionary Psychological Science*. Ed. Tood K. Shackelford and Viviana A. Weekes-Schackelford. Online.

Segerstråle, Ullica. 2000a. *Defenders of the Truth: The Sociology Debate*. Oxford: Oxford UP.

———, ed. 2000b. *Beyond the Science Wars: The Missing Discourse About Science and Society*. New York: SUNY P.

———. 2006. "Evolutionary Explanation: Between Science and Values." In Barkow, 121–47.

Slingerland, Edward, and Mark Collard, eds. 2012. *Creating Consilience: Integrating the Sciences and the Humanities*. Oxford: Oxford UP.

"Sociobiology." 2018 (2013). *Stanford Encyclopedia of Philosophy*. Online.

Tooby, John, and Leda Cosmides. 1990. "The Past Explains the Present: Emotional Adaptations and the Structure of Ancestral Environments." *Ethology and Sociobiology* 11: 375–424.

Wilson, Edward O. 1975. *Sociobiology: The New Synthesis*. Cambridge: Harvard UP.

———. 1978. *On Human Nature*. Cambridge: Harvard UP.

———. 1998. *Consilience: The Unity of Knowledge*. New York: Random House.

PART V

Contextualism—Changing the Paradigm in Literary and Performance Studies for the Twenty-First Century

A first question any analyst must ask: what is the context?

—David Gelernter

No simple answer, then, can be given to the question "What is context?"

—John Lyons

CHAPTER 14

Cognitive Literary Studies

> To put this more compactly: Interpretation is driven by the narrative context, not the code.
>
> —Paul Messaris

We are by no means the first to make many of the arguments of this book. It is only because we have had many specific examples of what can be done when certain aspects of cognitive science and biology form an explicit part of one's theoretical assumptions that we have been able to develop the program outlined in the previous chapters. Several good overviews, summaries, and updates to cognitive approaches to the study of literature exist. Rather than repeat them, our aim here is to concentrate (even if only briefly) on selected exemplary studies that provide a path for a humanities-based cognitive studies. This chapter is divided into the following sections: a brief look at the early and foundational work of two pioneers and some previously published overviews of the field; the rapidly growing and influential field of evolutionary approaches to the study of literature; cognitive poetics; reader-response and narrative theory; some additional cognitive approaches; and empirical studies in literature.

H. Mancing and J. Marston William, *Restoring the Human Context to Literary and Performance Studies*, Cognitive Studies in Literature and Performance,
https://doi.org/10.1007/978-3-030-89078-0_14

1 Pioneers and Overviews

As mentioned earlier, Norman N. Holland has been an influential pioneer in the field of literary studies with a cognitive-scientific orientation. After a long period of following a brand of reader-response theory centered on real human beings reading literary texts and grounded in Freudianism, Holland discovered neuroscience in the 1980s and substantially redefined his approach to literature. Although he never abandoned Freud, Holland was the first prominent literary scholar to take brain science seriously.

Holland's *The Brain of Robert Frost* (1988) presents an approach to reading literature that is as relevant today as it was when it was written, remaining a foundational book in the field. As Holland proudly notes, this is "the first book to bring to bear on literary criticism and theory the revolutionary discoveries of cognitive science and recent research into the brain" (v-vi). Holland's model of reading is consistent with the major ideas from biology, neuroscience, and cognitive psychology reviewed throughout this book. Holland underscores both what we have in common (our species-specific biological characteristics, our universal human nature) and what makes each of us unique (our individual circumstances, memories, experiences, and identity or sense of self), noting that this makes it inevitable "that we would simultaneously understand and misunderstand each other, because we are both different from and the same as our fellow humans" (172). In Sect. 4 of Chapter 4, we discussed Holland's use of the Lakoff and Johnson approach to metaphor for the experience of reading. (And a special endorsement from author Howard Mancing, who has been profoundly influenced by Holland's scholarship: I still believe that more than three decades later, it is one of the finest descriptions of what literature is all about and how it functions.)

In the new century's first decade, Holland completed a book that was long in the works and one of his last major publications before his death in 2017: *Literature and the Brain* (2009), the culmination of his work on neuroscience and literature. Rejecting that Freudian psychoanalytic theory "has no scientific standing" (16), Holland argues, "a good deal of psychoanalytic psychology can fairly claim the status of 'science,'" and states that he has drawn only on "psychoanalytic ideas that have received experimental confirmation" (16). Holland lays out his project as follows:

> The new discipline of "neuropsychoanalysis" looks at brain systems that might correspond to traditional psychoanalytic ideas like id, ego, and

> superego, repression, regression, impulsitivity, and compulsivity as well as psychiatric disorders like depression or schizophrenia. In an increasing number of articles and books, this group is tracing the ways brain systems correspond (approximately!) to long-applied and, sometimes, well-demonstrated psychoanalytic concepts. (18–19)

But notice the hedging here: "might correspond," "approximately!" and "sometimes." As with the discussion about the defense of psychoanalytic thought by Mark Solms in Sect. 3 of Chapter 8, many readers will not be convinced. We will not attempt to describe (or refute) Holland's many affirmations about the scientific validity of Freudian thought, but suffice it to say that his book consists of a mixture of often very valid and interesting discussions of both neuroscience and literature with a psychoanalytical theory that we find questionable. But as we try to avoid the binary thinking pitfall, we know that because we don't agree with every aspect of his scholarship, this does not mean he is automatically our rival! On the contrary, we are on the same side in promoting a biology and brain-oriented approach to literary studies.

We maintain that Holland's approach to cognitive studies of literature is often thought-provoking and original. Drawing on the work of scientists discussed throughout this book—Damasio, Gerrig, Nell, Pinker, and others, especially and at some length, Panksepp—Holland writes perceptively on attention, creativity, evaluation, memory, emotions, Theory of Mind and metarepresentation, and related subjects. Holland was an innovative leader in the movement to study literature within a framework of the cognitive sciences.

Mark Turner has arguably been the most prominent leader of the turn toward cognitive science in the study of literature. His first two books on cognitive metaphor theory (Turner 1987; Lakoff and Turner 1989) are still indispensable for anyone approaching that aspect of the cognition-literature connection. His *Reading Minds: The Study of English in the Age of Cognitive Science* (1991) both extends and deepens his earlier work and specifically addresses the relationships between the study of literature and cognitive science in a broader sense. More than anyone ever had, he specifically asks how it is possible to continue reading and writing about literature in a precognitive framework. Turner recognizes that his is "not a humble book" (viii). His proposal is radical: "I propose a revised concept of the humanities as the inquiry into what constitutes human beings and human acts. I take language and thought as constitutive of what is human.

On this view, the fundamental activity of the humanities is the discovery of the nature of human language and human thought" (47). We endorse his proposal wholeheartedly.

Next, Turner's brilliant *The Literary Mind* (1996), discussed previously in this book, sets forth the proposal that our normal cognitive processes in the real world are exactly as those we have always attributed to literary thinking:

> We imagine realities and construct meanings. The everyday mind performs these feats by means of mental processes that are literary and that have always been judged to be literary. Cultural meanings peculiar to a society often fail to migrate intact across anthropological or historical boundaries but the basic mental processes that make these meanings possible are universal. (11)

Combining insights from literature with those from neuroscience (Edelman, Damasio) and cognitive psychology and linguistics (Rosch, Lakoff), Turner's book makes an excellent case for the intimate—and inevitable and essential—relationship between literature and human embodied cognition. Finally, his work with Gilles Fauconnier, *The Way We Think* (Fauconnier and Turner 2002), and his later crossover book on the topic, *The Origin of Ideas* (2014), has made conceptual blending a central concept in contemporary thought.

With the exemplary and foundational work of Holland and Turner as a starting point in the 1980s, the field has grown rapidly. Since then, there has been a spate of introductions and overviews to the field of cognitive literary studies, or to its subfields. These overviews are very uneven (in general, the more recent ones offer a view of this rapidly expanding field that is both wider and deeper), but all of them introduce the reader to a number of specific works of theory and criticism; among the best in our opinions are Hart (2001), Richardson (2004, 2006), Jaén and Simon (2012a), and Mancing (2012). Almost all consider in some way THEORY's conceptual and theoretical inadequacy. As F. Elizabeth Hart has written,

> To varying degrees, cognitive literary and cultural critics either reject poststructuralism wholesale; or they seek to revise poststructuralist insights by calling into question the Saussurean roots of deconstruction while simultaneously confirming the theorized effects of socially constructive discourse and ideology. (2006, 34)

The following sections of this chapter illustrate the Rosch-Lakoff approach to categorization. Boundaries are fuzzy, structure is radial, and placement is somewhat arbitrary. It is rarely possible, for example, to discuss reader-response without simultaneous consideration of issues such as narrative theory, genre, or empirical studies. Innovative scholars are not prone to limiting their work to neat little niches. Thus, our organization here is rather arbitrary side, but not intended to be capricious.

2 Evolutionary Approaches

The most significant new approach to literature that has come from the disciplines discussed in this book is the one with explicit origins in evolution and biology, an approach often referred to as Literary Darwinism or Darwinian Literary Studies (DLS). Nothing is more diametrically opposed to the socially constructed subject than the evolved human animal. The leaders in the field have been Joseph Carroll, Jonathan Gottschall, and Brian Boyd.

Joseph Carroll's massive and groundbreaking *Evolution and Literary Theory* (1995) brought the concept of a Darwinian, or evolutionary, approach to the study to literature to the forefront. In laying out his program, Carroll stresses four points: (1) most importantly, the organism-environment relationship; (2) innate, evolved psychological structures; (3) the importance of ultimate as well as proximate causes; and (4) representation as a form of cognitive mapping (2). Carroll evokes Darwin, John Bowlby, and other evolutionary theorists and psychologists to argue that the "central purpose of literature … is to make sense of the world, to represent one's own sense of things" (426). At the same time, the book is a sustained, acerbic, and frontal attack on the assumptions of textualism and indeterminacy of hegemonic THEORY—particularly the versions promoted by Derrida and Foucault. Carroll has been the most prolific and influential critic/theorist writing from an evolutionary base. A number of his shorter essays, some previously published elsewhere, others original, are included in a second book, *Literary Darwinism* (2004). He updates his work, explores new areas, and continues his critique of THEORY in his collection *Reading Human Nature* (2011). Carroll is the biggest proponent of a Darwinian approach to literary study. His target article for a special issue of the journal *Style*, "An Evolutionary Paradigm for Literary Study" (2008a) together with the follow-up, "Rejoinder to the Responses" (2008b), features his overview of the field.

The most serious collaborative entry into the field of Darwinian approaches to literature is Jonathan Gottschall and David Sloan Wilson's anthology *The Literary Animal* (2005b), a collection of essays by literary scholars, social scientists, biologists, and even a distinguished novelist (Ian McEwan). Gottschall and Wilson (2005a) describe the difficulties they had in getting the book published because of resistance from the literary (THEORETICAL) establishment. The essays are uneven in quality, but some are penetrating interdisciplinary inquiries of the highest quality. This book is a good starting place for anyone interested in the relationships between biology/evolution and literary theory and criticism.

Gottschall joins Carroll as one of the most vocal proponents of an evolutionary approach to literature. He is also the scholar in the group who most advocates using quantitative methodologies in literary research, and this research area also fits within the section on empirical studies below. In *The Rape of Troy* (2008a), based on his controversial dissertation at SUNY-Binghamton, where he worked with Wilson, Gottschall lays out a case for reading Homer's epic poems as classic examples of some of the basic themes of Darwinian thought: "I argue that direct competition for women is a main cause of warfare and interpersonal male violence in the Homeric world. Like roads circling back to Rome, almost all of the main conflicts of the *Iliad* and *Odyssey* trace back to disputes over women" (58). Gottschall's position is that underlying all the proximate causes—such as the desire for honor and wealth—of the events that take place in the two poems lies the ultimate cause of sexual selection: "sexual selection has shaped men to compete for women *and* for concrete material resources *and* for intangible social resources because they are all *reproductive resources*" (49). As mentioned briefly in Sect. 6 of Chapter 8, Gottschall's *The Storytelling Animal* (2012) makes a case for the universality of narrative and stories.

The third leader in the growing field of Darwinian criticism is Brian Boyd, previously well established and highly regarded as a Nabokov scholar. He has proven himself to be as aggressive in his critique of THEORY as eloquent in his discussion of the ways that an understanding of the basics of evolution can inform literary criticism and theory. He makes one of the best side-by-side comparisons of THEORY and a "biocultural" approach to literature by comparing the legitimate (and compelling) concerns with both to such vital issues such as human nature, mind and self, power, ideology, sex and gender, individuality, intention and interpretation, and meaning (2008, 49–54). This presentation alone makes it clear that THEORY does not have, as it often implies, total and absolute claim to many of the most important concerns of humanist scholars in the twenty-first century, and that there are compelling reasons to address these and other questions from a position grounded in cognitive science.

In 2009, Boyd published his long-awaited book *On the Origin of Stories*, a work echoing Darwin's original title and encompassing a decade of his work in the field. In the book's first part, Boyd provides a solid introduction to many aspects of evolutionary theory. In the second and third parts, he makes a convincing case for art in general, and fiction particularly, as evolutionary adaptations. In the final two parts, he presents an analysis of two disparate works of narrative literature: Homer's *Odyssey* and Dr. Seuss's *Horton Hears a Who!* Along the way, Boyd maintains the strong critique of THEORY from his earlier studies. Boyd defines art as "a kind of cognitive *play*, the set of activities designed to engage human *attention* through their appeal to our preference for inferentially rich and therefore *patterned* information" (85). His take on fiction is similar: "Fiction, like art in general, can be explained in terms of cognitive play with pattern—in this case, with patterns of social information—and in terms of the unique importance of human shared attention" (130). Boyd's concluding chapter, "Conclusion: Retrospect and Prospects: Evolution, Literature, Criticism" (380–98) is the best place to see both his critique of THEORY and his case for "evocriticism" and is one of the best brief presentations of many aspects of the point of view we espouse throughout this book.

Boyd, Carroll, and Gottschall collaborated on a collection of the best essays in the field in a large book titled *Evolution, Literature, and Film: A Reader* (2010). The editors remind us, once more, as we maintained in the introductory chapter, that the assumptions behind classic THEORY did not go away in the twenty-first century:

> Many scholars working under the influence of "New Historicism" or "cultural studies" now claim that they are "post-theory" because they focus not on theories but on "empirical" historical data, especially data gleaned from archives. In reality, the archivalists have not left poststructuralist theory behind but have only internalized it. The categories they use derive chiefly from Foucauldian traditions: versions of Marxism and Freudianism filtered through deconstructive epistemology and cast in a programmatically oppositional mode. Simply eschewing explicit theoretical commitment does not invest findings with empirical validity or insulate them from theoretical

> critique. It has certainly done little to reverse the declining status of the humanities within the larger world of knowledge. (2010, 2)

Along with their critique, the editors provide a strong justification for incorporating an evolutionary approach to the study of literature. The book begins with a set of classic essays on evolution, evolutionary psychology, and evolutionary approaches to art and aesthetics by many of the scientists and others experts discussed in earlier chapters. These are followed by a stellar collection of essays (written between 1996 and 2008) by literature and film scholars. This is the place where students and scholars who desire to become familiar with evolutionary approaches to literature should begin.

Carroll, Gottschall, and Boyd may be the most prominent names in the field of evolutionary approaches to literature, but they are far from alone. The books by Storey (1996), Love (2003), Flesch (2007), Easterlin (2012), and Blair (2017) all deserve recognition. The edited volume of Craig and Linge (2017) offers a number of perspectives on Darwin's influence on multiple literary traditions around the turn of the twentieth century.

One of the richest and most innovative areas of non-THEORETICAL inquiry is the one grounded in evolutionary principles and values. When the evolved human mind-brain, human universals, Darwin's theory of sexual selection, theories of kin selection and reciprocal altruism, real male and female sex differences, homicide and jealousy, marriage and infidelity, parent-child relationships, metaphorical thought and conceptual integration, and Theory of Mind are taken into consideration, one clearly can make a myriad of observations and commentaries on literature that are not possible to make within a psychoanalytic, Marxist, or deconstructionist framework.

3 Cognitive Poetics

The second major area of research grounded in principles discussed throughout this book is cognitive poetics (also sometimes called cognitive stylistics or cognitive rhetoric, or simply cognitive literary analysis). As defined by Paula Leverage, one of the leaders in the field,

> [c]ognitive literary analysis then is a formal analysis of literary texts which applies cognitive psychology and linguistics to literary texts. The underlying assumption is that the cognitive processes through which we react to stimuli and situations in the non-literary world are the same as those with which we negotiate the everyday, and that literary texts then exploit these capacities for aesthetic effect. (Leverage 2010, 112)

This orientation has two points of origin. The first is the tradition initiated by Reuven Tsur, whose early work in cognitive poetics dates from the 1970s and is grounded, at least to some extent, in structuralism and Russian Formalism, as well as early work in the cognitive sciences. Tsur's work is best seen in his two books *What Is Cognitive Poetics?* (1983), a brief introduction to the new field of inquiry, and *Toward a Theory of Cognitive Poetics* (1992), in which he elaborates his approach to poetry, placing emphasis on metrics, sound patterns, and related matters. In some ways, the impressive work of prolific cognitive literary scholar Mark Bruhn (e.g., 2009, 2011a, 2011b, 2015, 2017a, 2017b, 2018a, 2018b, 2018c) follows in this tradition, while forging its own original path. In addition, Bruhn and Wehr's co-edited essay volume *Cognition, Literature, and History* (2013) is a must-read for anyone working in the field of cognitive poetics.

The second tradition is more explicitly derived from contemporary cognitive linguistics and has its starting point in Lakoff and Johnson's foundational *Metaphors We Live By* (1980), a book whose importance and influence we have mentioned before but can hardly be exaggerated (see Sect. 4 in Chapter 4). Important works coming out of the Lakoff-Johnson tradition include the books by Turner, and Lakoff and Turner, as discussed above. Peter Stockwell's text, *Cognitive Poetics* (2002), is a practical handbook and good introduction to the current state of the field. Among the best and most cited collections of essays, both theoretical and interpretive, are *Cognitive Stylistics* (2002), edited by Elena Semino and Jonathan Culpeper, and *Cognitive Poetics in Practice* (2003), conceived as a companion volume to Stockwell's book, edited by Joanna Gavins and Gerard Steen. The opening essay in the latter book by the editors (Gavins and Steen 2003) is an excellent place to start.

Another leader in the field is F. Elizabeth Hart, whose work in several essays (e.g., 1995, 2001, 2004, 2006, 2011) is exemplary. She consistently stresses a weak constructivist epistemology and embodiment as she calls for the recognition of the importance and relevance of cognitive studies: "The radical nature of the departure that second-generation cognitive science represents cannot be overstated because it amounts to a large-scale rejection of philosophical rationalism, on whose premise regarding the split of mind from body much of Western philosophy has subsequently been built" (2004, 95). All the while, she consistently points out the shortcomings and problems of some aspects of THEORY.

A second major figure is Mary Thomas Crane, whose *Shakespeare's Brain* (2001) is a study of Shakespeare's polysemous lexicon toward understanding the agency behind the works' language. Taking issue with THEORETICAL concepts of authorship, particularly Foucault's, Crane grounds her work in contemporary cognitive science (Damasio, Edelman, Lakoff, and others) in the conviction that "[a] focus on Shakespeare's brain allows us to attend to Shakespeare as author without losing the complexity offered by contemporary theory" (3). Among others whose work fits largely within the concept of cognitive poetics are Alan Richardon and Lisa Zunshine, whose work is discussed below.

Darwinian and cognitive rhetorical orientations are probably the two most important and most productive areas of non-THEORETICAL literary study. The two areas should complement each other, as cognitive linguistics is solidly grounded in neuroscience and biology. Yet the two approaches are not always in close harmony. Some of the practitioners of cognitive poetics seem to have little knowledge or awareness of evolutionary science or neuroscience; too often their work tends toward the abstractly rhetorical instead of acknowledging and incorporating embodied cognition. Some of the Darwinian scholars can at times be insensitive to matters of style, technique, and tone, placing too much emphasis on theme and character.

Joseph Carroll (2004, 130–36) is largely correct, if a bit harsh and over-generalizing, in his critique of cognitive rhetorical studies and his argument that such studies would gain from more explicit incorporation of biological and evolutionary concerns. For example, he oversimplifies and misrepresents the field when he writes that "cognitive rhetoricians tend to seek common ground with the discourse theory of poststructuralism, and they are uncomfortable with adaptationist claims that human nature consists in a highly structured set of motivational and cognitive dispositions that have evolved through an adaptive process" (xv). It is true that some do see their work as being within the poststructuralist tradition (e.g., Spolsky 2002; Zunshine 2010a), but others explicitly do not (e.g., Mancing 1999, 2003b; Hart 2006). Some may be uncomfortable with evolutionary approaches, while others are not; many deal with other (linguistic, stylistic) issues, and the question of their attitude toward adaptationist approaches is not raised. By the same token, some of those in the cognitive rhetoric group—e.g., Richardson and Zunshine—are critical of the Darwinian approach; see the report by journalist Britt Peterson (2008) and Zunshine's criticism (2010b, 313 n1). Perfect harmony among cognitive literary scholars with different focal points is neither possible nor necessarily desired: on the whole, scholars working with methods that align with modern cognitive and biological sciences have tended to respect each other's work and remain civil. There are healthy debates and critiques, but as yet, no outright "wars."

4 Reader-Response and Narrative Theory

Richard J. Gerrig is the cognitive psychologist who has done the most to further the understanding of literary reading. His *Experiencing Narrative Worlds* (1993) humanizes the reading of fiction by presenting it as an individually meaningful, contextualized *experience*. The guiding metaphors of readers are those of "*being transported* by a narrative by virtue of *performing* that narrative" (2). Gerrig deals with issues such as inference, participatory responses, the role of side participants and overhearers, authorial intentionality, anomalous suspense, the fiction vs. nonfiction debate ("toggle theories"), Spinoza's "willing creation of disbelief" (in place of Coleridge's "willing suspension of disbelief"), and other matters in ways that, if taken more frequently into account, could radically reorient the study of fictional narrative. Just as Gerrig's book deals with narrative theory and the act of reading, it is also a prime example of empirical approaches to literature discussed later in this chapter. His carefully crafted experiments, mostly published with a series of collaborators, are major contributions to our understanding of literary reading.

Real, embodied, and contextualized readers actively and imaginatively create their best understandings of the works they read. A novel or a poem does not determine how it is to be understood; it merely *affords* (referring back to Gibson's term) readers opportunities to create understandings. And the text also constrains our creative and sympathetic understandings. *Hamlet* is not about space exploration; *Candide* is not about baseball; *Beloved* is not about robber barons. It is only natural that each of us, with our I-language, sense of self, personal past, values, memories, feelings, expectations, and hopes will have (at least slightly and perhaps very) different understandings of the same text. We may be *attuned*—another Gibsonian term—to specific aspects of a text, such as the role of women, stylistic register, political orientation, consciousness of class relationships, considerations of genre or canon, racial prejudice, gender identities, structure of the work, patterns of metaphor, and so forth. Ideally, we attempt (again, using our ToM) to refine our unique understandings in dialogic exchange.

Since the major thrust of this book has been to direct attention to the embodied, contextualized reader of works of literature, it should be apparent that we see cognitive approaches to literary theory and criticism from the reader's (or audience's) point of view. Literary study grounded in cognitive science will, almost of necessity, be a form of reader-response criticism (and similar approaches to performance will concentrate on spectator experience; see Chapter 15). Recall Holland's notions on how readers craft, perform, and construct our literary experiences. Keith Oatley holds the same opinion and, simultaneously, uses a Gibson-like concept of the literary text as affordance and repeats his claim for reading as simulation: "The writer offers a kit of parts, or a set of cues. The reader does the construction, and makes the imagined dream start up and run.... Fictional narrative is a kind of simulation, but one that runs on minds rather than on computers" (2003, 166). We agree, and we take the work of literary scholars who place emphasis on the reading process and the reader's understanding of authorial intentions and characters' thought processes as being most representative of cognitive approaches to literature. Together, they form an area where more traditional theory has laid the foundation for cognitive approaches. It is not necessary to reject completely the THEORETICAL past to participate in the (cognitive) theoretical future.

David Herman's *Story Logic* (2002) is a major contribution to structuralist narratology with some contributions from other fields such as possible worlds theory and cognitive science (especially the concepts of scripts, frames, and schemata). Herman defines the title concept as "the logic by virtue of which people (including writers) know when, how, and why to use stories to enable themselves and others to find their way in the world" (24). Herman continues along these very interdisciplinary lines with highly engaging *Storytelling and the Sciences of Mind* (2013), in which he looks closely at various types of narrative in respect to what they can contribute to our understanding of intelligent life, relying in the process on research in psychology, philosophy, and linguistics. Herman has also edited two important collections of essays on cognitive approaches to narration and Theory of Mind. The first is *Narrative Theory and the Cognitive Sciences* (2003), a major work by all accounts, exploring aspects of narrative theory in light of CONTEXTUALISM and the cognitive sciences. As such, it breaks important new ground in the study of narratives, even as some of the essays struggle to reconcile new approaches and concepts with traditional structuralist narratology. The second is *The Emergence of Mind* (2011), essays on consciousness in English literature, with contributions by several of the scholars cited elsewhere in this chapter. It too is an original contribution to the field, albeit limited by the fact that it considers only English literature.

Marisa Bortolussi and Peter Dixon's *Psychonarratology* (2003) is a comprehensive work that makes significant contributions to multiple aspects of narrative theory, including characterization and reader-response theory, as well as to the empirical study of literature. Drawing on psychology, linguistics, discourse processing, and literary theory and criticism, the authors' goal is "to put methods of cognitive psychology at the service of understanding what we argue is the least understood dimension of narrative: its cognitive processing" (3). Their distinction between "textual features" and "reader constructions" (28) brings clarity to an area that has often been muddled, recognizing that the process of reading is at heart a matter of the reader's individual and contextualized, creative and imaginative, understanding of the objective reality (and the affordances) of the text: "What readers do will vary with the characteristics of the individual reader, the nature of the text, and the context in which the reading takes place" (35). Further, their proposal that the reader-narrator relationship is one of a conversation (29–30) humanizes the reading process in psychologically sound ways and recalls Bakhtin's dialogism. Their treatment of literary characters is exemplary: "literary characters are processed as if they were real people, and real people are processed in terms analogous to the categories brought to bear on the interpretation of literary characters" (140).

The authors provide a good, concise survey of some of the major approaches to reader-response criticism (5–10), perceptively calling attention to the theoretical inconsistencies and shortcomings of much of the most influential work done in the field. In particular, they contend—accurately—that many of the theories of Stanley Fish, Norman Holland,

Wolfgang Iser, and others are more intuitive than demonstrable, and that these theorists' work is largely uninformed by research in fields such as cognitive psychology and discourse processing. Inexplicably, however, while recognizing Holland's early and more subjective work, they fail to cite his later work that pioneered the very field they themselves claim. Similarly, they seriously undervalue, we believe, many of the most important contributions of Gerrig (1993) and completely ignore the pioneering work of Louise Rosenblatt (1994a, 1994b, 1995). It is frustrating to see how modern scholars (Holland, Bortolussi, and Dixon, for example) working within a cognitive paradigm keep reinventing the model that Rosenblatt so elegantly described in 1938 and repeated even more elegantly in 1978.

Blakey Vermeule's *Why Do We Care About Literary Characters?* (2010) is a major contribution to the cognitive study of narrative fiction. Recalling Robin Dunbar's work on grooming and gossip as discussed previously, her short answer to the question posed in her title is "gossip: we need to know what other people are like, not in the aggregate, but in the particular" (xii). She defends what she calls "an unlikely view: the reasons that we care about literary characters are finally not much different from the question of why we care about other people, especially people we have never met nor are ever likely to meet" (xiii). Vermeule deals prominently with the role of ToM in the understanding of fiction, although she uses the term Machiavellian intelligence more frequently, reasoning as follows: "if Machiavellian intelligence is the answer to why we are driven to make sense of other people, mind reading is the tool we use to put those abilities into play" (34–35). Vermeule thus agrees with virtually all others who study literature from a cognitive orientation that literary characters are best thought of as though they were real people. To do so, she writes, "is a habit lodged deep in the human psyche, and no amount of literary-critical sophistication is likely to cure people of it" (176). Holland makes the same point: "And even sophisticated literary critics treat fictional characters like real people" (2009, 107). Surely they are both right.

More recently, Vera Tobin (2018) has examined how our brains conspire with narration to help create the element of surprise. Through thoughtful close readings of disparate novels from Shakespeare's *Othello* to Richard Adams's *Watership Down* and Ian McEwan's *Atonement*, Tobin intriguingly and accessibly explores narrative unreliability and the "curse of knowledge" as well as the curse of empathy: "We care for literary characters in ways that are very like our attachments to other people in life," (282). She concludes that human brains are drawn to plot twists and eschew "spoilers," but ultimately crave the satisfaction of how simple and unsurprising everything seems when we look back on a text with hindsight.

As we have argued, narrative is one of the most important (arguably *the* most important) modes of epistemology; our thought is far more often narrative than it is logical. Narrative is also the literary art that most probably can be considered a part of our adapted minds; see the work of Scalise Sugiyama cited in Sect. 6 of Chapter 8, several of whose essays are also superb literary studies; a book-length collection/summary of her work is sorely needed. The human mind-brain functions by means of narrative as part of its neural structure. As Bakhtin (1981, 3–40) has suggested, the novel is the quintessential expression of the post-medieval human; it is the genre of modernity; it is the natural mode assumed by contextualized dialogism.

The biology and psychology of narrative may be one thing, but what about the techniques and rhetorical strategies of narrative texts? If we recognize that our concepts of authors, just like our concepts of ourselves and other people, are our own constructs, we have no need for a special term for the construct we have of the author. In other words, the pride of both formalist and structuralist narrative theory, Wayne Booth's "implied author," is redundant and unnecessary, and the same is true of the "implied reader." Narratives—novels and stories, biographies and autobiographies, memoirs and travel literature—like our memories and our sense of self—are always more-or-less fictional/factual versions of people and events. We know there is a teller, an author or a narrator, but we need not place any other formalist or structuralist ghosts in the text. See Richard Walsh (2007, 69–85) for a solid articulation of the position that the narrator of a story or novel is either a fictional narrator or the author; he argues solidly that concepts of implied author and implied reader are unnecessary. Mancing has argued, for example, that it is parsimonious to identify Cervantes himself, rather than an unnamed implied or inferred narrator, as the primary narrator of *Don Quixote* (2003a, 2003b, 2005). Our use of ToM to understand authors, narrators, characters, and ourselves and others as readers may turn out to be one of the most powerful pragmatic tools of our mind-brains as they read, understand, and write and teach about literature.

5 Other Cognitive Approaches[I]

Memory has become a growing field of inquiry in cognitive literary studies. Here the earliest important books are Nicolae Babuts's unjustly ignored *The Dynamics of the Metaphoric Field: A Cognitive View of Literature* (1992) and *Memory Metaphors, and Meaning* (2009). Both books present his idea of the "metaphoric field," which he defines as "a mnemonic space functionally organized to create meaning in terms of a dominant urgency…. Metaphors and dynamic patterns that gravitate around the urgency of that response are modulated and constitute a metaphoric field" (1992, 81). David C. Rubin's *Memory in Oral Traditions* (1995) is the first book-length study of orality from the point of view of our contemporary understanding of memory and mental imagery as described earlier in this book. Paul John Eakin's *How Our Lives Become Stories* (1999) significantly moves beyond his earlier work on self-writing that began in the 1970s and now deals with "recent research in neurology, cognitive science, memory studies, developmental psychology, and related fields to the task of rethinking the nature of self-experience" (xi).

Suzanne Nalbantian's *Memory in Literature* (2003) consists of penetrating studies of the work of writers like Jean-Jacques Rousseau, Marcel Proust, Virginia Woolf, André Breton, and several others. She draws upon the best work of some of the most outstanding memory specialists and neuroscientists, such as Schacter and Tulving, Berthoz, and Damasio, and she calls particular attention to the role of the amygdala and the hippocampus in memory processing. Evelyne Ender's *Architexts of Memory* (2005) deals, not surprisingly, with some of the same writers featured in Nalbantian's book: Proust and Woolf, along with others such as George Eliot and Gérard de Nerval. She, too, draws heavily on the work of major researchers and clinicians discussed throughout this book: Damasio, Sacks, Edelman, and Schacter (as well as the psychoanalytic approach of Freud and Lou Andreas-Salomé). Paula Leverage's *Reception and Memory* (2010) is the first book-length study of medieval literature grounded in cognitive science. Rather than focusing on the mode of production—the mnemonic techniques used by the singers/tellers of the *chansons de geste*—Leverage approaches the subject from the point of view of reader-response theory by discussing the role of the listening audience(s) in recalling repetition and division. Her study draws heavily on the approach to reading and memory by Gerrig and the idea of conceptual blending as discussed by Turner, combined with the rigorous work of neuroscientists like Edelman. To this she adds ideas from the empirical studies of David Miall and Zunshine's work on Theory of Mind.

Emotion and its relation to cognition have become a hot topic both in cognitive psychology and in cognitive literary studies. Keith M. Opdahl's *Emotion as Meaning* (2002) is a complex book, the scope of which includes a critique of semiotics and deals in some detail with emotion, imagination, mental imagery, memory, embodied cognition, the construction of meaning, and reader-response. Opdahl argues that we respond to and construct our understanding of a literary work as much by embodied emotion as by any other means: "One simply cannot do justice to a work of fiction without imagining it, and one cannot imagine a work of fiction without feeling it" (185).

Suzanne Keen's *Empathy and the Novel* (2007) was a strong contribution to the renewed interest in how readers perceive literary characters and the extent to which they experience feelings of empathy during their reading. Much like Palmer and Zunshine, discussed below, Keen affirms the stance that "[w]e often respond to characters as if they were human beings like us," which stands in opposition to "the professional convention of treating them as 'word masses' rather than as people" (101). Grounding her work in cognitive science (universal human emotions) and neuroscience (mirror neurons), as well as results of sociological studies of reader groups, Keen is concerned with the questions of how, when, and to what extent readers empathize with characters—if such empathy is more than the fervent hope of many literary scholars, philosophers, and psychologists. Importantly, Keen (146–47) argues that readers' understanding of and empathy for others are best developed when readers, especially when guided by an experienced teacher, engage in discussion—we would say Bakhtinian dialog—about the works read. One of the most moving examples of the empathy effects achieved by reading can be found in Azar Nafisi's beautiful book *Reading Lolita in Tehran* (2003); see Keen (2007, 166). See also Fritz Breithaupt's two provocative works on empathy, *Kulturen der Empathie* (Cultures of Empathy, 2009) and *Die dunklen Seiten der Empathie* (The Dark Sides of Empathy, 2017), as well as Frans de Waal's *The Age of Empathy* (2009) for an excellent and notably readable discussion of empathy's parallels between humans, primates, and other animals. Reynolds and Reason's essay collection *Kinesthetic Empathy in Creative and Cultural Practices* (2012) is a rich interdisciplinary contribution to empathy studies as relevant to the arts and humanities, covering embodied expression of performers and audiences in the realms of film, multimedia performance, dance, everyday actions, and even interactive sports equipment.

Empathy has continued as a popular topic of discussion and scholarly debate in recent years. Paul Bloom, who played a leading role in our deliberations above about the mind's tendencies toward categorization and essentialism (see Sect. 5 in Chapter 8), came out with his controversial *Against Empathy* trade book in 2016, in which he argues for "rational compassion" over a traditional understanding of empathy for guiding our decision-making. More recently, Muriel Cormican has pioneered an exciting new path in empathy and affect studies with her nuanced term "tender gaze" to describe the compassionate moments represented in literature, as well as on screen and on stage, eliciting perspective-taking by the reader (or spectator) that promote altruism and empathy while reducing implicit bias (Cormican 2021).

Novelist and psychologist Keith Oatley has written repeatedly on the role of emotions and empathy in the reading experience. His book *The Best Laid Schemes* (1992), a seminal psychological study of human emotions, ends with a discussion of what has become his signature concept: simulation. He describes art in general as a "simulation that runs on our minds rather than on a computer. The main value of any simulation is in testing the internal consistency of its assumptions, enabling them to be explored and to be improved" (398). More specifically, he says, a "play or novel allows us to compare an explicitly simulated set of actions and their results with emotions that are our own and that may move us considerably" (126). In his book *Such Stuff as Dreams* (2011), Oatley reshapes his simulation argument by claiming, as his title implies, that fiction "is a guided dream, a model that we readers and viewers construct in collaboration with the writer, which can enable us to see others and ourselves more clearly" (ix). Oatley takes advantage of his experience as both a psychologist and a novelist to make an important contribution to the field. Finally, he reminds us of an important truth: In the best fiction, he writes, "we are not told what to think and feel but are offered something about which we can think and feel" (46).

Ellen Spolsky's *Gaps in Nature* (1993) is, in many ways, a foundational document in cognitive approaches to the study of literature. Spolsky draws on the work of Eleanor Rosch and others on categorization, the modularity of mind as originally presented by Fodor (1983), and the computational approach to perception as developed by David Marr (1982), to study the theory and function of literary genres, a genre being almost a prototypical example of categorization. Her blend of fuzzy categorization, structure and function of the mind-brain, computational processes, feminist theory, and poststructuralist approaches to texts is highly original. Although Spolsky's attempts to find common ground between cognitive approaches to literature and some core assumptions of THEORY, especially deconstruction (see Spolsky 2002), are problematic, it is clear her book is one that future studies of literary genre will find indispensable. Spolsky has one of the broadest outlooks among the theorists and critics mentioned in this chapter, as her work ranges over areas of cognitive rhetoric, poststructuralism, and evolutionary theory.

Cognitive theories of schema and categorization (see Sect. 5 in Chapter 8) open the way to new approaches to the understanding of literary genres. A conception of genres as examples of prototype theory, existing with a radial structure, and subject to reconfiguration as our schemata change, liberates us from formalistic straightjackets and clichés about conventions and necessary and sufficient conditions to be met and from ideological determination. If the structures of a genre do not form a natural kind, but are perceived contextually by each individual reader, genre theory becomes more individualistic and psychological. In effect, genres are not natural categories; in Chomskyan terms, they are not E-genres, but I-genres. Recall Chomsky's (1995, 15) description of the I-language of a hypothetical person named Jones. We could rewrite it in terms of a literary genre: "The concept of genre is internal, in that it deals with an inner state of Jones's mind/brain, independent of other elements in the world. It is individual in that it deals with Jones, and with literary communities only derivatively, as groups of people with similar I-genres. It is intensional in the technical sense that the I-genre is a function specified in intension, not extension." That is, Jones's concept of a genre—say, the domestic novel, or the epic—is probably very similar to, but not exactly the same as, the concept of the same genre by someone else. Except for Spolsky's 1993 work and her later equally masterful book *The Contracts of Fiction* (2015), there is to date no major work on genre theory and cognitive concepts, but a series of shorter studies work toward such a position; see, for example, Mancing (2000), Hart (2004), and Sinding (2004). More recently, Margaret H. Freeman (2020) turns to the embodied schemata at the heart of conceptual metaphors, along with blending and other principles of "embodied subjectivity" (16) to explain the iconicity of poetry as the common ground of poet and reader, and to demonstrate how "[s]cientists and poets are not pursuing different realities: they are pursuing them from different perspectives" (1). Freeman incorporates an impressive number of cultural traditions and time periods into her study and includes a helpful glossary of relevant terms with less than conventional definitions, helping us to see seemingly obvious but deceptively complex concepts such as "aesthetic" and "cognition" in a nuanced light. Indeed, one of the book's contributions to scholarship is her suggestion that we rethink our use of the latter term, which Freeman proposes be used "not in adjectival but in nominalized form, as in poetic cognition, aesthetic cognition, and so on" (172).

Alan Palmer, in *Fictional Minds* (2004), winner of both the Perkins Prize and the MLA Prize for Independent Scholars in 2005, is very much a theoretical maverick. His fine, detailed, and extended critique of traditional structuralist narratology is exemplary as he proposes dealing with fictional characters in the same way we deal with people in our lives, employing the techniques and methods of cognitive psychology, philosophy of mind, psycholinguistics, and systems theory rather than those of semiotics, structuralism, or poststructuralism. Palmer takes characterization to be "an inference from an individual action, then, toward a supposed disposition or trait, and these are states of mind that extend over time" (40). His concept of both real human beings and literary characters as embodied, socially contextualized beings is quite consistent with the views of this book. He argues that "[j]ust as in real life the individual constructs the minds of others from their behavior, so the reader infers the workings of fictional minds and sees these minds in action from observation of characters' behavior and actions" (246). Cognitive literary scholars have expanded upon this idea in their scholarship in varying productive ways since then (see, e.g., Zunshine 2006; Keen 2007; Mancing 2017).

In *Social Minds in the Novel* (2010), Palmer extends his previous work by making the case that social minds are an essential aspect of *intermental thought*: "Such thinking is joint, group, shared, or collective, as opposed to *intramental*, or individual or private thought. It is also known as *socially distributed*, *situated*, or *extended cognition*, and also as *intersubjectivity*" (41, emphasis in the original). In effect, he is placing ToM in social contexts. Thus, Palmer makes yet another valuable contribution to narrative theory. Palmer's work ranks with Gerrig's *Experiencing Narrative Worlds* as one of the most original and influential studies of narrative fiction in recent decades. Its clarity, precision, and originality place him at the forefront of the movement to anchor the study of literature in the cognitive sciences.

Lisa Zunshine's brilliant *Why We Read Fiction* (2006) was the first book-length study of fictional characters grounded in ToM psychology. Her thesis is that without a ToM, we would not be able to understand what is going on in literature, particularly the novel:

> As a sustained representation of numerous interacting minds, the novel feeds the powerful, representation-hungry complex of cognitive adaptations whose very condition of being is a constant social simulation delivered either by direct interactions with other people or by imaginary approximation of such interactions. (10)

Why We Read Fiction has become one of the most influential books in the field, and rightly so. Later, Zunshine turned her attention to what she calls "conceptual hybrids" in a book titled *Strange Concepts and the Stories They Make Possible* (2008). Drawing from the work of several psychologists and anthropologists, Zunshine centers her work on our innate essentialist thinking. She defines psychological essentialism as "a hazy belief rather than well-thought-through theory, which influences our everyday thinking primarily about natural kinds (as contrasted with artifacts) and that can be reinforced or weakened by specific contexts" (11–12). Such thinking is "something that we as humans cannot help but do" (14). We naturally and automatically think differently about substances (things we can divide, measure, and carry around), artifacts (human-made objects designed for a purpose), animals (living organisms that move, reproduce, and can die), and human beings (purposeful agents with, among other things, a ToM). The problem comes in when we attempt to deal with entities that are a mixture of two or more of these categories, so most of Zunshine's book deals with the robots, androids, cyborgs, and other imagined human artifact, or human animal, hybrids that appear in science fiction, fantasy, and myth. She argues that our (largely nonconscious) essentialist thinking is as fundamental—as essential—to our comprehension of a literary work as our (also largely nonconscious) ToM. It is difficult to take issue with this position. This book solidifies Zunshine's role as one of the most influential leaders and versatile scholars in the general field of cognitive approaches to literature.

Another important contribution to the field is *Theory of Mind and Literature*, edited by Paula Leverage et al. (2011), which includes nineteen essays based on selected papers from the international "Theory of Mind and Literature" conference held at Purdue University in November 2007, featuring as keynote speakers Keith Oatley, Alan Palmer, Mark Turner, and Lisa Zunshine. No other work to date displays the range of application of the idea of Theory of Mind to literary studies, as is indicated by the five sections into which the essays are divided: (1) Theory of Mind Now and Then: Evolutionary and Historical Perspectives; (2) Mindreading and Literary Characterization; (3) Theory of Mind and Literary/Linguistic Structure; (4) Alternate States of Mind; and (5) Theoretical, Philosophical, Political Approaches.

Cognitive linguist Barbara Dancygier was a presenter at that pioneering ToM conference, and her outstanding book *The Language of Stories* (2012) furthers several of the concepts that were featured there. She discusses "literary texts as linguistic artifacts" (1), focusing on construction grammar and conceptual integration (blending) to increase our understanding of how literary interpretation happens. Dancygier provides an accessible explanation of blending theory as applied to narrative and, relying on numerous literary examples, presents viewpoint compression and decompression as concepts that greatly expand traditional understanding of narratorship. The book painstakingly and intriguingly outlines how lexical choices—from split-self-metaphors to deictic expressions and more—structure a text's viewpoint and its narrative spaces. Dancygier's work epitomizes the productive, synthesized result of cognitive-linguistically informed approaches to literary studies.

The first book to incorporate ToM, Machiavellian Intelligence, and CONTEXTUALISM into the study of Spanish Golden Age literature is Barbara Simerka's *Knowing Subjects* (2013). Simerka's chapters on the *Comedia* of the seventeenth century are especially original and insightful. In addition, she makes valuable contributions to the understanding of Cervantes's *Don Quixote*, the picaresque novel, and other works of fiction. She is particularly concerned with elucidating "the way that a specific literary form delineates the relationship between an urbanizing culture, unstable subject positions and hierarchies, and social anxieties about the relationship between cognition and cultural transformation" (22). This is in every way a groundbreaking book in Hispanic studies.

Several of the books and shorter essays previously mentioned are, in addition to the contributions they make to areas such as reader-response theory and narrative theory, valuable contributions to literary theory in general. The foundational books by Holland (1988, 1992) and Turner (1991, 1996) fill this bill, as does, for instance, Patrick Colm Hogan (2003a). Alan Richardson's *British Romanticism and the Science of the Mind* (2001) makes a unique contribution, examining the relationships between British Romantic writers (Coleridge, Wordsworth, Keats) and the studies of scientists (Charles Bell, Pierre-Jean-George Cabanis, Erasmus Darwin, Franz Joseph Gall) of the late eighteenth and early nineteenth centuries about the human brain and mind. With great originality, Richardson demonstrates that the brain science of the period was in many ways consistent with major themes of modern neuroscience and embodied cognition (Damasio, Mark Johnson, Pinker, Rosch). In *The Neural Sublime: Cognitive Theories and Romantic Texts* (2010), Richardson describes cognitive literary studies as now as "an established field with an international audience drawn from virtually every area within the professional study of literature" and notes that it "no longer needs to justify its very existence" (ix). Relating his work to cultural studies and New Historicism in this valuable contribution to the field, Richardson studies Romantic literature from the points of view of cognitive linguistics, ToM, and evolutionary biology.

There are several books that should be recognized as important statements regarding evolutionary/cognitive approaches to literary theory and criticism. Patrick Colm Hogan's *Cognitive Science, Literature, and the Arts* (2003a) is intended as a general introduction to cognitive science for humanists. He argues that "humanists must at least be among the leaders in the cognitive revolution," and, for this, he notes that humanist scholars "need to achieve familiarity with cognitive methods and principles" (3). He is right on both counts. In an effort to broaden his work's appeal, Hogan devotes much of his book to music and film, as well as to literature. His presentations of matters such as schema theory, memory, cognitive metaphor theory, conceptual blending, narrative epistemology, emotions, empathy, and others, are all fitting. A major problem with Hogan's approach, however, is his limited concept of what cognitive science is. He seems to be caught in the early stages of the cognitive revolution when he affirms that "the two most prominent schools of cognitive science today [are] representationalism and connectionism" (29). The lack of a perspective from evolution and the sort of embodied cognition described throughout this book, and the accompanying stress on the COGNITIVIST approach criticized in Chapter 11, severely limit his vision. Hogan's *The Mind and its Stories* (2003b) is a wide-ranging comparative study that involves an argument for literary universals and focuses on Sanskrit poetics, cognitive theories of emotions, and prototype theory to assert the primacy of three genres of tragicomedy: Romantic, heroic, and sacrificial. Interestingly, he draws primarily upon epic and dramatic literature to argue for narrative universals.

Jonathan Gottschall's bold *Literature, Science, and a New Humanities* (2008b), is divided into two parts. The first is a strong critique of the current state of literary scholarship and an argument for a "new humanities" that eschews the foundational bases of THEORY and proposes that "[l]iterary studies should move closer to the sciences in theory, method, and governing ethos" (3). Gottschall describes what he calls the "liberationist paradigm," the hegemonic form of THEORY and criticism, including Marxist, postcolonial, poststructuralist, new historical, queer, feminist, and psychoanalytic approaches. In spite of their minor internal squabbles, these studies are unified, he says, by a pursuit of a radical political agenda, strong social constructionism, and poststructuralist antifoundationalism (5–6). In place of this paradigm, Gottschall argues for "an approach that is based in scientific theory and grounded, ultimately, in the bedrock of evolution by natural selection" (13). Important to his argument is the idea of consilience, E. O. Wilson's (1998) term to describe how the work in any discipline should not contradict the findings of other disciplines. Gottschall makes a case for the use of empirical and quantitative methods in the study of literature (43–65), and the second half of his book consists of reprints or adaptations of some of his previously published shorter essays, all employing quantitative studies of folk literature.

Edward Slingerland's *What Science Offers the Humanities* (2008) bolsters the fundamental points we have made throughout this book, namely

- that the mind–body dualism at the heart of postmodernism—with its blank slate concept of mind, linguistic determinism, and strong social constructionism—is profoundly flawed and inadequate as an epistemology;

- that an integration of the arts and sciences is now more possible—and needed—than ever before;
- that human cognition involves metaphorical thought, blended spaces, and Theory of Mind in an essential way; and
- that literature can best be read and understood within the concept of embodied cognition.

Paul B. Armstrong's *How Literature Plays with the Brain* (2013) draws on neuroscience to illuminate the aesthetic process. His thesis is that "literature plays with the brain through experiences of harmony and dissonance that set in motion and help to negotiate oppositions that are fundamental to the neurobiology of mental functioning" (ix). His chapters on learning to read and the neuroscience of the hermeneutic process are especially valuable as a basis for future studies.

Fortunately, innovative cognitive approaches to literary studies continue to emerge, marking this section of our book only as a starting point. A final mention here goes to Ellen Spolsky's aforementioned 2015 work *The Contracts of Fiction*, in which she addresses in an original way and with rich literary cultural examples many of the principles discussed already in this book. Spolsky broadly considers genre reception in the context of cognition and in terms of similarities as well as differences between fictions and real life:

> As with ToM, so with much of the cognitive work described by the contracts of fiction: whatever describes normal cognitive processing is usually good enough to describe how we process the imaginative stuff. But there will need to be extensions of the rules. Rules about transforming tokens into types, for example, something we do normally and easily, is with no great difficulty expanded to make sense of Robert Frost's claim that "something there is that doesn't love a wall." The contracts, for those whose contexts allow it, offer to put some extra pressure on fictions, getting them to work harder by labeling and framing them as worthy of extended considerations. They describe the ways the different kinds of fictions allow and protect that extra attention in generic niches, such as lyric poetry, cubist painting, Italian opera. The specifics of the works reinforce or complicate the clues the genre gives. (xxi)

Spolsky's highly recommended work highlights the importance of context as well as the utility of cognitively oriented approaches not only to literature, but also to the arts and humanities more generally.[2]

6 Empirical Studies

Empirical approaches to literature involve attempts to study literary experiences by using a variety of testable means, data gathering, and carefully designed experiments. The study of actual readers in the act of reading and the analysis of their reactions and understandings is a potentially enormous area of investigation clearly related to cognition and therefore of interest in the context of this book. The work in this field is carried out by psychologists, sociologists, and humanities scholars. Related areas of investigation involve reader-response theories, studies of discourse processing, and literacy studies, and sometimes reading research fades into linguistic studies. Boundaries, in other words, are not clear or fixed. Overall, relatively little work has been done in this area by literary scholars in the U.S. and Great Britain, while Germany and the Netherlands have been the site of a great deal of activity. Empirical studies offer the possibility of understanding actual reading practices and real literary understanding, as opposed to more abstract theoretical approaches (e.g., psychoanalytic theory or poststructuralism) that may have little or nothing to do with real reading practices. Thus, empirical studies offer a reality-based alternative to sterile theoreticism, the valuing of theory over reality.

On the other hand, empirical studies frequently deliver much less than they seem to promise. Small-scale studies of what a group of readers say or write about their reactions to bits of text or a short story may conclude that certain feelings, thoughts, ideas, or memories were evoked. Jèmeljan (Frank) Hakemulder, whose own research (2000) is discussed below, notes the frustration one encounters when attempting to understand the effects of literary reading on actual individuals: Too many factors are involved to be sure just what effect the reading itself has in specific cases. To remedy this, he suggests, "future research may consider using texts that are easier to manipulate" (144), and notes that "[e]mpirical research obviously tends to reduce complex realities (e.g., the value or effects of literature) to measurable, seemingly superficial variables" (163). But if the reductionist approach of using simpler, more manipulable, less authentic texts might lead to more reliable data, such a procedure might also fail to replicate what happens with reading sophisticated works of literature. As Hakemulder suggests, "it seems more likely that longer, and perhaps therefore even more complex narratives (e.g., novels) have a stronger effect on self-concept than short, experimenter-generated ones"

(144).[3] This statement encapsulates one of the most serious challenges faced by empirical studies of literature.

Another obvious caveat is that the laboratory or classroom setting of these studies may contaminate the authenticity of the reading experience. Few large-scale neural imaging studies (fMRI, CAT, etc.) of actual brain activities of readers have been carried out, and to date have not produced significant results.[4] Furthermore, if—as Rosenblatt, Holland, and others maintain—every act of literary reading is a unique and non-repeatable aesthetic experience, how can literary reading be reduced to data that would be significant? Nevertheless, some of the work in this promising area is more interesting than others. The three names that have been most prominent since the 1980s are Siegfried J. Schmidt, David S. Miall, and the previously mentioned Hakemulder. We will limit our comments primarily to the work of these three leading scholars. Much needed still is a detailed book-length study surveying the history and accomplishments of the field.

An early introduction to empirical studies may be found in some of the essays in *Foundation for the Empirical Study of Literature* (1982), ed. Siegfried J. Schmidt, and also in Schmidt's shorter studies (e.g., 1983, 1984, 1985, 1992). Noteworthy in Schmidt's approach is his recognition that human beings are constructivist autopoietic systems whose understandings (of worlds and texts) always stem from their unique perspective within the total social literary system. His work tends increasingly toward the study of media in general. As part of this biologically based concept of literature, Schmidt rejects code theories of language and proposes that language is a behavioral system, which even more squarely positions him within a view compatible with what we propose.

If Schmidt's empirical program is largely a theoretical approach, that of David S. Miall is much more research oriented. For a number of years, Miall has been writing essays based on studies of small groups of specific readers. Some of the best and most recent of these essays, along with a new one, have been collected in his *Literary Reading* (2006). From this book, the chapter titled "The Empirical Approach: A Survey and Analysis" (89–117) provides what is perhaps the best introduction to the field. Miall's overview (like his book in general), however, is not without its own problems. In his review of reader-response theories, for example, he mentions Rosenblatt only once, briefly and in passing, and limits his comments on Holland to his early work of the 1960s, completely ignoring Holland's innovative approach since the 1980s based on cognitive science. He makes little use of the important work on reading from cognitive scientists like Gerrig, Gibbs, Smith, and others. He neglects, undervalues, or underestimates the significance of the work of other scholars on feelings and emotions in reading, a subject in which he has particular interest and which he repeatedly insists has never been sufficiently studied. Nonetheless, Miall's work is essential reading for those interested in empirical approaches to the reading process.

One of the most balanced and nuanced books in the field is Hakemulder's *The Moral Laboratory* (2000). Hakemulder explains the metaphor of his title by proposing that "literature can be considered a Moral Laboratory, in which plausible implications of human conduct and ideas can be studied in a relatively controlled and safe way" (150). He critically reviews theoretical approaches to the idea that reading can contribute to morality and ethics, evaluating many of the empirical studies that others have carried out. He then describes a series of experiments he conducted to test some of the theories and confirm (or disconfirm) some of the earlier research. Hakemulder is keenly sensitive to the complexity of the questions involved and the difficulty of designing test instruments that can produce reliable results. He tentatively concludes that reading literature can lead to increased empathy, self-knowledge, and self-discipline, but he acknowledges the difficulties of finding definitive proof for these conclusions. In the end, he calls for further research, particularly for collaborative interdisciplinary efforts involving both psychologists and literary scholars (164).

The rise of literary theory and criticism grounded in the cognitive sciences in the last quarter of a century has been impressive, even spectacular. Important anthologies of criticism have appeared in the field over the past decade: the collection of evolutionary literary studies by Boyd et al. (2010); the essays on cognitive cultural studies in Zunshine (2010b); the volume of essays on Theory of Mind edited by Leverage et al. (2011); Aldama's edited volume exploring possibilities for "a cognitive theory of narrative acts" (2010); the essays on consciousness in English literature by Herman (2011); the collection of literary studies by Jaén and Simon (2012b; see their overview, 2012a) as well as their collection of cognitive approaches to early modern Spanish literature (2016); the essays on history and literature edited by Bruhn and Wehrs (2013); the studies in narrative theory by Bernaerts et al. (2013); essays on aesthetics and cognition by Danta and Groth (2014); and a collection of studies on embodied cognition in Shakespeare by Johnson, Sutton, and Tribble (2014). The fact that the field, in its various manifestations, has gelled to the extent that so many major collections of studies can be published attests to the fact that the relationship between the cognitive sciences and the study of literature has become a recognized area of inquiry.

The works reviewed in this chapter make manifest a wide variety of approaches to literature that are characterized by the concepts and values from contemporary linguistics, biology, and psychology described in previous chapters. Taken together, these contextualized approaches open up new, exciting, and sometimes conflicting ways to conceive of the study of literature. In their wake, literary analysis and criticism in the twenty-first century will be substantially different from what it was in the final half of the twentieth—and as the next chapter demonstrates, for performance studies as well.

Notes

1. A few important and outstanding studies mentioned in earlier chapters are worth mentioning again here: On mental imagery, these include Esrock's *The Reader's Eye* (1994) and Elaine Scarry's *Dreaming by the Book* (1999) discussed in Sect. 6 of Chapter 8; and on relevance theory and its relation to literature, Ian MacKenzie's *Paradigms of Reading: Relevance Theory and Deconstruction* (2002), as discussed in Sect. 4 of Chapter 4.
2. For an array of essays on embodied approaches to the humanities more generally, see, e.g., Garratt (2016).
3. Hakemulder has continued to wrestle with these issues productively, for example in the co-authored textbook (2012) with Willie van Peer and Sonia Zyngier, *Scientific Methods for the Humanities*, also a useful resource for linguists and film scholars (van Peer et al. 2012).
4. However, it should be noted that at this writing, the field of NeuroHumanities and its subfield of "literary neuroscience" is still relatively new and still emerging. One important researcher in this field is Natalie Phillips, an English professor at Michigan State University who has done imaging studies involving the reading of Jane Austen. See "Literary Neuroscience" 2015.

References

Aldama, Frederick Luis, ed. *Toward a Cognitive Theory of Narrative Acts*. Austin: U of Texas P, 2010.

Armstrong, Paul B. 2013. *How Literature Plays with the Brain: The Neuroscience of Reading and Art*. Baltimore: Johns Hopkins UP.

Babuts, Nicolae. 1992. *The Dynamics of the Metaphoric Field: A Cognitive View of Literature*. Newark: U of Delaware P.

———. 2009. *Memory, Metaphors, and Meaning: Reading Literary Texts*. New Brunswick: Transaction Publishers.

Bakhtin, Mikhail M. 1981. *The Dialogic Imagination: Four Essays*. Ed. Michael Holquist. Trans. Caryl Emerson and Michael Holquist. Austin: U of Texas P.

Bernaerts, Lars, Dirk De Geest, Luc Herman, and Bart Vervaeck, eds. 2013. *Stories and Minds: Cognitive Approaches to Literary Narrative*. Lincoln: U of Nebraska P.

Blair, Linda Nicole. 2017. *Virginia Woolf and the Power of Story: A Literary Darwinist Reading of Six Novels*. Jefferson: McFarland.

Bloom, Paul. 2016. *Against Empathy: The Case for Rational Compassion*. New York: Ecco.

Bortolussi, Marisa, and Peter Dixon. 2003. *Psychonarratology: Foundations for the Empirical Study of Literary Response*. Cambridge: Cambridge UP.

Boyd, Brian. 2008. "Art and Evolution: The Avant-Garde as Test Case: Spiegelman's *The Narrative Corpse*." *Philosophy and Literature* 32: 31–57.

———. 2009. *On the Origin of Stories: Evolution, Cognition, and Fiction*. Cambridge: Belknap P of Harvard UP.

———, Joseph Carroll, and Jonathan Gottschall, eds. 2010. *Evolution, Literature, and Film: A Reader*. New York: Columbia UP.

Breithaupt, Fritz. 2009. *Kulturen der Empathie*. Berlin: Suhrkamp.

———. 2017. *Die dunklen Seiten der Empathie*. Berlin: Suhrkamp.

Bruhn, Mark, 2009. "Romanticism and the Cognitive Science of Imagination." *Studies in Romanticism* 48.4: 543.

———, ed. 2011a. "Exchange Values: Poetics and Cognitive Science (I)." Special issue of *Poetics Today: International Journal for Theory and Analysis of Literature and Communication* 32.3.

———. 2011b. "Introduction: Exchange Values: Poetics and Cognitive Science." *Poetics Today* 32.3: 403–60.

———. 2015. "A Mirror on the Mind: Stevens, Chiasmus, and Autism Spectrum Disorder." *The Wallace Stevens Journal* 39.2: 182–206.

———. 2017a. "Ambiguity in Affect: The Modernity of Wordsworth's Lyrical Ballads." In *Romantic Ambiguities: Abodes of the Modern*. Ed. Sebastian Domsch, Christoph Reinfandt, and Katharina Rennhak. Trier: Wissenschaftlicher Verlag Trier. 21–36.

———. 2017b. "'The History and Science of Feeling': Wordsworth's Affective Poetics, Then and Now." In *The Palgrave Handbook of Affect Studies and Textual Criticism*. Ed. Donald R. Wehrs and Thomas Blake. Cham: Springer. 671–93.

———. 2018a. "Intentionality and Constraint in Conceptual Blending." In *Expressive Minds and Artistic Creations: Studies in Cognitive Poetics*. Ed. Szilvia Czábi. New York: Oxford UP. 79–100.

———. 2018b. "Target First: On 'Bidirectionality and Metaphor.'" *Poetics Today* 39.4: 703–33.

———. 2018c. "Citation Analysis: An Empirical Approach to Professional Literary Interpretation." *Scientific Study of Literature* 8.1: 76–112.

———, and Donald Wehrs, eds. 2013. *Cognition, Literature, and History*. New York: Routledge.

Carroll, Joseph. 1995. *Evolution and Literary Theory*. Columbia: U of Missouri P.

———. 2004. *Literary Darwinism: Evolution, Human Nature, and Literature*. New York: Routledge.

———. 2008a. "An Evolutionary Paradigm for Literary Study." *Style* 42.2–3: 103–43.

———. 2008b. "Rejoinder to the Responses." *Style* 42.2–3: 308–70.

———. 2011. *Reading Human Nature: Literary Darwinism in Theory and Practice*. Albany: SUNY P.

Chomsky, Noam. 1995. *The Minimalist Program*. Cambridge: MIT P.

Cormican, Muriel. 2021. "Toward a Theory of the Tender Gaze: Affect, Critical Insight, and Empathy in Contemporary German Cinema." In Cormican and William, 17–39.

———, and Jennifer Marston William, eds. 2021. *The Tender Gaze: Compassionate Encounters on the German Screen, Page, and Stage*. Rochester: Camden House.

Craig, Robert, and Ina Linge, eds. 2017. *The Language of Science and Literature Around 1900*. Oxford: Peter Lang.

Crane, Mary Thomas. 2001. *Shakespeare's Brain: Reading with Cognitive Theory*. Princeton: Princeton UP.

Dancygier, Barbara. 2012. *The Language of Stories: A Cognitive Approach*. Cambridge: Cambridge UP.

Danta, Chris, and Helen Groth, eds. 2014. *Mindful Aesthetics: Literature and the Science of Mind*. New York: Bloomsbury.

de Waal, Frans. 2009. *The Age of Empathy: Nature's Lessons for a Kinder Society*. New York: Harmony Books.

Eakin, Paul John. 1999. *How Our Lives Become Stories: Making Selves*. Ithaca: Cornell UP.

Easterlin, Nancy. 2012. *A Biocultural Approach to Literary Theory and Interpretation*. Baltimore: Johns Hopkins UP.

Ender, Evelyne. 2005. *Architexts of Memory: Literature, Science, and Autobiography*. Ann Arbor: U of Michigan P.

Esrock, Ellen J. 1994. *The Reader's Eye: Visual Imaging as Reader Response*. Baltimore: Johns Hopkins UP.

Fauconnier, Gilles, and Mark Turner. 2002. *The Way We Think: Conceptual Blending and the Mind's Hidden Complexities*. New York: Basic Books.

Flesch, William. 2007. *Comeuppance: Costly Signaling, Altruistic Punishment, and Other Biological Components of Fiction*. Cambridge: Harvard UP.

Fodor, Jerry A. 1983. *The Modularity of Mind*. Cambridge: MIT P.

Freeman, Margaret H. 2020. *The Poem as Icon: A Study in Aesthetic Cognition*. Oxford: Oxford UP.

Garratt, Peter, ed. 2016. *The Cognitive Humanities: Embodied Mind in Literature and Culture*. London: Palgrave Macmillan.

Gavins, Joanna, and Gerard Steen, eds. 2003. *Cognitive Poetics in Practice*. London: Routledge.

Gerrig, Richard J. 1993. *Experiencing Narrative Worlds: On the Psychological Activities of Reading*. New Haven: Yale UP.

Gottschall, Jonathan. 2008a. *The Rape of Troy: Evolution, Violence, and the World of Homer*. Cambridge: Cambridge UP.

———. 2008b. *Literature, Science, and a New Humanities*. New York: Palgrave Macmillan.

———. 2012. *The Storytelling Animal: How Stories Make Us Human*. Boston: Houghton Mifflin Harcourt.

———, and David Sloan Wilson. 2005a. "Introduction: Literature—A Last Frontier in Human Evolutionary Studies." In Gottschall and Wilson, xvii–xxvi.

———, eds. 2005b. *The Literary Animal: Evolution and the Nature of Narrative*. Evanston: Northwestern UP.

Hakemulder, Jèmeljan. 2000. *The Moral Laboratory: Experiments Examining the Effects of Reading Literature on Social Perception and Moral Self-Concept*. Amsterdam: John Benjamins.

Hart, F. Elizabeth. 1995. "Cognitive Linguistics: The Experiential Dynamics of Metaphor." *Mosaic* 28: 1–23.

———. 2001. "The Epistemology of Cognitive Literary Studies." *Philosophy and Literature* 25: 314–34.

———. 2004. "Embodied Literature: A Cognitive-Poststructuralist Approach to Genre." In Richardson and Spolsky, 85–106.

———. 2006. "Performance, Phenomenology, and the Cognitive Turn." In McConachie and Hart, 29–51.

———. 2011. "Reading, Consciousness, and Romance in the Sixteenth Century." In Herman, 103–31.

Herman, David. 2002. *Story Logic: Problems and Possibilities of Narrative*. Lincoln: U of Nebraska P.

———, ed. 2003. *Narrative Theory and the Cognitive Sciences*. Stanford: CSLI Publications.

———, ed. 2011. *The Emergence of Mind: Representations of Consciousness in Narrative Discourse in English*. Lincoln: U of Nebraska P.

———. 2013. *Storytelling and the Sciences of Mind*. Cambridge: MIT Press.

Hogan, Patrick Colm. 2003a. *Cognitive Science, Literature, and the Arts: A Guide for Humanists*. New York: Routledge.

———. 2003b. *The Mind and Its Stories: Narrative Universals and Human Emotion*. Cambridge: Cambridge UP.

Holland, Norman N. 1988. *The Brain of Robert Frost: A Cognitive Approach to Literature*. New York: Routledge.

———. 1992. *The Critical I*. New York: Columbia UP.

———. 2009. *Literature and the Brain*. Gainesville: The PsyArt Foundation.

Jaén, Isabel, and Julien Simon. 2012a. "An Overview of Recent Developments in Cognitive Literary Studies." In Jaén and Simon, 13–21.

———, eds. 2012b. *Cognitive Literary Studies: Current Themes and New Directions*. Austin: U of Texas P.

———, eds. 2016. *Cognitive Approaches to Early Modern Spanish Literature*. Oxford: Oxford UP.

Keen, Suzanne. 2007. *Empathy and the Novel*. Oxford: Oxford UP.

Johnson, Laurie, John Sutton, and Evelyn Tribble, eds. 2014. *Embodied Cognition and Shakespeare's Theater: The Early Modern Body-Mind*. New York: Routledge.

Lakoff, George, and Mark Turner. 1989. *More than Cool Reason: A Field Guide to Poetic Metaphor*. Chicago: U of Chicago P.

———, and Mark Johnson. 1980. *Metaphors We Live By*. Chicago: U of Chicago P.

Leverage, Paula. 2010. *Reception and Memory: A Cognitive Approach to the Chansons de Geste*. Amsterdam: Rodopi.

———, Howard Mancing, Richard Schweickert, and Jennifer Marston William, eds. 2011. *Theory of Mind and Literature*. West Lafayette: Purdue UP.

"Literary Neuroscience and the History of Attention: An fMRI Study of Reading Jane Austen." 2015. *The Oxford Handbook for Cognitive Approaches to Literature*. Ed. Lisa Zunshine. Oxford: Oxford UP.

Love, Glen A. 2003. *Practical Ecocriticism: Literature, Biology, and the Environment*. Charlottesville: U of Virginia P.

MacKenzie, Ian. 2002. *Paradigms of Reading: Relevance Theory and Deconstruction*. Houndmills: Palgrave Macmillan.

McConachie, Bruce, and F. Elizabeth Hart, eds. 2006. *Performance and Cognition: Theatre Studies and the Cognitive Turn*. London: Routledge.

Mancing, Howard. 1999. "Against Dualisms: A Response to Henry Sullivan." *Cervantes: Bulletin of the Cervantes Society of America* 19.1: 158–76.

———. 2000. "Prototypes of Genre in Cervantes's *Novelas ejemplares*." *Cervantes* 20.2: 127–50.

———. 2003a. "Cervantes as Narrator of *Don Quijote*." *Cervantes: Bulletin of the Cervantes Society of America* 23.1: 117–40.

———. 2003b. "Rastier Revisited: Paradigms in Conflict." *Semiotica* 145: 139–49.

———. 2005. "Response to 'On Narration and Theory." *Cervantes: Bulletin of the Cervantes Society of America* 24.2: 135–54.

———. 2012. "Embodied Cognitive Science and the Study of Literature." In *Cognitive Cervantes*, Special Issue of *Cervantes: Bulletin of the Cervantes Society of America*. Ed. Howard Mancing, Barbara Simerka, and Julien Simon. 32.1: 25–69.

———. 2017. "Don Quixote's Affective Thoughts." In *The Palgrave Handbook of Affect Studies and Textual Criticism*. Ed. Don Wehrs and Thomas Blake. New York: Palgrave Macmillan. 623–48.

Marr, David. 1982. *Vision: A Computational Investigation into the Human Representation and Processing of Visual Information*. Cambridge: MIT P.

Miall, David S. 2006. *Literary Reading: Empirical and Theoretical Studies*. New York: Peter Lang.

Nafisi, Azar. 2003. *Reading Lolita in Tehran*. New York: Random House.

Nalbantian, Suzanne. 2003. *Memory in Literature: From Rousseau to Neuroscience*. Houndmills: Palgrave Macmillan.

Oatley, Keith. 1992. *Best Laid Schemes: The Psychology of Emotions*. Cambridge: Cambridge UP.

———. 2003. "Writingandreading: The Future of Cognitive Poetics." In Gavins and Steen, 161–73.

———. 2011. *Such Stuff as Dreams: The Psychology of Fiction*. Malden: Wiley-Blackwell.

Opdahl, Keith M. 2002. *Emotion as Meaning: The Literary Case for How We Imagine*. Lewisburg: Bucknell UP.

Palmer, Alan. 2004. *Fictional Minds*. Lincoln: U of Nebraska P.

———. 2010. *Social Minds in the Novel*. Columbus: Ohio State UP.

Peterson, Britt. 2008. "Darwin to the Rescue." *The Chronicle of Higher Education* (August 1): B7.

Reynolds, Dee, and Matthew Reason, eds. 2012. *Kinesthetic Empathy in Creative and Cultural Practices*. Bristol/Chicago: Intellect.

Richardson, Alan. 2001. *British Romanticism and the Science of the Mind*. Cambridge: Cambridge UP.

———. 2004. "Studies in Literature and Cognition: A Field Map (Introduction)." In Richardson and Spolsky, 1–29.

———. 2006. "Cognitive Literary Criticism." In *Literary Theory and Criticism: An Oxford Guide*. Ed. Patricia Waugh. Oxford: Oxford UP. 544–56.
———. 2010. *The Neural Sublime: Cognitive Theories and Romantic Texts*. Baltimore Johns Hopkins UP.
———, and Ellen Spolsky, eds. 2004. *The Work of Fiction: Cognition, Culture and Complexity*. Aldershot: Ashgate.
Rosenblatt, Louise M. 1994a (1978). *The Reader, the Text, the Poem: The Transactional Theory of the Literary Work*. Carbondale: Southern Illinois UP.
———. 1994b. "The Transactional Theory of Reading and Writing." In *Theoretical Models and Processes of Reading*. Eds. Robert B. Ruddell, Martha Rapp Ruddell, and Harry Singer. Newark: International Reading Association. 1057–92.
———. 1995 (1938). *Literature as Exploration*. New York: Modern Language Association.
Rubin, David C. 1995. *Memory in Oral Traditions: The Cognitive Psychology of Epic, Ballads, and Counting-out Rhymes*. New York: Oxford UP.
Scarry, Elaine. 1999. *Dreaming by the Book*. New York: Farrar, Straus and Giroux.
Schmidt, Siegfried J., ed. 1982. *Foundation for the Empirical Study of Literature*. Amsterdam: John Benjamins.
———. 1983. "The Empirical Science of Literature ESL: A New Paradigm." *Poetics* 12: 19–34.
———. 1984. "The Fiction Is that Reality Exists: A Constructivist Model of Reality, Fiction, and Literature." *Poetics Today* 5.2: 253–74.
———. 1985. "Foundations of a Constructivist Empirical Study of Literature." *Reader* 14: 5–22.
———. 1992. "Media, Culture: Media Culture: A Constructivist Offer of Conversation." *Poetics* 21: 191–210.
Semino, Elena, and Jonathan Culpeper, eds. 2002. *Cognitive Stylistics: Language and Cognition in Text Analysis*. Amsterdam: John Benjamins.
Simerka, Barbara. 2013. *Knowing Subjects: Cognitive Cultural Studies and Early Modern Spanish Literature*. West Lafayette: Purdue UP.
Sinding, Michael. 2004. "Beyond Essence (or, Getting over 'There'): Cognitive and Dialectical Theories of Genre." *Semiotica* 149: 37–95.
Slingerland, Edward. 2008. *What Science Offers the Humanities: Integrating Body and Culture*. Cambridge: Cambridge UP.
Spolsky, Ellen. 1993. *Gaps in Nature: Literary Interpretation and the Modular Mind*. Albany: State U of New York P.
———. 2002. "Darwin and Derrida: Cognitive Literary Theory as a Species of Poststructuralism." *Poetics Today* 23.1: 43–62.
———. 2015. *The Contracts of Fiction: Cognition, Culture, Community*. Oxford: Oxford UP.
Stockwell, Peter. 2002. *Cognitive Poetics: An Introduction*. London: Routledge.

Storey, Robert. 1996. *Mimesis and the Human Animal: On the Biogenetic Foundations of Literary Representation*. Evanston: Northwestern UP.

Tobin, Vera. 2018. *Elements of Surprise: Our Mental Limits and the Satisfactions of Plot*. Cambridge: Harvard UP.

Tsur, Reuven. 1983. *What Is Cognitive Poetics?* Tel Aviv: Katz Research Institute for Hebrew Literature.

———. 1992. *Toward a Theory of Cognitive Poetics*. Amsterdam: North-Holland.

Turner, Mark. 1987. *Death Is the Mother of Beauty: Mind, Metaphor, Criticism*. Chicago: U of Chicago P.

———. 1991. *Reading Minds: The Study of English in the Age of Cognitive Science*. Princeton: Princeton UP.

———. 1996. *The Literary Mind*. New York: Oxford UP.

———. 2014. *The Origin of Ideas. Blending, Creativity, and the Human Spark*. New York: Oxford UP.

van Peer, Willie, Frank Hakemulder, and Sonia Zyngier. 2012. *Scientific Methods for the Humanities*. Amsterdam: John Benjamins.

Vermeule, Blakey. 2010. *Why Do We Care About Literary Characters?* Baltimore: Johns Hopkins UP.

Walsh, Richard. 2007. *The Rhetoric of Fictionality: Narrative Theory and the Idea of Fiction*. Columbus: Ohio State UP.

Wilson, Edward O. 1998. *Consilience: The Unity of Knowledge*. New York: Random House.

Zunshine, Lisa. 2006. *Why We Read Fiction: Theory of Mind and the Novel*. Columbus: Ohio State UP.

———. 2008. *Strange Concepts and the Stories They Make Possible: Cognition, Culture, Narrative*. Baltimore: Johns Hopkins UP.

———. 2010a. "Introduction: What is Cognitive Cultural Studies?" In Zunshine, 1–33.

———, ed. 2010b. *Introduction to Cognitive Cultural Studies*. Baltimore: Johns Hopkins UP.

CHAPTER 15

Cognitive Approaches to Performance Studies

You know, taken out of context
I must seem so strange.
—Singer/songwriter Ani DiFranco
From the song "Firedoor"

Cognitive approaches have been leading the way to new understandings of nonliterary art (i.e., multi-media, performance, perceptual genres). Indeed, theater studies, film studies, and dance/movement studies may well turn out to be the scholarly areas where work grounded in contemporary embodied cognitive science has the most to offer. At the same time, the foundation is being prepared for the next generation of cognitive studies, with intriguing cognitive analyses involving multimedia works, video games, and AI. New research on "4E" (embodied, embedded, enactive, and extended) cognition holds tremendous promise for interdisciplinary collaborations and possibilities for reciprocity between the humanities and the experimental sciences. Taking into consideration the dynamic and contextual nature of cognition, 4E research illuminates the way we understand the many forms of performance in terms of both production and reception. The leading reference work on this approach, *The Oxford Handbook of 4E Cognition* (2018) addresses questions like "What is the nature of literature (or the theatrical

H. Mancing and J. Marston William, *Restoring the Human Context to Literary and Performance Studies*, Cognitive Studies in Literature and Performance,
https://doi.org/10.1007/978-3-030-89078-0_15

play, or a film) if it enacts meaning or a world only when the reader (or audience engages with it?") (Newen et al. 13). Bakhtin's dialogic processes underpin performances of all types, which are often created in response to sociopolitical circumstances and events, and which carry out a dialog with their audiences, both through verbal language and body movements.

1 Theater

Bruce McConachie, the leading theorist of cognitive approaches to theater history and performance, uses cognitive metaphor theory and especially the image schema of CONTAINMENT[1]—which implies an inside, an outside, and the boundary between them—as the primary theoretical context for his book *American Theater in the Culture of the Cold War* (2003). He takes his work to a higher level in *Engaging Audiences* (2008), in which he also continues and expands his critique of poststructuralism and other types of THEORY. Calling psychoanalytic and other approaches "ascientific theories," McConachie asks the following questions:

> What happens, though, when theories deriving from good science come into conflict with critical theories that have no basis in scientific evidence or logic? Are scholarly conclusions based on theories that are no longer valid or that claim to be beyond empirical evidence and falsifiability just as reliable as conclusions rooted in scientific understanding? (12)

His answer is: "I believe that they are not" (12). Drawing then on ideas discussed in this book as well—conceptual blending, mirror neuron systems, cognitive anthropology, and more—McConachie describes a sophisticated way of understanding and discussing audience members' ways of engaging with theatrical performance. The foundational anthology titled *Performance and Cognition* (2006), edited by McConachie and F. Elizabeth Hart, contains essays that address a variety of aspects of theory and performance. At the same time, it clearly outlines the differences between a cognitive approach to the subject and more traditional approaches of THEORY. McConachie later published the likewise commendable *Evolution, Cognition, and Performance* (2015), in which he uses the scientific concept of enactment to explore the biological and cultural aspects underlying performance and spectatorship in the theater and beyond.

Because the body is so obviously physically and visually present in theatrical performance, studies in embodied cognition are particularly important in theater and performance studies. Following Mary Thomas Crane's pioneering book on Shakespeare (2001) cited in the previous chapter, and McConachie's groundbreaking work, there has been an impressive spate of cognitively oriented monographs dealing with theater and performance studies so far in the twenty-first century, among them being: Matthews and McQuain (2003), Blair (2008), Cook (2010), Stevenson (2010), Lutterbie (2011), Lyne (2011), Rokotnitz (2011), Tribble (2011), Paavolainen (2012), Shaughnessy (2012, 2013), Booth (2017), Murphy (2019), and Helms (2019). In addition, the past decade has seen the emergence of further anthologies on these topics, for example, three that all happened to appear in 2016: Blair and Cook; Budra and Werier; and Falletti et al. Special journal issues or sections on the topic have also appeared, such as the five essays on cognitive studies, theater, and performance in the *Journal of Dramatic Theory and Criticism* (Spring 2011), and *Theatre Journal*'s issue dedicated to Artificial Intelligence (2021).

Drama (like film and other visual media) is not literature per se. But if a play and a novel are different media, comprehended differently (recall Gibson's distinction between direct perception and mediated perception), they are also related, insofar as both involve comparable creative interpretive acts on the part of the viewer/reader. And if a play's text is read, and if the work is not seen as a performance—as usually happens in classroom situations—it is indeed a literary act, comparable to the reading of a novel or poem, even for the reader who attempts to visualize a performance while reading. When viewing a play, just as when reading a poem or a story, it is important to understand that the language of performance is a Chomskyan I-language, a Bakhtinian dialogic exchange, and not an independent Saussurean social force that constructs our subjectivity. The relationship between theater and literature is an intimate one, and so much of what is said about cognitive acts involved in the study of literature is also true of cognitive acts involved in theatrical performance. We understand the words and actions of performers using the same ToM we use in reading; we understand the feelings and emotions of the characters on stage as we do those of people in real life.

It is, of course, not just the minds on stage that audiences relate to and connect to during a theater performance, but also the bodies. In her comprehensive view of cognitive studies as related to the performing arts, Naomi Rokotnitz (2018) shows with various theatrical examples "just how indispensable our bodies are to the process of knowledge acquisition and processing, and also how participation in dramatic production can elicit and focus such learning and become instrumental in instigating profound physical, intellectual, and spiritual developments" (471). Importantly, Rokotnitz reminds us how the cognitive humanities "fosters cross-fertilization between theorists and practitioners from the humanities and empirical experimenters in the sciences" (472). Although experimental scientists are doing fascinating work in measuring the neurophysiological responses of readers and viewers, like EEG and eye-gaze metrics,[2] Rokotnitz reminds us that "laboratories are not sufficiently equipped to test the full range of human cognition. Multiple aspects of human existence and interaction are, in fact, far better observed, experienced, and analyzed through various forms of cultural production and participation" (473). The performing arts are a laboratory in themselves,[3] and their entertainment value and sometimes diversionary functions should not detract from the contributions that theatrical events can make to cognitive science, and to our developing understanding of how a production's meaning is collectively created and embodied within the theater space (see William 2021).

2 Film and Other Media

Since film can be, and very often is, brought into the literature classroom in a way that theater generally cannot be (except when filmed or read as a dramatic text), and since film is taught along with, and often as a part of, the study of literature, it is even harder to separate film studies from literary studies. Both fields require an understanding of linguistics and pragmatics, biology and evolution, neuroscience, and cognitive psychology.

David Bordwell is a pioneer in approaching the study of film from cognitive science. His first effort along these lines was *Narration in the Fiction Film* (1985), in which he assumes a constructivist and schema-theory approach by which perception consists of "active, goal-oriented processes.... The organism *constructs* a perceptual judgment on the basis of nonconscious *inferences*" (31). Unlike in literature, Bordwell maintains, there normally is no narrator in a film and the experience is more like seeing the events play out as if in reality: "a narrative film works quite directly on the limits of the spectator's perceptual-cognitive abilities" (74). In his *Making Meaning* (1989), Bordwell again studies film from a constructivist stance. He draws heavily on schema theory, along with prototype categorization, arguing against a semiotic signification model of meaning (a film is not a text; we do not read a film) and in favor of a poetics, or critical practice, that "requires the interpreter to draw upon schemata and procedures built up in the context of situations outside this film and, indeed, outside any film" (135). The reception of film, like the reception of literature, is contextual and relies in that way on a dialogic exchange between the work and its recipient. The meaning of a film is dynamic and fluid, not predetermined but instead is always dependent on what the spectator brings to it.

The first critical anthology to make a frontal attack on THEORY (specifically poststructuralist film theory) largely from a cognitive perspective was the massive *Post-Theory* (1996), edited by David Bordwell and Noël Carroll. Several of the essays are superb examples of theoretical critique and lay out a cognitive alternative to the psychoanalytic-Marxist stances that inform the hegemonic THEORETICAL approach to the discipline. This work was essential in establishing the contextualist stance we have taken throughout this book.

Philosopher Noël Carroll joins Bordwell as an early and prolific critic of THEORY in film studies. His major theoretical statement is *Mystifying Movies* (1988), a strong critique of the establishment's "contemporary film theory," which, he describes as consisting of "an amalgam of Anglo-French vintage, most often comprised, at least, of Althusserian-derived Marxism, Bathesian textual criticism, and, most importantly, of Lacanian psychoanalysis" (226), with its consistent emphasis on "absence," "suture," and "mirror," etc. In place of this *top-down approach* (i.e., starting with the theory and interpreting everything through its lens), he proposes a *bottom-up approach* of drawing on a variety of disciplines (including cognitive science) in piecemeal fashion as appropriate in specific cases. He continues his critique in a long series of shorter essays, many of which are collected in the book *Theorizing the Moving Image* (1996). To a large extent, Carroll is less a theorist than a pragmatist whose stance is largely anti-theoretic. He calls himself a "cognitivist" (not in the way we use the term in small caps in this book) to describe his "fixed opinion that many of our questions about film—especially concerning comprehension and reception—can be answered without resorting to psychoanalysis" (xviii).

Film director Joseph D. Anderson authored a breakthrough book on cognition and film theory, *The Reality of Illusion: An Ecological Approach to Cognitive Film Theory* (1996). Anderson's approach is grounded in cognitive science in general and Gibson's ecological theory of perception in particular. It is both a solid critique of semiotic film theory and a reminder of the simple truth that the cognitive processes involved in seeing and understanding a film are very different from those involved in understanding a literary work consisting of nothing more than symbolic language: "Because motion pictures can be constructed of the stuff of everyday experience, they can function as a surrogate for the physical world, not in the way that arbitrary symbols such as words may stand for physical objects, but as an actual substitute for the thing itself" (164). Visual media are, by definition, not literature. They are perceived directly and not understood only symbolically as are texts consisting of language alone. The processes of watching film, television, video and video games, live theater, computer animation, and other types of visual media are closer to actual visual perception than they are to reading. Recognizing that visual media are not texts, and that we perceive these media directly through our embodied perceptual systems, places the study of them in an area quite different from those assumed in THEORY.

With these foundational assertions as their backbone, recent film scholarship has applied cognitive-psychological principles such as conceptual blending and perspective-taking to the experience of intercultural spectatorship, for example Sung-Ae Lee's essay "Fairy-Tale Scripts and Intercultural Conceptual Blending in Modern Korean and Television Drama" (2014) and William's monograph *Cognitive Approaches to German Historical Film* (2017). Twentieth and twenty-first century approaches to world cinema in respect to empathy (cognitive, neurocinematic, as well as phenomenological/philosophical takes on the subject) are summarized well in Jane Stadler's chapter "Empathy and Film" in the *Routledge Handbook of Philosophy of Empathy* (2016).

The study of film and other media may be revolutionized even more than literary studies by our realization and appreciation of embodied cognition. The insightful publications in the past few decades of Laura U. Marks (2000) on intercultural cinema and embodiment, Roger Cook

(2015) on embodied simulation and social cognition, and Arthur Shimamura (2013) on "psychocinematics" point in this direction, as do edited volumes such as Maarten Coëgnarts and Peter Kravanja's *Embodied Cognition and Cinema* (2015). A Berghahn journal has even been established to cover this topic exclusively: *Screen Bodies: The Journal of Embodiment, Media Arts, and Technology* (since 2016).

While it does an injustice to the topic of non-film media to lump them into one "other" category, space constraints allow us to mention here only briefly some fascinating research being conducted in these areas, in the hope that we will spark interest in our readers to investigate further. Kenny K. N. Chow and D. Fox Harrell, digital media scholars at the Georgia Institute of Technology, have written about conceptual blending in relation to animated images (2009), examining how imaginative blends are crucial to making sense of animation. Chow and Harrell argue that their study and its implications "collectively form a step toward an embodied cognition approach to animation phenomena and toward recentralizing understanding of artistic and humanistic production in cognitive research" (2). Film and media expert Kathrin Fahlenbrach's edited volume *Embodied Metaphors in Film, Television, and Video Games: Cognitive Approaches* (2015) focuses heavily on cognitive metaphor theory (see our discussion in Sect. 4 of Chapter 4). In addition to considering films in this context, the volume's contributors examine other interactive experiences such as embodied video-game avatars and artgames (i.e., video games conceptualized as esthetic experiences or works of art in themselves), making this book an excellent starting point for anyone wanting to learn more about embodied cognition in the fascinating realm of so-called "other media."[4]

3 Dance and Movement Studies

Although often conveniently grouped in with theater due to conforming to the live stage-performance prototype, dance and movement studies deserve their own fanfare when it comes to embodiment and Bakhtinian dialogic expression. Indeed, recent scholarship in this field has turned to Bakhtin and his Circle, most notably Marilia Amorim (2020), whose work's aim is "to transpose the concept of discursive genre from Bakhtinian texts and those from the Circle to a language of dance" (72). Amorim's hypothesis is as follows:

> the active understanding of the audience and the work of producing meaning is based on elements of the socio-cultural reality shared between it and the artist creator-interpreter of the dance. However, this does not mean that the dance discourse copies or imitates the shared reality. It is through the genre that the relation with reality illuminates the work and allows us to understand it and not just through the iconicity of its signs. (73)

Amorim analyzes discursive fragments from ballet, hip-hop, and street dance, affirming with her study the dialogic meaning-making process during a dance performance and the establishment of "one valued discursive reality that is shared between artists and their public" (98). In other words, meaning is neither arbitrary nor predetermined, but arises organically as a co-created product of each performance in its unique context.[5] Ann Cooper Albright (2015) has also examined dance and other live theatrical performances that demonstrate a shared creation of meaning by performers and audience, as well as "the power of the exchange between performers and the importance of the audience's role as witnesses" (33).

Other dance and movement scholarship over the past decade has concentrated on neuroscience and kinesthetics in intriguing ways. In a special neuroscientifically oriented issue of *The Journal of the Society for Dance Research*, Jola et al. (2011) presented their research on measured audience motor simulation in which they observed spectators' brain activity while watching dancers in action that is similar to what it would be if the movements were actually being performed by the spectator (see our discussion of mirror neurons in Sect. 3 of Chapter 10).[6] As the authors summarize, the study showed that:

> 1) real-life conditions matter because they change the results and call into question the validity of laboratory based experimental conditions for research which is making claims about lived experience outside the laboratory and
> 2) new hypotheses emerged about the effects that the length of time spent watching dance can have on cortical excitability and action observation responses. (378)

Their research, as it indicated variability in brain activity depending on the length of time spent watching a dance performance, called into question prior understanding of our mirror neuron system's automaticity.

More recent scholarship has continued to reference mirror neurons in relation to movement (not always restricted to dance), such as Tracy E. Bersley (2018), who argues in her article for improved actor training through the embodied exploration of an actor's character. Bersley presents a very readable introductory overview of some of the issues that are (or should be) at the forefront of movement studies, such as touch receptors and the way performers "touch" space, as well as the roles of "flinch cells" and mirror neurons. Stanton B. Garner's impressively interdisciplinary monograph *Kinesthetic Spectatorship in the Theatre* (2018) also concentrates on movement and mirroring in dance as well as theater and even puppet performances, making a noteworthy contribution to empathy studies as well as disability studies as it considers the dialogically oriented phenomena of kinesthetic resonance and movement perception: "What does it mean to say that we resonate with the movements of others? Do we actually inhabit their movements in an experiential sense, or do we vicariously activate our own motor repertoires and experiences? How do we engage with sensorimotor performances that are radically different from ours? " (11). The answers to these questions are never clear-cut, but the exploration of them is both fascinating and crucial to a performance studies that is concilient with current science. Also in 2018, Kathryn Banks and Timothy Chesters published their volume *Movement in Renaissance Literature*, which looks at movement in terms of a "kinesic intelligence" that, as the editors argue, informed Renaissance texts and shaped responses to them by readers and audience members. While this volume would fit well with the theater studies works that address embodied cognition as mentioned above in Sect. 1, the focus on physical receptive response as something to which Renaissance writers and playwrights were apparently already attuned merits mention here as a work with implications impacting movement studies more generally.

While published a few years earlier than the studies mentioned above, the volume *Dance Dramaturgy: Modes of Agency, Awareness and Engagement*, edited by Pil Hansen and Darcey Callison (2015), provides a notable bridge between the various approaches to dance and movement described in this section as well as between these art forms and theatrical or film productions. As Hansen describes the project in the Introduction, the volume's essays point to the nuances of dramaturgical agency, which is not held exclusively by the dance dramaturg but is shared with collaborators and spectators. The traditional notion of the authoritative dramaturg is replaced by a "call for broad and differentiated articulation of the modes of engagement and relationships that dramaturgs embody or promote in different contexts: collaboration, participation, friendship, mentorship, critique, support, memory, dialogue, navigation, and many more" (2). In other words, the dialogic nature of dance extends beyond one key person and leads to the creation of meaning that is greater than the sum of a production's parts. We can extend these principles of distributed agency back quite easily to directorship of films and theater productions—a fitting conclusion to this chapter on cognitive approaches to performance studies of various types, even as the dialogs on this topic are still at their beginning. The concluding chapter of this book makes the final case for cognitive approaches as the paradigm that will help literary and performance studies continue to thrive.

Notes

1. See Mark Johnson's (1987) description of this important image schema.
2. See for instance Christoforou et al. (2017).
3. See the as yet unpublished research of Devlin et al., experimental psychologists at University College London, who found that audience members' heartbeats became synchronized while watching a live theater performance ("Audience members' hearts beat together at the theatre," 2017).
4. For previous research on video games and embodiment in particular, see for example Gee (2008) and Harvey (2009).
5. See also Gonçalves et al. (2019) on the notion of a Bakhtinian "danced-utterance." The authors take a Bakhtinian perspective on performative education: "In understanding the communication of the body as a materiality of the individual consciousness that enunciates infinite dialogic chains through dancing, it is essential to conceive movement as a sign text, filled with voices that can be read and interpreted" (136).
6. Interestingly, the researchers conducted their study during dress rehearsals of the Scottish Ballet's *Sleeping Beauty* due to "ethical reasons of avoiding potential disturbance to paying spectators in a full performance" (381).

References

Albright, Ann Cooper. 2015. "Split Intimacies. Corporeality in Contemporary Theater and Dance." In *The Oxford Handbook of Dance and Theater*. Ed. Nadine George-Graves. Oxford: Oxford UP. 19–34.

Amorim, Marilia. 2020. "Dance Discourse and the Concept of Genre – Some Interpretive Elements." *Bakhtiniana: Revista de Estudos do Discurso* 15.2: 67–102.

Anderson, Joseph. 1996. *The Reality of Illusion: An Ecological Approach to Cognitive Film Theory*. Carbondale: Southern Illinois UP.

"Audience Members' Hearts Beat Together at the Theatre." 2017. University College London, Psychology and Language Sciences website. Online.

Banks, Kathryn and Timothy Chesters, eds. 2018. *Movement in Renaissance Literature*. New York: Palgrave Macmillan.

Bersley, Tracy E. 2018. "The Body's Brain: Neurology in Theatrical Practice." *Critical Stages/Scènes Critiques* 17: online.

Blair, Rhonda. 2008. *The Actor, Image, and Action: Acting and Cognitive Neuroscience*. London: Routledge.

———, and Amy Cook, eds. 2016. *Theatre, Performance and Cognition. Language, Bodies and Ecologies*. London / New York: Bloomsbury Methuen Drama.

Booth, Michael. 2017. *Shakespeare and Conceptual Blending: Cognition, Creativity, Criticism*. New York: Palgrave Macmillan.

Bordwell, David. 1985. *Narration in the Fiction Film*. Madison: U of Wisconsin P.

———. 1989. *Making Meaning: Inference and Rhetoric in the Interpretation of Cinema*. Cambridge: Harvard UP.

———, and Noël Carroll, eds. 1996. *Post-Theory: Reconstructing Film Studies*. Madison: U of Wisconsin P.

Budra, Paul and Clifford Werier, eds. 2016. *Shakespeare and Consciousness*. New York: Palgrave Macmillan.

Carroll, Noël. 1988. *Mystifying Movies: Fads and Fallacies in Contemporary Film Theory*. New York: Columbia UP.

———. 1996. *Theorizing the Moving Image*. Cambridge: Cambridge UP.

Chow, Kenny K.N., and D. Fox Harrell. 2009. "Material-Based Imagination: Embodied Cognition in Animated Images." *UC Irvine: Digital Arts and Culture 2009*. Online.

Christoforou, Christoforos, et al. 2017. "Your Brain on the Movies: A Computational Approach for Predicting Box-Office Performance from Viewer's Brain Responses to Movie Trailers." *Frontiers in Neuroinformatics* 19 December. Online.

Coëgnarts, Maarten, and Peter Kravanja. 2015. *Embodied Cognition and Cinema*. Leuven: Leuven UP.

Cook, Amy. 2010. *Shakespearean Neuroplay: Reinvigorating the Study of Dramatic Texts and Performance through Cognitive Science*. New York: Palgrave Macmillan.

Cook, Roger. 2015. "Embodied Simulation, Empathy, and Social Cognition: Berlin School Lessons for Film Theory." *Screen* 56.2: 153–71.

Crane, Mary Thomas. 2001. *Shakespeare's Brain: Reading with Cognitive Theory*. Princeton: Princeton UP.

Fahlenbrach, Kathrin. 2015. *Embodied Metaphors in Film, Television, and Video Games: Cognitive Approaches*. London: Routledge.

Falletti, Clelia, Gabriel Sofia, and Victor Jacono, eds. 2016. *Theatre and Cognitive Neuroscience*. London/New York: Bloomsbury Methuen Drama.

Garner, Stanton B. 2018. *Kinesthetic Spectatorship in the Theatre: Phenomenology, Cognition, Movement*. New York: Palgrave Macmillan.

Gee, James Paul. 2008. "Video Games and Embodiment." *Games and Culture* 3.3: 253–63.

Gonçalves, Michelle Bocchi, Thais Castilho, and Jair Mario Gabardo Junior. 2019. "Dancing Bodies at School: Dialogues between Performative Education and the Bakhtinian Perspective." *Bakhtiniana: Revista de Estudos do Discurso* 14.3: 136–55.

Hansen, Pil, and Darcey Callison, eds. 2015. *Dance Dramaturgy: Modes of Agency, Awareness and Engagement*. New York: Palgrave Macmillan.

Harvey, Alison. 2009. "Seeking the Embodied Mind in Video Game Theory: Embodiment in Cybernetics, Flow, and Rule Structures." *Journal of the Canadian Game Studies Association* 3.4: online.

Helms, Nicholas R. 2019. *Cognition, Mindreading, and Shakespeare's Characters*. New York: Palgrave Macmillan.

Jola, Corinne, Frank Earl Pollick, and Marie-Helene Grosbras. 2011. "Arousal Decrease in Sleeping Beauty: Audiences' Neurophysiological Correlates to Watching a Narrative Dance Performance of Two-and-a-Half Hours." *The Journal of the Society for Dance Research* 29 (Special On-line Supplement on Dance and Neuroscience): 378–403.

Johnson, Mark. 1987. *The Body in the Mind: The Bodily Basis of Meaning, Imagination, and Reason*. Chicago: U of Chicago P.

Lee, Sung-Ae. 2014. "Fairy-Tale Scripts and Intercultural Conceptual Blending in Modern Korean and Television Drama." In *Grimms' Tales Around the Globe*. Ed. Vanessa Joosen and Gillian Lathey. Detroit: Wayne State UP. 275–94.

Lutterbie, John Harry. 2011. *Toward a General Theory of Acting: Cognitive Science and Peformance*. New York: Palgrave Macmillan.

Lyne, Raphael. 2011. *Shakespeare, Rhetoric and Cognition*. Cambridge: Cambridge UP.

McConachie, Bruce. 2003. *American Theater in the Culture of the Cold War: Producing and Contesting Containment, 1947–1962*. Iowa City: U of Iowa P.

———. 2008. *Engaging Audiences: A Cognitive Approach to Spectating in the Theatre*. New York: Palgrave Macmillan.

———. 2015. *Evolution, Cognition, and Performance*. Cambridge, UK: Cambridge UP.

———, and F. Elizabeth Hart, eds. 2006. *Performance and Cognition: Theatre Studies and the Cognitive Turn*. London: Routledge.

Marks, Laura U. 2000. *The Skin of the Film: Intercultural Cinema, Embodiment, and the Senses*. Durham: Duke UP.

Matthews, Paul M., and Jeffrey McQuain. 2003. *The Bard on the Brain: Understanding the Mind through the Art of Shakespeare and the Science of Brain Imaging*. Washington, DC: Dana Press.

Murphy, Maiya. 2019. *Enacting Lecoq: Movement in Theatre, Cognition, and Life*. New York: Palgrave Macmillan.

Newen, Albert, Leon De Bruin and Shaun Gallagher, eds. 2018. *The Oxford Handbook of 4E Cognition*. Oxford: Oxford UP.

Paavolainen, Teemu. 2012. *Theatre/Ecology/Cognition: Theorizing Performer-Object Interaction in Grotowski, Kantor, and Meyerhold*. New York: Palgrave Macmillan.

Rokotnitz, Naomi. 2011. *Trusting Performance: A Cognitive Approach to Embodiment in Drama*. New York: Palgrave Macmillan.

———. 2018. "Performance and Cognition: How the Performing Arts Contribute to the Science of Mind." *Interdisciplinary Literary Studies* 20.4: 470–85.

Shaughnessy, Nicola. 2012. *Applying Performance: Live Art, Socially Engaged Theatre and Affective Practice*. New York: Palgrave.

———. 2013. *Affective Performance and Cognitive Science: Body, Brain and Being*. London: Methuen.

Shimamura, Arthur P. 2013. *Psychocinematics: Exploring Cognition at the Movies*. Oxford: Oxford UP.

Stadler, Jane. 2016. "Empathy and Film." In *Routledge Handbook of Philosophy of Empathy*. Ed. Heidi L. Maibom. New York: Routledge. 317–26.

Stevenson, Jill. 2010. *Performance, Cognitive Theory, and Devotional Culture: Sensual Piety in Late Medieval York*. New York: Palgrave Macmillan.

Tribble, Evelyn. 2011. *Cognition in the Globe: Attention and Memory in Shakespeare's Theatre*. New York: Palgrave Macmillan.

William, Jennifer Marston. 2017. *Cognitive Approaches to German Historical Film: Seeing is Not Believing*. New York: Palgrave Macmillan.

———. 2021. "The Tender Gaze, Embodied Politics, and Perspective-Taking in Contemporary German Theater." In *The Tender Gaze: Compassionate Encounters on the German Screen, Page, and Stage*. Ed. Muriel Cormican and Jennifer Marston William. Rochester: Camden House. 40–59.

CHAPTER 16

Conclusion: The Bridging Function of Contextualism and the Cognitive Paradigm

For example, to understand the meaning of a literary text, one needs to establish the multiple contexts of its words and phrases.

—Fritjof Capra

In this brief concluding chapter, we recap the spectrum of cognitive approaches to literary studies, speculate briefly on how cognitive literary and performance studies provides a new context for some previous practices, and assess the relationship between cognitive studies and THEORY.

1 A New Direction: Which Path(s) to Take?

Toward the beginning of this century, theater historian Bruce McConachie expressed his dissatisfaction with the reigning THEORETICAL approaches to the understanding and appreciation of theatrical performance in the following terms:

> Beyond semiotics and some warmed-over Freudianism, our current critical and theoretical discourses have very little to say about theatrical pleasure and social experience. And besides, I did not really trust Ferdinand de Saussure or Jacques Lacan to tell me much about enjoyment or experience. Do people really find much enjoyment in reading signs? Perhaps,

H. Mancing and J. Marston William, *Restoring the Human Context to Literary and Performance Studies*, Cognitive Studies in Literature and Performance,
https://doi.org/10.1007/978-3-030-89078-0_16

> but it has always seemed like thin gruel to me. And why take the trouble to go to the theater for something you can do anywhere, for free, by yourself? Lacan's theories, by linking human development to language and desire, can help to answer questions about enjoyment and experience, but only if one believes in the Freudian premises of his arguments. Remove those, and the Lacanian house of signs and stages comes tumbling down. Besides, why should the historian attempt to encompass the experience of the theater within such a narrow understanding of enjoyment? More significantly, perhaps, is there any experimental validity to Lacan's conclusions? Or, for that matter, to the psychology of semiosis? Most psychologists work out a tradition of empirical experimentation that Lacan and other Continental philosophers simply reject out of hand. (2003, vii–viii)

It was this sense of dissatisfaction and frustration that led McConachie to turn to the experiential realism of Lakoff and Johnson and to find within their cognitive metaphor theory a much more satisfactory way of describing his own experiences, and then to move on to broaden his theoretical base with work from other cognitive sciences. A similar sense of growing dissatisfaction with THEORY has led an increasing number of literary scholars to turn to cognitive science and biology in search of an alternative framework. Certainly this is true in our own cases.

Literary theory must be rethought from the ground up to secure a meaningful place in today's intellectual world. There is no such thing as a Saussurean *langue*, an independent social entity that is a closed, self-referential system consisting only of differences, within which signifiers signify. Neither the word nor the sign is the primary element in linguistic communication, and a text or an utterance is not merely the sum of its structural elements. Human symbolic communication does not consist of a process of telementation with an active sender and a passive receiver, nor do all members of a linguistic community have identical dictionaries in their heads. Coding and decoding of arbitrary signs is a basic but minor technical part of the process of linguistic communication and should rarely be the primary (let alone the exclusive) focus of scholarly attention. Dealing with language in Saussurean terms is not a valid enterprise. The modifications made to Saussurean concepts by leading figures in THEORY are even less valid. Proponents of THEORY conceived of language as an active agent that works through people and texts, inscribing itself on the blank slates of our socially constructed subjectivity. In this approach, strings of unmotivated signifiers float free and leave only traces and aporias, and the world itself is but a text and must be read, even though all reading is misreading.

In place of this, we should conceive of language as a Chomskyan I-language, a matter of species biology and individual psychology and as a cognitive tool that is used pragmatically in specific sociocultural contexts for specific purposes. We should recognize that we use language figuratively and creatively, just as our cognitive processes are most often figurative and imaginative rather than literal and logical. Rather than a copy machine that creates identical meanings in our mental lexicons, language is an inexact tool that makes mutual understanding possible, at least to some degree, but that never guarantees precise understanding. Meanings and understandings are constructed, negotiated, and contingent—but real.

We do not simply manipulate representations, translate codes, or decipher signs. As Bakhtin eloquently argues, we engage in dialogic relationships with other once-occurrent human beings in specific social contexts. Listening and understanding are as much an active process as speaking and writing. Who speaks, when, how, where, and why are all important, necessary considerations. We infer meanings, paying attention to what is relevant to us in specific contexts, because of our experiential and sympathetic understanding of speakers' and authors' intentions. Linguistic communication is a joint action in which speakers and writers are conscious of the interests and experiences of their addressees, as well as the context within which the utterance takes place or the text is read. Listeners/readers are active participants in the construction of utterances (and texts), and they are active constructors of contextual meaning. Approaches to language consistent with contextualized and embodied cognition provide a basis for literary theory that can participate in viable, convergent research in all the physical and biological sciences, social sciences, and humanities in the twenty-first century.

Just as literary theory should be based on a realistic, modern linguistics, it should also be based on current concepts of biology and psychology and not the discredited versions of Cartesian dualism and behaviorism. The "subject" described by and discussed in THEORY bears no resemblance to actual human beings. We are active, embodied agents who bring forth our realities within a complex social context. The human sense of self emerges, grows, and changes throughout a lifetime, as we negotiate and construct our realities and narrate our life stories to ourselves and others. We are never separate from our physical, biological, and social contexts, but form part of and co-define those contexts. The human mind-brain is not a debiologized blank slate that is determined by social, ideological, and linguistic forces. It is, rather, a complex result of interacting and mutually defining biological, personal, and social factors.

To do literary criticism and theory within a modern linguistic-biological-psychological framework does not mean all discourse about literature must explicitly be about linguistics, biology, or psychology. Literary criticism and theory should not conflict with what is known about linguistics, biology, and psychology, just as it should not conflict with what is known about physics, chemistry, geology, or astronomy, even when we do not foreground any of these subjects in our discussion of literature. To write under the assumption, for example, that the flat earth is the center of the solar system disqualifies one from taking a meaningful role in contemporary discourse. The same is true of an assumption or specific statement about the alchemy of turning lead into gold or the influence of the stars and planets on human events. And the same should be true of the following obsolete assumptions:

- that all communication consists of nothing but sign systems;
- that speakers are active and listeners are passive;
- that language is an active agent in culture;
- that texts are sequences of floating signifiers and the infinite free play of language;
- that all reading is misreading;
- that there exists a tidy division between mind and body;
- that the human mind is a blank slate; or
- that people are socially constructed and debiologized subjects.

An approach to literature (or anything else in life) that is properly grounded in the vast interdisciplinary, convergent, contextualized paradigm of the cognitive sciences does not determine *what* you will say as much as it frees you from repeating truisms from the tired, intellectually discredited, and superseded discourse of THEORY.

We see the paradigm described in this book above all as liberating: it frees us from modes of discourse and modes of cognition that have limited our activities and have distanced literary and performance study from the rest of the scholarly world. We agree with Pinker that

> [t]oday we may be seeing a new convergence of explorations of the human condition by artists and scientists—not because scientists are trying to take over the humanities, but because artists and humanists are beginning to look to the sciences, or at least to the scientific mindset that sees us as a species with a complex psychological endowment. (2002, 418)

The linguistic, biological, and psychological concepts described in this book will, as Zunshine's elegant work of 2006 illustrates, enable us to change the way we conceive of literary texts and our pedagogical and scholarly practices. Literary texts are not self-referential codes inscribed by language (which uses author-functions as the vehicles of its own agency) that determine how they are to be understood by the socially constructed subjects that always already misread them, but products of individually situated creative authors that are read and understood uniquely by individually situated creative and imaginative readers. Literary genres are not merely conventions that determine how we understand texts, but radial categories constructed by individuals. Novels, stories, poems, and essays are not self- and/or inter-referential closed systems, but pragmatic tools or prostheses that refer to, and afford the possibility of understanding, at least to some degree, the embodied mind-brains of their authors, readers, and characters and the contextual, ecological, and cultural realities within which these mind-brains function.

The best studies of literature and performance emerge when we as scholars recognize that we are embodied, contextualized, unique individuals who use the cognitive tool of language to understand that authors, narrators, characters, and readers are (or are like) entities similar to ourselves. As addressed below, and as we have hoped to have made clear throughout the book, there is not one single scholarly path to pursue as we keep these truths in mind. The plethora of interpretive focal points, all grounded in the principles of embodied cognition, may seem overwhelming at first. But once we inform ourselves about some fundamental principles of modern-day biology and psychology, and consider carefully their relevance to what we have been studying all along, the contextualist orientation becomes nothing short of liberating for literary and performance scholars.

2 Literary and Performance Studies in Context

It is impossible to "do cog. lit. crit." the way some graduate students in recent decades were taught to stamp out cookie-cutter Marxist, deconstructive, or Lacanian readings. More than anything else, embodied cognition provides a worldview and a set of values and assumptions. Approaching literature and performance from the cognitive science standpoint allows you to ask interesting questions, but how you answer them is up to you. Several areas of literary studies will be directly affected by a change to a linguistically, biologically, and psychologically valid approach to literature. Since recent literary and cultural THEORY has been dominated by an amorphous combination of psychoanalytic, sociological, and political theories, all derived from and dependent upon Saussurean linguistics, we will comment very briefly on how the cognitive framework described in this book might suggest a reconceptualization of some of these approaches.

As we have argued at some length, modern biology, neuroscience, and cognitive and developmental psychology have devastated Freudian psychoanalytic theory. Nearly every aspect of Freud's complex program has been called into question: the unconscious, repressed memory, dream theory, the Oedipal complex, and all the rest. Freud, who, ironically, began his career as a brain scientist, lived at a time when his fanciful ideas about how the brain was structured and how it functioned could not be disproved. What he proposed made as much sense as anything that had preceded it. Clearly, it caught on and became internalized in Western societies.

Today, in the era of the advanced study of neuroanatomy, noninvasive brain monitoring techniques, neural mapping, advances in genetics, new revelations about the process of evolution, comparative psychology, cognitive psychology, empirical research in sleep and dreams, and the study of patients with all sorts of brain injuries, there is little if any scientific support for Freud's concept of the brain and the theories based on it. Psychoanalytic literary theory and criticism, which has been so rich a vein, begins to look more vacuous and irrelevant every day as we see how it depends on a biology and a psychology that are as up-to-date as a horse and buggy. Only to the degree that any theory of the psyche, subjectivity, self, and related concepts is consistent with evolution, modern biology, neuroscience, and the main contemporary trends in cognitive and developmental psychology, can it claim legitimacy. Any attempt to bring

Freudian theory into the twenty-first century needs to acknowledge and respond thoroughly to the criticism that comes from scientists such as those discussed in this book.

And if Freud is in trouble, Lacan's case is hopeless. Take away from Lacan any kind of *langue* with signifiers and signifieds, any kind of dynamic unconscious supposedly structured like a Saussurean E-language, and the mirror stage, and there is very little left to his theory. Since none of these concepts has any status whatsoever in modern biology and cognitive science, it should not surprise anyone to learn that while Freud is still debated somewhat, Lacan is mostly forgotten.

If Freud is in big trouble in cognitive psychology, Marx's problems are more of a political nature. Many feel that the fall of the Berlin Wall and the breakup of the Soviet Union and its satellites discredited Marx and Marxist theories once and for all. Maybe or maybe not; Marx has always been subtler and more nuanced than either his most fervent followers or the self-appointed political leaders who proclaim themselves Marxists have wanted to recognize. Marxist theory need not be so thoroughly called into question as is Freudian theory by what we know of evolution, the mind-brain, and human cognition. Unlike Freud's, Marx's original theory was, after all, grounded in convincing (and still relevant) social realities.

But Marxism will have some difficulty with cognitive science in at least three ways. First, to the degree that Marxism is proclaimed—e.g., by Jameson (1981)—to be the one and only fully explanatory theory in all circumstances, it has no place in a world of pragmatics, constructivism, or embodied cognition. If part of human cognition is to create or construct our worlds based on our personal experiences and our sociohistorical context, there can be no totalizing theory, Marxist or otherwise. Second, to the extent that Marxists have bought into Saussurean linguistics and postmodern subjectivity, as it has in the Althusserian model, it becomes less useful. Ideology does not hail and interpolate us within texts; such an idea should be dismissed out of hand, as should all the other cant accompanying this brand of Marxist thought. And third, the standard Marxist position that the mind is in no way influenced by biology and that historical, economic, and class issues alone determine the human condition, must be abandoned. Karl Marx, who so admired Charles Darwin, would have no difficulty incorporating the realities of human cognition into his theories.

Surely Marx today would gladly recognize that we now know more about how the embodied mind-brain operates. First, it should be a part of the recognition of an individual's contextualized life that her class-consciousness is an important formative part of her construction of her reality. Second, schema theory and non-essentialist concepts of categorization can help clarify Marxist values. And third, the pragmatic insights of Marx and other comparable theorists can make advantageous use of concepts of relevance, joint action, and integration in the study of language in actual use by situated human agents. Marxism may begin to sound a little different in a cognitive paradigm, but it is still an important voice to be reckoned with.

Feminism has offered the most original and vital force in literary study of the past few decades. We endorse the following statement by Frank B. Farrell:

> It is quite possible that much of the work in literary criticism over the last three decades will come to seem empty since it is determined by ideological drives little controlled by considerations of evidence and inference. But the case with feminism is different. It is a genuine revolution in the history of thought, and we do not yet know how narrow and limited our previous conceptions of literature will come in the future to appear. (2004, 148)

Feminism, even more than psychoanalytic theory and Marxist theory, is not a single entity, and so the effect that cognitive science will have on it will depend on certain specific factors. For example, the strand of feminist thought still couched in Lacanian language and operating under the assumptions of a Saussurean concept of language will be in trouble. Similarly, certain kinds of essentialist feminism, stressing necessary differences between the way men and women conceive and do things will not fare well unless the embodied, evolved mind-brain is taken into consideration along with the oppressive sociocultural forces that continue to disadvantage women (intersectional feminism, for example, duly recognizes the different extents and types of disadvantages across demographic categories). Because equity for women has, unfortunately, not yet been achieved, feminism should be fortified with support from science whenever possible toward that goal.

Biology, evolution, neuroscience, and psychology (cognitive, developmental, and comparative) can contribute significantly to supporting and developing feminist thought. These fields converge on the necessity of radically revising all the easy deterministic explanations for everything, including sexuality, gender, and patriarchy. Presenting these issues in a contextualized way with the backing of current legitimate science allows room to expand theoretical premises to include not just those who identify as women and men, but also nonbinary and gender-fluid people. In addition, moving away from a poststructuralist rejection of gender categories makes room for new progressive developments that move beyond binary thinking, such as the budding field of trans linguistics, which seeks to explore gender (both in society and in grammar) in the context of how trans people experience and use language in reference to their identity (see Zimman 2019, 2020).

Feminist theory that stresses situatedness, context, and dialogic relationships can only be reinforced by considerations of embodied cognition. Bakhtinian metalinguistics and modern pragmatics provide positive support for feminist thought that eschews facile binaries (male-female, self-other, mind-body, self-environment), and simplistic determinism (we are completely socially constructed). Women are not determined solely by language, ideology, patriarchy, or any other single thing. Rather, women (and men and trans people and gender-fluid people) are embodied agents, individually located at specific places and in specific times, with individual memories and experiences, who typically use language as their most important cognitive tool in developing and maintaining their sense of self and in conducting their relationships with other people.

Since what we posit in this book is, above all, a conceptual framework for work in literary and performance studies, and other humanistic and social science endeavors, it has the same relevance for cultural studies as for literature. Cultural studies are sometimes understood to have arisen in opposition to THEORY, yet many foundational tenets of THEORY are fully integrated into cultural studies and are implicitly accepted as givens. Nearly everything said above about the relevance of cognitive psychology and biological/evolutionary science to psychoanalytic, Marxist, and feminist theory also applies to cultural studies. One could make the argument that an embodied cognitive approach to literature—and film, drama, and other areas of the arts—is itself a kind of cultural studies.

3 Bakhtin for the Twenty-First Century

We have referred frequently throughout this book to Mikhail Bakhtin. Bakhtin's contextualism and dialogism bridge various cognitive approaches from Chomsky to Turner to more recent 4E cognition, and other schools of thought as outlined in the previous chapters. Bakhtin is an accessible starting point for literary and performance scholars who may not have much previous experience in the domains of cognitive linguistics and language philosophers. His work aids us conceptually in moving beyond cognitivism. Chomsky, Lakoff, and later cognitive linguists and language philosophers carried further—in diverse ways—the foundations that Bakhtin had laid for considerations of the utterance in context. Once we are listening, Bakhtin's voice reverberates frequently in works of cognitive studies, for instance in Maturana and Varela's "imagined dialogue" as present in every thought and reflection (1992), Sperber and Wilson's listener who is an "active interpreter" of a speaker's intentions (1995), Tomasello's "dialogic engagement with other minds" that allows for linguistic communication (1999), Thompson's "dialogical dynamic" that lays the foundation for empathy (2007), and many more that are relevant to scholarly considerations of literature and performance.

Reading Bakhtin opens our eyes to the many ways that we can return to the human, embodied context in analyzing and appreciating the fruits of human creativity. The paradigm we advocate does not rely solely on Bakhtin, however. The title of this book is "Restoring the Human Context to Literary and Performance Studies" because our fields of study were shaped for so long by poststructuralism and constructivist theories that, in our view, there needs to be a deliberate recognition and reckoning process so that we can move forward fully with updated approaches that are compatible with and adaptable to newly emerging scientific developments surrounding mind and body. Bakhtin is conceived of as a guide in this exciting journey, not as the sole path.

4 Assessment and Final Remarks

THEORISTS often assume that one is either with them or against them, buying into their values and methods or dogmatically clinging to the ways of old. This position is not merely one of the early days of THEORY but continues in some form today. However, there have also been some efforts from the structuralist-semiotic-poststructuralist camp to promote rapprochement with cognitive science, albeit without a focus on embodied cognition; see Schleifer et al. (1992), Rastier (1997), Barry (1999), Broden (1999), and Jackson (2000); and see the critiques of Merrell (1994) and Mancing (2003). More common than attempts to acknowledge or incorporate cognitive science into theoretical discourse, though, is dismissiveness by THEORISTS (see e.g., Habib 2008).

A prime example of this stance is an essay by Slavoj Žižek (2000) in which this leading promoter of Lacanian psychoanalytic theory offers a critique of cognitive approaches to literature (he continues his strong defense of Lacan in his 2007 book *How to Read Lacan*). Apparently, Žižek lives by the conceptual metaphor that ARGUMENT IS WAR, since he opens his essay by claiming to witness "the struggle for intellectual hegemony ... the fight for ideological hegemony" between theoretical cultural studies and what he calls "cognitivists" (9). Either/or, mutually exclusive binaries, here locked in a (dare we say Darwinian?) struggle for survival: intellectual warfare red in tooth and claw. According to Žižek, the evil hegemonic cognitive establishment aims to "get rid of" cultural studies, "a foreign body unable to fit fully into the existing academy" (19). Throughout his essay, however, Žižek grossly distorts a supposedly ideology-obsessed cognitive science by reducing it all to a simplistic naive realism and contrasting it with an equally simplistic supposedly mature relativism (see the sharp and compelling critique by Hart 2001; see also McConachie 2006).

Not surprisingly, Žižek proposes Lacan as the middle-ground option between the two extremes. One is tempted to accuse Žižek of either ignorance or dishonesty in his discussion of cognitivism (although this term is used by theorists like Noël Carroll, for many—including the present authors—it carries with it the negative connotations of AI); but, unlike Žižek, we do not want to depend on facile and misleading either/or binaries. Rather, we suspect that Žižek misunderstands CONTEXTUALISM and embodied cognitive science, not realizing how they differ from the COGNITIVIST approach. We surmise that there could be some degree of conscious distortion of his opponents' position in order (like the critics of sociobiology) to make his political point; and probably his view of the world is narrowed to some extent by his THEORETICAL blinders that render him incapable of understanding discourse in a radically different paradigm. (Note: This is a ToM approach to how Žižek's mind is working in this essay; that is, this is how we think Žižek might have been thinking as he wrote the essay.) The end result, however, is an attack on a nonexistent straw man; in other words, a useless exercise in rhetoric.

A similar stance is adopted in the Lacanian James M. Mellard's sometimes sneering critique of Darwinian approaches to literature (2007), in which he uses Žižek and psychoanalytic theory to critique innovative scholars like Joseph Carroll and the contributors to Gottschall and

Wilson's (2005) volume, and even the novelist Ian McEwan. Although we applaud the efforts of those who would attempt to build bridges between a cognitive approach and the Saussurean-derived ones of THEORY, we are pessimistic that it can be carried off successfully.

The works mentioned at times throughout this book and then reviewed briefly in the previous two chapters point the way to a new and exciting context for literary theory and criticism. But the work done in cognitive literary and performance studies is (as in any field, particularly a newer one) uneven in merit and often only partially informed by the overall approach to cognition described in this book. A number of these studies, especially some of the earlier ones from the 1980s and 1990s, reveal a lack of familiarity with Chomskyan and post-Chomskyan linguistics and at times even with the contributions of Bakhtin, pragmatics, and related fields. Several of these works, as indicated, attempt to reconcile a cognitive approach with Saussurean linguistics, still struggling to articulate a modern psychology in terms of signifiers and signifieds, codes, and the conduit theory of communication. Efforts to reconcile cognitive approaches with structuralism, semiotics, and/or poststructuralism have not been very successful. At times, especially in some of the earlier cognitive essays, the authors seem not to grasp the full import of embodied cognition and misrepresent some of the main tenets of cognitive science. Often they perpetuate the THEORETICAL ignorance and/or denial of the role of evolution and biology (sometimes even jumping on the anti-evolutionary psychology bandwagon), and too frequently they fall into the trap of understanding the mind-brain as a computer and write of information processing and internal representations. Our hope is that this book will alleviate some of the misinformation and lack of adequate knowledge about basic principles of and developments in cognitive science so that humanities-based cognitive studies going forward will be that much richer.

The lingering influence of Freud, a simplistic brand of Marxism, and a radical strong social constructionist stance can still be found at times in cognitive-based writing. At other times, there is a clear personal animosity toward feminist theories and even toward women in general. Sometimes one gets an impression of a conservative, even reactionary, political stance. Some arguments are made more forcefully and more convincingly than others. In other words, many of the early entries into cognitive studies are tentative efforts, reaching toward a new paradigm but not yet free of old ones. Such is to be expected from pioneers. Just because a scholar is writing within a cognitive paradigm does not mean that they are always right. What is important, however, in every one of these works is that a conscious attempt is made to conceive of the enterprise of literary and performance study in a new way, and to contribute to a new critical orientation.

The program we present here is in no way prescriptive. While we respect his work overall, we don't agree with Joseph Carroll's stance that a Darwinian approach to literature, such as the one he proposes, should (or could) aim "to establish a new alignment among the disciplines and ultimately to subsume all other possible approaches to literary study" (2008). Employing explicit evolutionary terminology to describe every work of literature is unwarranted. While nothing we write about literature should contradict present-day evolutionary theory, subjecting everything to Darwinian imperialism is unnecessary. Jonathan Gottschall (2008) also gives us pause when he seems to argue that all literary criticism be based on quantitative methodology. There is no single way to do (read, discuss, teach, criticize, theorize) literature from a cognitive (or evolutionary) perspective. We are looking to promote a new paradigm, not a new cookie cutter.

Three groups of scholars have the most to gain from the many interdisciplinary manifestations of literary study grounded in biology and cognition:

- First, students and particularly graduate students who are about to enter a scholarly world that is evolving away from the increasing irrelevance of THEORY and toward new theoretical bases.[1] One of the worst fates that can befall someone about to begin a professional career is to be at the forefront of the rear guard.
- Second, scholars whose approaches to literary and cultural studies have been grounded largely in the disciplines of history, sociology, and politics, or feminist, minority, or developing nation studies, as well as those who have already been active in reader-response study, reception theory, phenomenology, and hermeneutics. Such scholars will find that their critical assumptions are already close to and can only be enriched by the advances in cognition.
- And finally, scholars who have been working within the structuralist-semiotic-deconstructive paradigms of THEORY but who have been

> less than fully satisfied with the linguistic and psychological assumptions of those approaches. Such scholars will face the large task of retooling, but that task is not impossible.

As alluded to in the Introduction, while we are harsh toward constructivist theories that are used unreflectively, it doesn't mean that we think all scholars who refer to such in their work are "uptight, dogmatic professionals who take resentful pleasure in ignoring texts' clear meanings, denying the pleasures they afford, and denouncing anyone who dares think otherwise" as English literature scholar David Kurnick (2021) has described some recent representations of literary critics by, well, other literary critics. Some critics will be dogmatic, while others will find that a certain aspect of THEORY seems to support the argument they would like to make about a text. There's nothing inherently wrong with this, if what is referred to does not contradict current science. THEORY has influenced many literary writers over time, and thus it makes sense sometimes to read those writers' texts through that lens, while also keeping at the forefront the contextualized, dialogic nature of the cultural artifact that results from human embodied cognition. We are all interested in discerning the meaning of texts, and debates about how to do it are part of the fun. What we put forth here isn't meant to be the one definitively "right" way, but a path to eliminating approaches that have no grounding in—and indeed that conflict with—the realities of contemporary biological and psychological sciences.

In summary, there is life after THEORY, and a good, robust, and exciting one at that. Literary scholars and cultural critics in general can benefit greatly from investigating and strengthening the links between literature and linguistics (including post-Chomskyan linguistics, Bakhtinian metalinguistics, and pragmatics), biology (including evolution and neuroscience), and psychology (including cognitive, developmental, comparative, and evolutionary psychology). Literary theory has long been an interdisciplinary activity, but now it is time to try a new mix of disciplines. It is time to replace formal, structural, semiotic, and explicitly or implicitly poststructural approaches to literature with more attention to real flesh-and-blood human beings. The alternative offered by the integrated study of literature, performance, bodies, and cognition is one that holds out this possibility of rehumanizing literature, in a way that may seem counterintuitive to some at first—via updated science and empiricism.

And, finally: we are often asked when we begin to talk about cognitive science and literature, "Yes, but how do you *do* it?" The question, if asked by someone open-minded enough to get past knee-jerk skepticism, still reveals a deep-seated fear: show me—fast!—how it works so that I can apply it quickly. Quite honestly, we are getting a little weary of the question and need to call for backup: in a famous short essay entitled "My Religion" (1974; originally 1907), Spanish philosopher Miguel de Unamuno refuses to explain exactly what he believes, preferring to explore some basic principles and leave conclusions up to his individual readers. His self-defined role is to challenge his readers intellectually and emotionally, and not to give answers. Near the conclusion of his essay, he writes that

> the questioners and curious who chance to read these words of mine will ask me again: "All right, now, but what solutions do you offer?" And I, in conclusion, will tell them that if they want solutions, to step over to the stand across the street, for I do not carry any such line of goods. My endeavor has been, still is, and will always be to make those who read me think and meditate on fundamentals. I have always sought to agitate and, at most, to suggest rather than instruct. If I start to sell bread, it will not be bread, but yeast and leavening. (216)

Cognitive approaches are something you can "take" but not necessarily something you "do" in a certain way. In this book, we haven't prescribed a method or singular system for doing literary studies. There is no such thing as *the* cognitive method. The scholarly work described throughout the previous chapters illustrates the wide range of approaches, methodologies, and critical assumptions that fall within what we call the cognitive paradigm. We have described a manner of thinking and talking, and a context, that will enable creative embodied mind-brains to discover their own ways of doing literary analysis and, most importantly, in a way that does not contradict modern science. What to do and exactly how to do it will be figured out by each individual scholar, an exciting prospect that bodes well for the future flourishing of creative, cognitively oriented approaches to literature and performance as well as to other humanities fields.

Note

1. For scholarship on the pedagogy of cognitive literary studies, see Jaén (2014) and William (2017).

References

Barry, Jackson. 1999. *Art, Culture, and the Semiotics of Meaning: Culture's Changing Signs of Life in Poetry, Drama, Painting, and Sculpture*. New York: St. Martin's P.

Broden, Thomas F. 1999. "Linguistic Semantics for Literature and the Human Sciences Today." *Semiotica* 124: 81–127.

Carroll, Joseph. 2008. "An Evolutionary Paradigm for Literary Study." *Style* 42.2–3: 103–43.

Farrell, Frank B. 2004. *Why Does Literature Matter?* Ithaca: Cornell UP.

Gottschall, Jonathan. 2008. *Literature, Science, and a New Humanities*. New York: Palgrave Macmillan.

———, and David Sloan Wilson, eds. 2005. *The Literary Animal: Evolution and the Nature of Narrative*. Forewords by E. O. Wilson and Frederick Crews. Evanston: Northwestern UP.

Habib, M. A. R. 2008. *Modern Literary Criticism and Theory: A History*. Oxford: Blackwell.

Hart, F. Elizabeth. 2001. "The Epistemology of Cognitive Literary Studies." *Philosophy and Literature* 25: 314–34.

Jackson, Tony. 2000. "Questioning Interdisciplinarity: Cognitive Science, Evolutionary Psychology, and Literary Criticism." *Poetics Today* 21: 319–47.

Jaén, Isabel. 2014. "Teaching Cervantes's *Don Quixote* from a Cognitive Historicist Perspective." *Cognition in the Classroom. Interdisciplinary Literary Studies* 16.1: 110–26.

Jameson, Fredric. 1981. *The Political Unconscious: Narrative as a Socially Symbolic Act*. Ithaca: Cornell UP.

Kurnick, David. 2021. "Queer Theory and Literary Criticism's Melodramas. Recent Debates Suffer from a Strange Amnesia." *Chronicle of Higher Education Review*, 16 February (Online).

Mancing, Howard. 2003. "Rastier Revisited: Paradigms in Conflict." *Semiotica* 145: 139–49.

Maturana, Humberto R., and Francisco J. Varela. 1992 (1984). *The Tree of Knowledge: The Biological Roots of Human Understanding*. Trans. Robert Paolucci. Boston: Shambhala.

McConachie, Bruce. 2003. *American Theater in the Culture of the Cold War: Producing and Contesting Containment, 1947–1962*. Iowa City: U of Iowa P.

———. 2006. “Cognitive Studies and Epistemic Competence in Cultural History: Moving beyond Freud and Lacan.” In *Performance and Cognition: Theatre Studies and the Cognitive Turn*. Ed. Bruce McConachie and F. Elizabeth Hart. London: Routledge. 52–75.

Mellard, James M. 2007. “‘No Ideas But in Things’: Fiction, Criticism, and the New Darwinism.” *Style* 41.1: 1–29.

Merrell, Floyd. 1994. “On Bifurcating *Semiosis*: Or, How to Stop Worrying about Those Elusive Signs and Learn to Live with Them.” *Semiotica* 99.1–2: 101–25.

Pinker, Steven. 2002. *The Blank Slate: The Modern Denial of Human Nature*. New York: Viking.

Rastier, François. 1997 (1989). *Meaning and Textuality*. Toronto: U of Toronto P.

Schleifer, Ronald, Robert Con Davis, and Nancy Mergler. 1992. *Culture and Cognition: The Boundaries of Literary and Scientific Inquiry*. Ithaca: Cornell UP.

Sperber, Dan, and Deirdre Wilson. 1995 (1986). *Relevance: Communication and Cognition*. Cambridge: Harvard UP.

Thompson, Evan. 2007. *Mind in Life: Biology, Phenomenology, and the Sciences of Mind*. Cambridge: Belknap P of Harvard UP.

Tomasello, Michael. 1999. *The Cultural Origins of Human Cognition*. Cambridge: Harvard UP.

Unamuno, Miguel de. 1974 (1907). “My Religion.” In *The Selected Works of Miguel de Unamuno*, Vol. 5: *The Agony of Christianity and Essays on Faith*. Trans. Anthony Kerrigan. Annotated by Martin Nozick and Anthony Kerrigan. Princeton: Princeton UP. 209–17.

William, Jennifer Marston. 2017. “The Case for Teaching Cognitive-Literary Studies: Approaches, Challenges, and Benefits.” In *Self, Other, and Context in Early Modern Spain (Studies in Honor of Howard Mancing)*. Ed. Carolyn Nadeau, Isabel Jaén Portillo, and Julien J. Simon. Newark: Juan de la Cuesta Hispanic Monographs, LinguaText. 29–42.

Zimman, Lal. 2019. “Trans Self-identification and the Language of Neoliberal Selfhood: Agency, Power, and the Limits of Monologic Discourse.” *International Journal of the Sociology of Language* 256: 147–75.

———. 2020. “Transgender Language, Transgender Moment: Toward a Trans Linguistics.” In *The Oxford Handbook of Language and Sexuality*. Ed. K. Hall and R. Barrett. Online.

Žižek, Slavoj. 2000. “Lacan between Cultural Studies and Cognitivism.” *Umbr(a): A Journal of the Unconscious* 9: 9–32.

———. 2007. *How to Read Lacan*. New York: W.W. Norton and Company.

Zunshine, Lisa. 2006. *Why We Read Fiction: Theory of Mind and the Novel*. Columbus: Ohio State UP.

Index

H. Mancing and J. Marston William, *Restoring the Human Context to Literary and Performance Studies*, Cognitive Studies in Literature and Performance,
https://doi.org/10.1007/978-3-030-89078-0

Printed in the USA
CPSIA information can be obtained
at www.ICGtesting.com
LVHW011133041023
760095LV00007B/377

9 783030 890773